Fracture Mechanics

Fracture mechanics studies the development and spread of cracks in materials. It uses methods of analytical solid mechanics to calculate the driving force on a crack and those of experimental solid mechanics to characterize the material's resistance to fracture. The subject has relevance to the design of machines and structures in application areas including aerospace, automobiles, power sector, chemical industry, oil industry, shipping, atomic energy and defense.

This book presents the gradual development in the fundamental understanding of the subject and in numerical methods that have facilitated its applications. The subject can be studied from the viewpoint of material science and mechanics; the focus here is on the latter.

The book, consisting of nine chapters, introduces readers to topics like linear elastic fracture mechanics (LEFM), yielding fracture mechanics, mixed mode fracture and computational aspects of linear elastic fracture mechanics. It also discusses the calculation of theoretical cohesive strength of materials and the Griffith theory of brittle crack propagation and its Irwin and Orowan modification. Explaining analytical determination of crack tip stress field, it also provides an introduction to the airy stress function approach of two dimensional elasticity and Kolosoff-Mukheslishvili potential formulation based on analytic functions. In addition a chapter deals with the characteristics of fracture in terms of crack opening displacement (COD) and J integral and the interpretation of J as potential energy release rate for linear elastic materials. Other relevant topics discussed in the book include stress intensity factor (SIF); factors that affect cyclic crack growth rate and Elber's crack closure effect; fundamentals of elastic plastic fracture mechanics (EPFM) and the experimental measurements of fracture toughness parameters KIC, JIC, crack opening displacement (COD), K-resistance curve etc.

Surjya Kumar Maiti is Professor at the Department of Mechanical Engineering, Indian Institute of Technology Bombay. He received his Ph.D. from Indian Institute of Technology, Bombay, and he worked as post-doctoral assistant at the University of Cambridge for two years (1981–1983). For over thirty-five years Maiti has been teaching courses on solid mechanics, strength of materials, stress analysis and pressure vessel design at both undergraduate and postgraduate levels. His research spans fracture mechanics, finite and boundary element methods, structural health monitoring and stress corrosion cracking.

FRACTURE MECHANICS

Fundamentals and Applications

Surjya Kumar Maiti

CAMBRIDGE
UNIVERSITY PRESS

CAMBRIDGE
UNIVERSITY PRESS

4843/24, 2nd Floor, Ansari Road, Daryaganj, Delhi - 110002, India

Cambridge University Press is part of the University of Cambridge.

It furthers the University's mission by disseminating knowledge in the pursuit of education, learning and research at the highest international levels of excellence.

www.cambridge.org
Information on this title: www.cambridge.org/9781107096769

© Surjya Kumar Maiti 2015

This publication is in copyright. Subject to statutory exception and to the provisions of relevant collective licensing agreements, no reproduction of any part may take place without the written permission of Cambridge University Press.

First published 2015

Printed in India by Thomson Press India Ltd., New Delhi 110001

A catalogue record for this publication is available from the British Library

Library of Congress Cataloguing in Publication data
Maiti, Surjya Kumar.
 Fracture mechanics: fundamentals and applications / Surjya Kumar Maitii.
   pages cm
 Includes bibliographical references and index.
 Summary: "Deals with the characteristics of fracture in terms of crack opening displacement (COD) and J integral and the interpretation of J as potential energy release rate for linear elastic materials"- - Provided by publisher.
 ISBN 978-1-107-09676-9 (hardback)
1. Fracture mechanics. I. Title.
 TA409.M336 2015
 620.1'126–dc23
 2015004526

ISBN 978-1-107-09676-9 Hardback

Cambridge University Press has no responsibility for the persistence or accuracy of URLs for external or third-party internet websites referred to in this publication, and does not guarantee that any content on such websites is, or will remain, accurate or appropriate.

To my parents, family (Baruna, Rakhi and Raman), brothers (Bijoy and Ajoy), sisters (Dulali and Malati), teachers, and students

Contents

List of Figures	*xi*
List of Tables	*xvii*
Preface	*xix*

1 Introduction — 1
   1.1 Introduction — 1
   1.2 Linear Elastic Fracture Mechanics — 1
   1.3 Elastic Plastic or Yielding Fracture Mechanics — 2
   1.4 Mixed Mode Fracture — 2
   1.5 Fatigue Crack Growth — 3
   1.6 Computational Fracture Mechanics — 4
   1.7 Scope of the Book — 4
   References — 5

2 Linear Elastic Fracture Mechanics — 6
   2.1 Introduction — 6
   2.2 Calculation of Theoretical Strength — 6
   2.3 Griffith's Explanation Based on Stress Concentration — 8
   2.4 Griffith's Theory of Brittle Fracture — 10
       2.4.1 Irwin–Orowan Modification — 12
   2.5 Stress Intensity Factor (SIF) Approach — 13
       2.5.1 Relationship between G and K — 17
   2.6 Concepts of Strain Energy and Potential Energy Release Rates — 22
       2.6.1 Crack Extension Under Load Control (Soft Loading) — 22
       2.6.2 Crack Extension Under Displacement Control (Hard Loading) — 23
   2.7 Irwin Plastic Zone Size Correction — 24
   2.8 Dugdale–Barenblatt Model for Plastic Zone Size — 26
   2.9 Crack-Tip Plastic Zone Shape — 27
       2.9.1 Mode I Plastic Zone — 28
       2.9.2 Plane Strain Constraint — 30
       2.9.3 Mode II and Mode III Plastic Zones — 30
   2.10 Triaxiality at Crack Front — 31
   2.11 Thickness Dependence of Fracture Toughness K_C — 33

2.12 Design Applications	34
Appendix 2.1 SIFs for Various Configurations	48
Exercise	54
References	60

3 Determination of Crack-Tip Stress Field — 65

3.1 Introduction	65
3.2 Airy Stress Function Approach	65
3.3 Kolosoff–Muskhelishvili Potential Formulation	68
3.4 Examples on Analytic and Stress Functions	68
3.5 Westergaard Stress Function Approach	69
3.5.1 Mode I Crack-Tip Field	71
3.5.2 Mode II Crack-Tip Field	75
3.6 Mode III Solution	77
3.7 Williams' Eigenfunction Expansion for Mode I	80
3.8 Williams' Eigenfunction Expansion for Mode II and Mixed Mode	83
Exercise	84
References	84

4 Crack Opening Displacement, J Integral, and Resistance Curve — 86

4.1 Introduction	86
4.2 Crack Opening Displacement	87
4.3 Special Integrals	89
4.4 Rice's Path-Independent Integral J	91
4.5 J As Potential Energy Release Rate	92
4.6 Graphical Representation of J for Non-linear Elastic Case	94
4.7 Resistance Curve	95
4.8 Stability of Crack Growth	97
Exercise	100
References	101

5 Determination of Stress Intensity Factors — 102

5.1 Introduction	102
5.2 Analytical Methods	103

	5.2.1 Boundary Collocation Method	108
	5.2.2 Green's Function Approach	108
	5.2.3 Method of Superposition	109
	5.2.4 Weight Function Method	110
5.3	Numerical Technique: Finite Element Method	113
	5.3.1 Displacement and Stress-based Methods for Extraction of SIFs	118
	5.3.2 Energy-based Methods for Determination of SIFs	120
5.4	FEM-Based Calculation of G Associated with Kinking of Crack	138
5.5	Other Numerical Methods	139
5.6	Experimental Methods	140
	5.6.1 Strain Gauge Technique	140
	5.6.2 Photoelasticity	142
Exercise		143
References		146

6 Mixed Mode Brittle Fracture — 152

- 6.1 Introduction — 152
- 6.2 Theory based on Potential Energy Release Rates — 153
- 6.3 Maximum Tangential Stress Criterion — 154
- 6.4 Maximum Tangential Principal Stress Criterion — 157
- 6.5 Strain Energy Density Criterion — 159
- Exercise — 163
- References — 166

7 Fatigue Crack Growth — 168

- 7.1 Introduction — 168
- 7.2 Fatigue Crack Growth Rate under Constant Amplitude Loading — 170
- 7.3 Factors Affecting Fatigue Crack Propagation — 174
- 7.4 Crack Closure — 174
- 7.5 Life Estimation Using Paris Law — 176
- 7.6 Retardation of Crack Growth Due to Overloads — 178
- 7.7 Variable Amplitude Cyclic Loading — 182
  - 7.7.1 Rainflow Cycle Counting — 190
- 7.8 Closure — 192

Appendix 7.1 Fortran Program for Crack Growth Calculations — 192
Exercise — 196
References — 199

8 Elastic Plastic Fracture Mechanics — 202

8.1 Introduction — 202
8.2 Briefs on Plasticity — 202
    8.2.1 Incremental Theories of Plasticity — 207
8.3 Crack Opening Displacement Criterion — 208
8.4 Mode III Crack-Tip Field for Elastic-Perfectly-Plastic Materials — 209
8.5 Relationship between J and COD — 212
8.6 Fracture Assessment Diagram and R-6 Curve — 213
8.7 Mode I Crack-Tip Field — 216
    8.7.1 Rice–Rosengren Analysis — 216
    8.7.2 Hutchinson's Analysis — 221
8.8 Experimental Determination of J — 226
8.9 Alternative Methods for Measuring J — 227
8.10 Crack-Tip Constraints: T Stress and Q Factor — 230
8.11 Crack Propagation and Crack Growth Stability — 233
8.12 Engineering Estimates of J — 237
8.13 Closure — 245
Exercise — 254
References — 254

9 Experimental Measurement of Fracture Tougness Data — 257

9.1 Introduction — 257
9.2 Measurement of Plane Strain Fracture Toughness K_{IC} — 257
9.3 Measurement of J_{IC} — 263
9.4 Measurement of Critical COD δ_C — 267
9.5 Measurement of K-Resistance Curve — 269
    9.5.1 Linear Elastic Material — 269
    9.5.2 Elastic Plastic Material — 270
References — 272

Index — 273

List of Figures

Fig. 1.1	Brittle fractures of plates.	2
Fig. 1.2	Crack extensions attendant with elastic plastic deformation.	3
Fig. 1.3	Mixed mode crack extensions in plate and hollow shaft.	3
Fig. 1.4	Cracks resulting from fatigue loading at the root of the gear tooth, window corner of an aircraft and step of a shaft.	3
Fig. 1.5	Constant amplitude fatigue loading with overload cycle and random cyclic fatigue loading.	4
Fig. 2.1	Atomic-level modelling of cleavage fracture. (a) Schematic representation of atomic interactions. (b) Variation of inter-atomic forces with spacing. (c) New surfaces created after fracture.	7
Fig. 2.2	Stress concentration at the tip of crack-like defect in plate subjected to tension.	9
Fig. 2.3	(a) Griffith crack. (b) Centre crack under uniaxial tension.	10
Fig. 2.4	Plastically deformed material adjoining crack edges.	13
Fig. 2.5	Three modes of loading of crack. (a) Mode I, (b) Mode II and (c) Mode III.	15
Fig. 2.6	Variation of stress σ_y ahead of tip of original crack and crack opening displacements $2v$ behind the tip of extended crack.	17
Fig. 2.7	Cracks subjected to more than one opening modes of loading.	21
Fig. 2.8	More than one mode of loading on crack. (a) Crack in beam. (b) Angled crack in plate.	21
Fig. 2.9	Three modes of loading on circumferential crack in shaft.	22
Fig. 2.10	(a) Soft and (b) hard loading arrangements.	23
Fig. 2.11	Variation of load with displacement.	24
Fig. 2.12	Crack-tip plastic zone and equivalent crack size. (a) Stress field ahead of given crack a. (b) Tensile stress–strain property of material considered. (c) Stress field ahead of equivalent crack $(a + r_y)$.	25
Fig. 2.13	Strip-yield plastic zone.	27
Fig. 2.14	Mode I crack-tip plastic zone shape according to (a) Tresca criterion and (b) Mises criterion.	29
Fig. 2.15	Plastic zone shape according to Mises criterion. (a) Mode II and (b) Mode III.	30

Fig. 2.16	Plastic zone size variation along thickness. (a) Specimen. (b) Plastic zone shape around crack front. (c) Stress state at points J and K. (d) Stress state at point I.	31
Fig. 2.17	Shear planes.	32
Fig. 2.18	Variation of fracture toughness and percentage flat fracture with specimen thickness.	33
Fig. 2.19	C-frame.	34
Fig. 2.20	Spherical pressure vessel with crack.	35
Fig. 2.21	Plate with hole and crack.	37
Fig. 2.22	A proving ring with internal crack.	38
Fig. 2.23	Gear tooth with crack located at base location.	39
Fig. A2.1.1	Internal crack in finite plate. (a) Mode I and (b) Mode II.	48
Fig. A2.1.2	(a) Single and (b) double edge crack in finite plate.	49
Fig. A2.1.3	Beam segment with crack under bending moment.	49
Fig. A2.1.4	Crack in beam under three point bending load.	49
Fig. A2.1.5	Compact tension specimen.	50
Fig. A2.1.6	Thumbnail crack in plate.	50
Fig. A2.1.7	Internal crack in infinite plate under (a) pressure load and (b) point load.	50
Fig. A2.1.8	Crack at hole edge in infinite plate. (a) two cracks and (b) single crack.	51
Fig. A2.1.9	Penny-shaped crack in infinite three-dimensional body. (a) point load at A; (b) pressure loading on annular area $r = b$ to $r = a$.	51
Fig. A2.1.10	Penny-shaped crack in shaft under tension and bending loads.	52
Fig. A2.1.11	Mode III SIF for circumferential crack in shaft under torsion.	53
Fig. A2.1.12	Round bar with circumferential crack under tension.	53
Fig. A2.1.13	Crack edge loading.	54
Fig. A2.1.14	Disc-shaped compact tension specimen.	54
Fig. 3.1	Stresses at a point.	66
Fig. 3.2	Boundary forces over span AB.	67
Fig. 3.3	Mode I crack under remote tension.	70
Fig. 3.4	Mode I crack in infinite plate under remote equi-biaxial tension σ.	72
Fig. 3.5	Mode II crack in infinite plate under remote shear τ.	75
Fig. 3.6	Mode III crack in infinite plate under remote out-of-plane shear τ.	78

Fig. 3.7	Mixed mode crack.	84
Fig. 4.1	Crack-tip blunting and crack opening displacement δ for centre crack under remote Mode I loading.	88
Fig. 4.2	Contour for movement of W.	89
Fig. 4.3	Close contour for line integral in stress field.	90
Fig. 4.4	Contour for J integral.	91
Fig. 4.5	Rectangular plate with mixed mode loading.	93
Fig. 4.6	Crack extension under constant (a) displacement and (b) load.	94
Fig. 4.7	Resistance curve for brittle material.	95
Fig. 4.8	Resistance curve for materials with small-scale plastic deformation at crack-tip.	96
Fig. 4.9	Resistance curves under load and displacement controls.	98
Fig. 5.1	Plate with angled crack and crack-tip coordinates.	103
Fig. 5.2	Point loading on crack edges.	104
Fig. 5.3	Crack edge loading.	106
Fig. 5.4	Loading on edges of infinite plate.	107
Fig. 5.5	Uniform pressure loading on crack edges.	109
Fig. 5.6	(a) Crack-free plate. (b) and (c) Two cases of complementary crack loadings.	110
Fig. 5.7	Edge crack under opening load.	110
Fig. 5.8	Internal crack with two different loadings but similar symmetry. (a) Uniform pressure on crack edges. (b) Uniform tension on plate edges.	112
Fig. 5.9	Edge crack geometry.	112
Fig. 5.10	(a) Typical FE discretization and (b) u-displacement surface.	115
Fig. 5.11	Extrapolation method.	119
Fig. 5.12	Shifting of crack-tip to accommodate small crack extension.	121
Fig. 5.13	Crack-tip shift to accommodate small Δa and contour for J integral.	122
Fig. 5.14	Contour for domain integration.	123
Fig. 5.15	Symmetric and anti-symmetric displacements and stresses about crack plane.	126
Fig. 5.16	Typical arrangement of crack-tip elements for evaluation of domain integral and shape of typical element in mapping plane $\xi - \eta$.	127

Fig. 5.17	Typical crack-tip discretization.	131
Fig. 5.18	(a) Quadratic and (b) quarter point singular elements around crack-tip.	133
Fig. 5.19	Crack-tip element arrangements. (a) Quarter point singularity elements. (b) Quadratic elements.	134
Fig. 5.20	(a) Thermo-elastic problem and (b) crack-tip discretization.	137
Fig. 5.21	(a) Angled crack. (b) Discretization around tip and knee.	138
Fig. 5.22	Strain gauge arrangement for (a) Mode I and (b) mixed mode.	141
Fig. 5.23	Typical isochromatic fringes around crack-tip.	143
Fig. 6.1	Cracks loaded in mixed mode.	152
Fig. 6.2	(a) Loading. (b) Typical variation of G with θ.	153
Fig. 6.3	Stresses in polar coordinates.	154
Fig. 6.4	Mixed mode crack and failure locus. (a) Angled crack problem. (b) Failure locus.	157
Fig. 6.5	Unstable crack path.	157
Fig. 6.6	Directions θ_{c1} and θ_{c2} corresponding to maximum tangential stress and maximum tangential principal stress.	158
Fig. 6.7	Graphical determination of θ_{c1} and θ_{c2}.	159
Fig. 6.8	Circumferential variation of S.	160
Fig. 7.1	Constant amplitude fatigue loading with overload cycle and random fatigue loading.	168
Fig. 7.2	Different types of fatigue loading. (a) Time-varying loadings with well defined cycles. (b) Randomly time-varying load without well defined cycles.	169
Fig. 7.3	Cracks resulting from fatigue loading at the root of a gear tooth, window corner of an aircraft fuselage, and step of a shaft.	170
Fig. 7.4	Plot of crack growth rate with stress intensity factor range.	171
Fig. 7.5	Crack closure during cyclic loading. (a) Load-displacement variation. (b) Locations for displacement monitoring. (c) Opening SIF K_{op}	175
Fig. 7.6	(a) Crack-tip plastic zone. (b) Roughness-induced closure. (c) Debris or metal oxide induced closure.	175
Fig. 7.7	Internal crack in plate of constant thickness.	177
Fig. 7.8	(a) Loading. (b) Retardation of crack growth immediately following application of overload.	179

Fig. 7.9	(a) Overload and instantaneous plastic zones. (b) Instantaneous plastic zone touching overload plastic zone boundary. (c) New overload plastic zone spreading beyond first overload plastic zone boundary.	180
Fig. 7.10	Reduction of cyclic stress intensity factors due to residual stresses after overload cycle.	182
Fig. 7.11	Variable cyclic loading. (a) Actual loading. (b) Load sequencing for life calculation. (c) Crack growth due to two loading sequences.	183
Fig. 7.12	Rayleigh-type distribution of range of loading. (a) Frequency variation with stress range. (b) Load range increasing in a block. (c) Load range decreasing in a block. (d) Increasing–decreasing load range in a block.	184
Fig. 7.13	(a) Rainflow counting from load record. (b) Cycle counting considering kinematic hardening material properties.	191
Fig. 8.1	(a) Tensile test specimen. (b) Stress–strain plot under uniaxial tension/compression.	204
Fig. 8.2	(a) Two yield loci in $\sigma_1 - \sigma_2$ plane. (b) von Mises yield locus in $\tau - \sigma$ plane.	205
Fig. 8.3	(a) Isotropic hardening. (b) Kinematic hardening.	206
Fig. 8.4	Plasticity modulus H'.	208
Fig. 8.5	Crack opening displacement.	209
Fig. 8.6	(a) Mode III loading and coordinates. (b) Crack-tip elastic plastic field.	210
Fig. 8.7	Mode I crack with strip yield zones.	212
Fig. 8.8	Variation of σ_C with a.	214
Fig. 8.9	Fracture assessment diagram.	215
Fig. 8.10	Crack-tip polar coordinates.	217
Fig. 8.11	Radial variation of plastic zone size.	220
Fig. 8.12	Crack-tip opening displacement according to Tracy (1976).	224
Fig. 8.13	Near-tip fracture process zone and extent of J field.	226
Fig. 8.14	Experimental determination of J. (a) Monitoring P and u. Plots of (b) P versus u for constant crack sizes, (c) U versus a for constant displacements and (d) J versus u for constant crack sizes.	227
Fig. 8.15	Bending deformation of three point bend specimen.	228

Fig. 8.16	Areas involved in J calculation.	229
Fig. 8.17	Variation of fracture toughness with constraint parameter Q.	232
Fig. 8.18	Elastic unloading and fracture process zones embedded in J field.	233
Fig. 8.19	J-resistance curve.	236
Fig. 8.20	Geometry associated with Tables 8.2 and 8.3.	246
Fig. 8.21	Geometry associated with Tables 8.4 and 8.5.	247
Fig. 8.22	Geometry associated with Tables 8.6 and 8.7.	249
Fig. 8.23	Geometry associated with Tables 8.8 and 8.9.	251
Fig. 9.1	Different test specimen identifications depending on loading direction and crack plane orientation.	258
Fig. 9.2	Compact tension specimen.	258
Fig. 9.3	Three point bend specimen.	259
Fig. 9.4	Crack opening displacement or clip gauge.	260
Fig. 9.5	Typical load-displacement records.	260
Fig. 9.6	Illustration of crack front at different stages.	261
Fig. 9.7	Standard geometries for (a) compact tension and (b) three point bend specimens for J_{IC} testing.	263
Fig. 9.8	Loading and unloading curves associated with measurement of J.	266
Fig. 9.9	Experimental resistance curve.	266
Fig. 9.10	Components of total displacement.	268
Fig. 9.11	Displacements associated with crack opening displacement measurement.	268
Fig. 9.12	Three different forms of load versus CMOD records. (a) Load reaches level P_c then drops suddenly. (b) Load reaches slowly to highest level P_u then drops suddenly. (c) Load reaches maximum level P_{max} followed by slow drop and then sudden drop.	269
Fig. 9.13	(a) Centre-cracked tensile specimen. (b) Compact tension specimen. (c) Wedge loaded specimen.	270
Fig. 9.14	Typical load-displacement curve of elastic material.	271
Fig. 9.15	Typical load-displacement curve of elastic plastic material.	271

List of Tables

Table 2.1	Room temperature fracture toughness of structural materials.	41
Table 7.1	Threshold stress intensity factors.	171
Table 7.2	Typical Paris law constants.	173
Table 7.3	Paris law constants for some more materials.	174
Table 8.1	Dependence of I_n on n and state of stress.	223
Table 8.2	Compact tension specimen in plane strain.	246
Table 8.3	Compact tension specimen in plane stress.	247
Table 8.4	Centre crack tension specimen under remote tension and in plane strain.	248
Table 8.5	Centre crack tension specimen under remote tension and in plane stress.	249
Table 8.6	h_1, h_2 and h_3 for SECP in plane strain under three point bending.	250
Table 8.7	h_1, h_2 and h_3 for SECP in plane stress under three point bending.	251
Table 8.8	h_1, h_2, h_3 and h_5 for SECP in plane strain under remote uniform tension.	252
Table 8.9	h_1, h_2, h_3 and h_5 for SECP in plane stress under remote uniform tension.	253

List of Tables

Preface

Currently, there are a large number of books available on 'Fracture Mechanics', varying widely in coverage and levels of difficulty. After having taught the subject for more than twenty-five years to senior undergraduate and graduate students, I felt the need for a book that can help students supplement their understanding after classroom exposure, and also be of use to others especially self-learners, interested in learning the fundamentals of the subject.

The subject has become very important in design and safety analysis of critical components of machines and structures in aerospace, space explorations, automobiles, power plants, chemical plants, oil exploration, shipping, defense, civil applications, and so on. Because of its wide relevance, there is a need for large manpower with requisite knowledge of the subject and, hence, for books on the subject to suit a variety of maturity and capability levels.

This book has been written with the intent of expounding major fundamental concepts and their mathematical foundations from the point of view of mechanics; that can be covered in a course of about one semester. It is designed to be helpful to senior undergraduates, postgraduates, researchers, practising engineers, faculty members and self-learners. The material has been presented in a fashion that helps the reader grasp the ideas easily and see through all stages of their mathematical development. It is envisaged as a starting text and should provide readers the foundation to appreciate and grasp advanced books and research publications.

The book has nine chapters. In each chapter, basic ideas have been adequately explained. Their applications have been further illustrated by solved numerical examples. These will foster practical applications of the theoretical concepts. Wherever possible, unsolved problems are included at the end of a chapter.

I am thankful to the reviewers for their suggestions. I am also thankful to my colleagues Dr Dnyanesh Pawaskar, Dr Salil S Kulkarni, Dr Krishna N. Jonnalagadda and Dr Tanmay K. Bhandakkar, each of whom has gone through a part of the whole write-up and given useful inputs/comments.

I will feel satisfied if the readers find the book useful.

1

Introduction

1.1 Introduction

Fracture mechanics provides the basis for designing machine and structural components with materials containing defects such as crack, gives rational approach for assessing degree of safety or reliability of an in-service degraded machine component, and helps to calculate the life of a component with crack subjected to cyclically fluctuating load, corrosion, creep, or a combination of all these. A crack is a discontinuity, internal or external (Figs 1.1 and 1.2), in the material with zero tip radius. The development of the subject has been driven by the stringent safety requirements of the aerospace industry, nuclear power plants and other safety-critical applications. The advancement in the understanding of the subject coupled with developments in material science, experimental methods, and numerical techniques such as finite element, boundary element, and meshless methods, has facilitated optimum design and minimization of material usage for an application.

This book presents the gradual development in the fundamental understanding of the subject and in numerical methods that have facilitated its applications. Though the subject can be studied from the viewpoint of material science and mechanics, the focus here is on the latter.

1.2 Linear Elastic Fracture Mechanics

Development of the subject originated with the work of Griffith (1921), who propounded the condition of unstable extension of an existing crack in a brittle material within the framework of global energy balance or the First Law of Thermodynamics. The shortcomings of the approach were eliminated by Irwin (1948), who classified the three fundamental modes of crack extension and presented the condition of fracture in terms of a parameter associated with the stress–strain field in the close neighbourhood of the crack-tip. He also showed the link between the crack-tip field parameter, the stress intensity factor (SIF), and the energy release rate parameter introduced by Griffith. These parameters have proved useful in characterizing the fracture of brittle materials and have helped in practical design applications. Brittle materials fracture without showing any plastic deformation before the onset of crack extension or during crack propagation. This type of failure is distinguished by the fact that the fractured parts can be put together to get the original geometry almost reconstructed (Fig. 1.1).

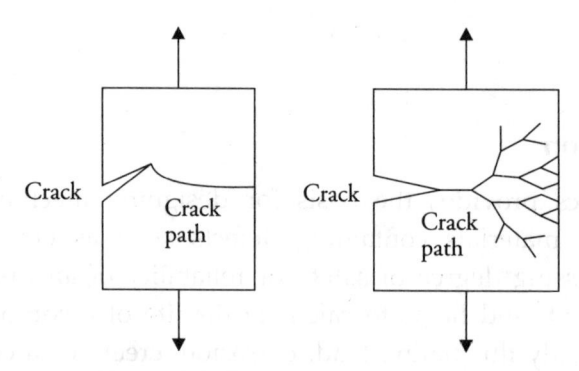

Figure 1.1 Brittle fractures of plates.

1.3 Elastic Plastic or Yielding Fracture Mechanics

Most materials that are used in engineering constructions and machine building are metallic and show plastic deformation around the crack-tip prior to crack extension and during crack extension (Fig. 1.2). They fracture showing features, combined to a varying degree, of normal stress driven brittle and shear stress driven ductile fractures. The plastic deformation preceding fracture was brought within the scope of energy balance approach and SIF through some minor amendments. A substantial increase in the scope was possible through an introduction of crack-tip field characterizing parameters such as the crack opening displacement (COD) by Wells (1961) and J integral by Rice (1968). While the energy release rate and the SIF concepts helped to lay the foundation of the linear elastic fracture mechanics (LEFM), the latter two provided much of the basis for yielding or elastic plastic fracture mechanics (YFM or EPFM).

1.4 Mixed Mode Fracture

Although Irwin classified the crack extensions into three fundamental modes, practical problems very often involve mixed mode of crack extensions (Fig. 1.3).

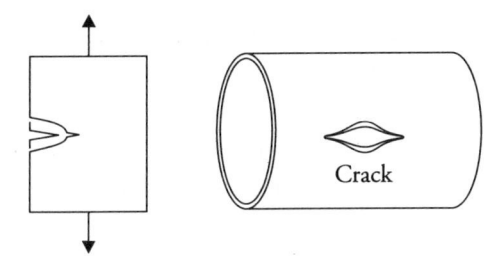

Figure 1.2 Crack extensions attendant with elastic plastic deformation.

Independent developments have subsequently taken place which explain such fractures within the scope of LEFM.

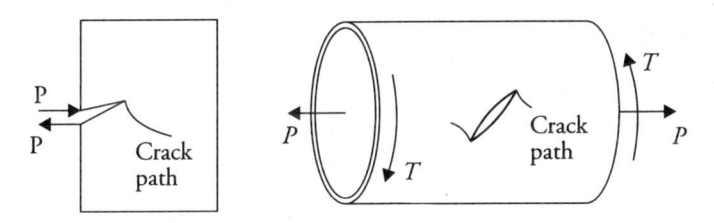

Figure 1.3 Mixed mode crack extensions in plate and hollow shaft.

1.5 Fatigue Crack Growth

A majority of machine and structural components (Fig. 1.4) are subjected to cyclically fluctuating load (Fig. 1.5).

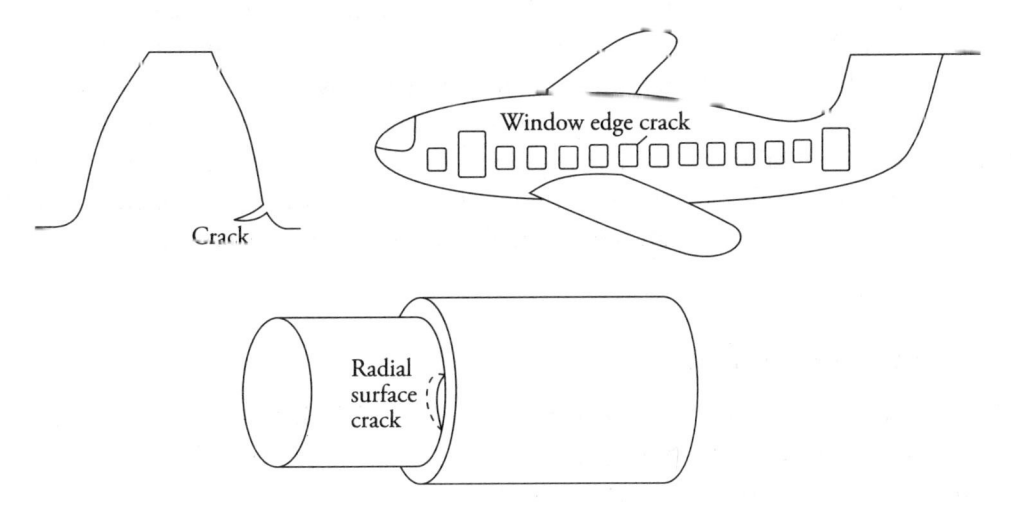

Figure 1.4 Cracks resulting from fatigue loading at the root of the gear tooth, window corner of an aircraft, and step of a shaft.

Based on an experimental study, Paris (1963) showed that the cyclic crack growth rate under fatigue loading can be characterized in terms of cyclic SIF range. This paved the way for an estimation of cyclic life of components. Paris law was subsequently enlarged in scope to accommodate variable and random amplitude cyclic loading, effects of occasional overloads, and so on.

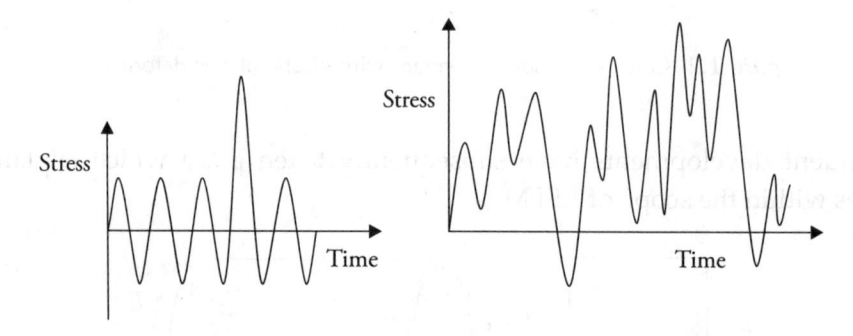

Figure 1.5 Constant amplitude fatigue loading with overload cycle and random cyclic fatigue loading.

Gradually the scope of LEFM has enlarged to take care of crack growth under stress corrosion, creep, and their combinations.

1.6 Computational Fracture Mechanics

The development of fracture mechanics has been driven by safety critical applications in defence, aerospace, power plants, oil industry, transport, etc. The applications of the principles of fracture mechanics in these areas have been facilitated by developments in numerical methods such as finite element method (FEM), boundary element method (BEM), and meshless method. Although many problems can be solved analytically and practical geometries can be handled by FEM or experimental methods, but problems of EPFM could rarely be handled by the analytical methods; solutions were obtained only through numerical methods such as FEM. Thus, computational methods have become part and parcel of fracture mechanics.

1.7 Scope of the Book

The book deals with the fundamentals of LEFM, EPFM, fatigue, computational issues, and mixed mode fracture. It also covers experimental methods and applications of fracture mechanics in design.

Developments in computational methods and experimental techniques have facilitated the study of fracture under impact loading conditions, high strain rates of deformation, large deformation processes such as metal forming, and so on. In addition methods have been developed for the study of fracture of composites. However these issues are beyond the scope of the book.

References

1.1 Griffith, A.A. 1921. 'The Phenomena of Flow and Rupture in Solids.' *Philosophical Transaction of the Royal Society, London, Series A*221: 163–97.

1.2 Irwin, G.R. 1948. 'Fracture Dynamics.' In *Fracturing of Metals*, 147–66. Cleveland: American Society for Metals.

1.3 Paris, P.C. and F. Erdogan. 1963. 'A Critical Analysis of Crack Propagation Laws.' *Journal of Basic Engineering, Transactions of ASME*85: 528–34.

1.4 Rice, J.R. 1968. 'A Path Independent Integrals and the Approximate Analysis of Strain Concentration by Notches and Cracks.' *Journal of Applied Mechanics, Transactions of ASME*35: 379–86.

1.5 Wells, A.A. 1961. 'Unstable crack propagation in metals: cleavage and fast fracture', Vol. 1, 210–30. *Proceedings of the Crack Propagation Symposium*, College of Aeronautics, Cranfield.

2

Linear Elastic Fracture Mechanics

2.1 Introduction

The foundation for the understanding of brittle fracture originating from a crack in a component was laid by Griffith (1921), who considered the phenomenon to occur within the framework of its global energy balance. He gives the condition for unstable crack extension in terms of a critical strain energy release rate (SERR) per unit crack extension. The next phase of development, which is due to Irwin (1957a and b), is based on the crack-tip local stress–strain field and its characterization in terms of stress intensity factor (SIF). The condition of fracture is given in terms of the SIF reaching a critical value, and the parameter is shown to be related to the critical energy release rate given by Griffith. Later, the scope of the SIF approach was amended to take care of small-scale plastic deformation ahead of the crack-tip. Most of the present applications of the principles of linear elastic fracture mechanics (LEFM) for design or safety analysis have been based on this SIF.

This chapter presents the gradual developments that have taken place to advance the understanding of fracture of brittle materials and other materials that give rise to small-scale plastic deformation before the onset of crack extension. Examples are presented to illustrate the applications of LEFM to design.

2.2 Calculation of Theoretical Strength

A fracture occurs at the atomic level when the bonds between atoms are broken across a fracture plane, giving rise to new surfaces. This can occur by breaking the bonds perpendicular to the fracture plane, a process called cleavage, or by shearing bonds along a fracture plane, a process called shear. The theoretical tensile strength of a material will therefore be associated with the cleavage phenomenon (Tetelman and McEvily 1967; Knott 1973).

Generally, atoms of a body at no load will be at a fixed distance apart, that is, the equilibrium spacing a_0 (Fig. 2.1). When the external forces are applied to break the atomic bonds, the required force/stress (σ) increases with distance (a or x) till the theoretical strength σ_c is reached. Further displacement of the atoms can occur under a decreasing applied stress. The variation can be represented approximately by a sinusoidal variation as follows.

$$\sigma = \sigma_c \sin \frac{2\pi x}{\lambda} \tag{2.1}$$

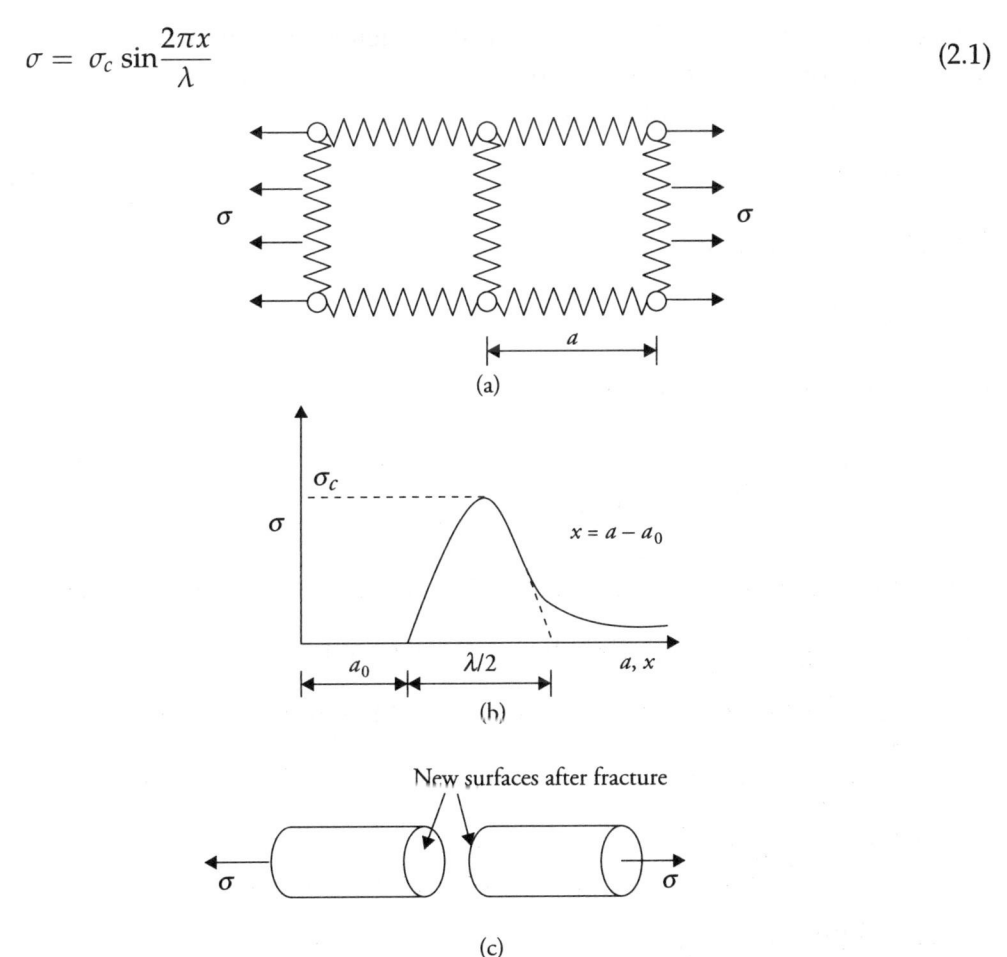

Figure 2.1 Atomic-level modelling of cleavage fracture. (a) Schematic representation of atomic interactions. (b) Variation of inter-atomic forces with spacing. (c) New surfaces created after fracture.

where λ is the wavelength of the load variation. The work done over half-cycle or span $\lambda/2$ is given by

$$W = \int_0^{\lambda/2} \sigma dx = \int_0^{\lambda/2} \sigma_c \sin \frac{2\pi x}{\lambda} dx = \sigma_c \frac{\lambda}{\pi} \tag{2.2}$$

If a cylinder of unit cross-sectional area breaks upon the application of load variation as shown in Fig. 2.1(b), two new surfaces of unit area each are created. To create each of these areas, specific surface energy γ_s, which is a material property, is needed. The energy stored at the two surfaces W_s is therefore given by

$$W_s = 2\gamma_s \tag{2.3}$$

This energy comes from the work done W in deforming the material before the separation. For conservation of energy $W = W_s$. For small displacements x, stress σ can be written using Hooke's law

$$\sigma = E\epsilon = E\,x/a_0 \tag{2.4}$$

Eqs. (2.1) and (2.4) give

$$\lambda = \sigma_c \frac{2\pi a_0}{E} \tag{2.5}$$

Combining Eqs. (2.2), (2.3), and (2.5)

$$\sigma_c^2 = \frac{E\gamma_s}{a_0} \quad \text{or} \quad \sigma_c = \sqrt{\frac{E\gamma_s}{a_0}} \tag{2.6}$$

For many materials $\gamma_s = \frac{Ea_0}{100}$ (Knott 1973), hence, $\sigma_c = \frac{E}{10}$. Actual strength, the ultimate strength σ_{ult}, measured during the tensile test lies in the range $\frac{E}{1000}$ to $\frac{E}{100}$. For example, for steel, $\sigma_{ult} = 400 - 800$ MPa and $E = 200$ GPa; therefore, $\sigma_c = \frac{E}{500}$ to $\frac{E}{250}$. Similarly, for aluminum, $\sigma_{ult} = 100 - 500$ MPa and $E = 70$ GPa; therefore, $\sigma_c = \frac{E}{700}$ to $\frac{E}{140}$.

The observed ultimate strength for most engineering metals is much below this theoretical prediction $\frac{E}{10}$ for σ_c. The observed strength is, in some cases, of the order of $\frac{E}{1000}$. The atomic model failed to explain the observed reduction in strength of a material.

2.3 Griffith's Explanation Based on Stress Concentration

Griffith(1921) attempted to explain the discrepancy between the theoretical and the actual strengths based on stress concentration. He suggested that a material, although apparently homogeneous, contains small defects such as cracks, which act as stress concentrators. The stress at the tip (A or B in Fig. 2.2) of such a crack reaches very high value σ_{max}, which may be comparable to the theoretical strength σ_c, although the applied stress σ is low. Using Inglis's (1913) solution, he was able to provide some justification for the reduction in the theoretical strength.

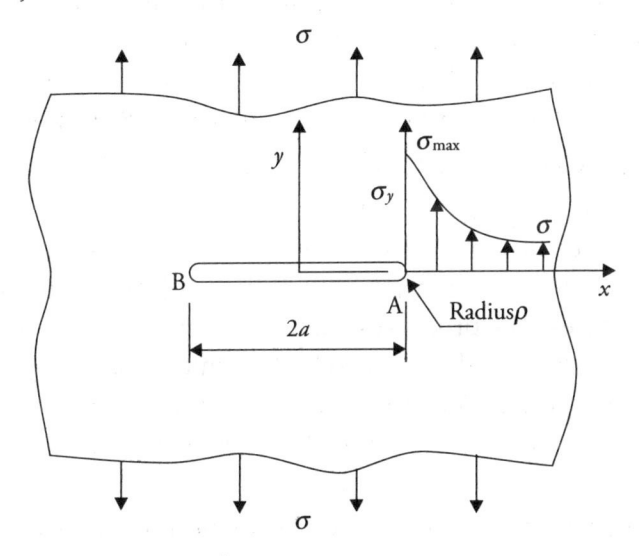

Figure 2.2 Stress concentration at the tip of crack-like defect in plate subjected to tension.

For a uniformly loaded tensile panel (Fig. 2.2), the maximum stress due to concentration at the tip A (Timoshenko and Goodier 1970) is given by

$$\sigma_{max} = \sigma \left(1 + 2\sqrt{\frac{a}{\rho}} \right) \tag{2.7}$$

where $2a$ is the crack size and ρ is the tip radius. Assuming $\rho \ll a$, and equating σ_{max} with σ_c of Eq. (2.6)

$$\sigma = \frac{1}{2}\sqrt{\frac{E\gamma_s}{a}\frac{\rho}{a_0}} \tag{2.8}$$

Considering $\sigma = E/1000$, $\gamma_s = 0.01 E a_0$, and $\rho \approx a_0$ from Eq. (2.8), $2a = 5000 a_0$. This means that the measured strength σ is two orders less than the theoretical strength $E/10$ in the presence of a crack size 5000 times the inter-atomic spacing a_0 at no load.

Such a method of establishing the influence of defects on the theoretical strength suffers from a drawback (Knott 1973). It is based on the correlation of two expressions, which are valid at two different dimension levels; the atomic bond strength model is valid at a level where the dimensions are comparable to an inter-atomic spacing and the Inglis's solution is valid at a macroscopic level.

2.4 Griffith's Theory of Brittle Fracture

Griffith(1921) provided a theory within the framework of thermodynamic energy balance for fracture of brittle materials and to calculate the fracture strength of a material with a crack. He assumed the material to be brittle and linearly elastic till fracture. He argued that when a crack in a stressed body extends, there are two forms of energy at play: the strain energy (W_e) and the surface energy (W_s). Further, at the onset of crack extension, the rate of change of potential energy (π) with crack length, where $\pi = W_e - W_s$, is zero, that is, $\frac{\partial \pi}{\partial a} = \frac{\partial(W_e - W_s)}{\partial a} = 0$. This relation can be interpreted in a different manner. The rate of release of strain energy is equal to the rate of increase of surface energy. Alternatively, the release in strain energy per unit crack extension gets converted into the surface energy of the newly created surfaces. The process of energy conversion is irreversible. In a sense, 'source' for energy supply is the strained body; 'sink' is the newly created surfaces. Hence, the unstable crack propagation takes place if the strain energy released rate associated with a crack extension is more than the corresponding energy absorbed in creating new surfaces.

Griffith followed a rigorous method of calculation for changes in strain energy due to an internal crack symmetrically located in an infinite plate of uniform thickness under equi-biaxial tensions (Fig. 2.3(a)) at its outer boundary using Inglis's solution. Here, an approximate method of calculation is presented to help in understanding, considering a centre crack in an infinite plate of uniform thickness subjected to uniaxial tension in the direction perpendicular to the crack

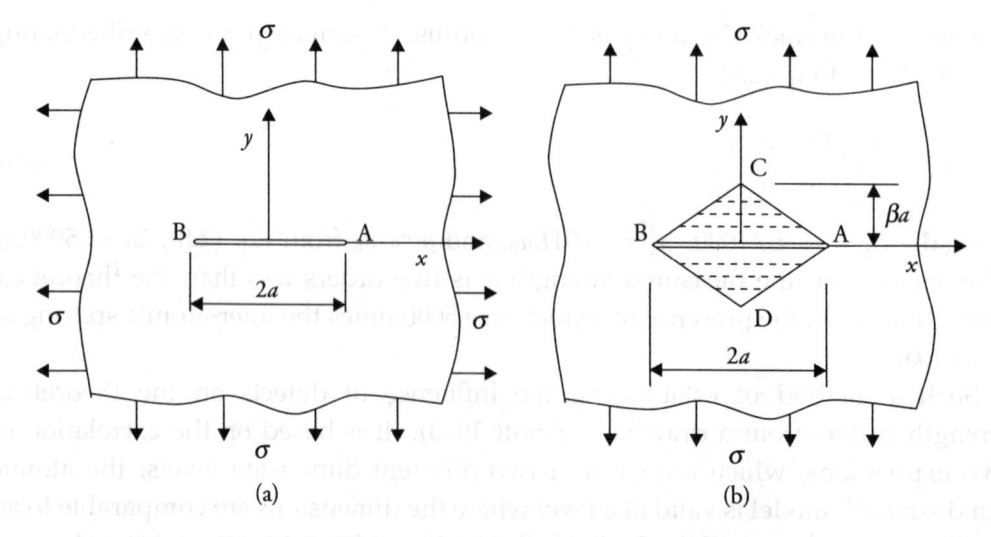

Figure 2.3 (a) Griffith crack. (b) Centre crack under uniaxial tension.

(Fig. 2.3(b)). The loading tries to open the crack edges. As the crack extends from 0 to $2a$, approximately two triangular regions (ABC and ABD) of materials immediately above and below the crack become stress-free. Hence, the reduction in strain energy W_e as the geometry changes from the plate with no crack to one with a crack of length $2a$ is equal to the strain energy of the material enclosed by ACBD approximately in the original crack-free configuration. Therefore,

$$W_e = a\, 2\beta\, a\, \frac{\sigma^2}{2E} = \frac{\beta \sigma^2 a^2}{E} \tag{2.9}$$

where E is the modulus of elasticity, thickness of the plate is taken as unity, and β is a constant. Using the rigorous solution of Inglis, β was obtained as π. The increase in surface energy W_s of the plate due to the crack extension is given by

$$W_s = 2(2a\gamma_s) \tag{2.10}$$

where γ_s is surface energy per unit area. The release of strain energy ΔW_e for an infinitesimal crack extension Δa is given by

$$\Delta W_e = \frac{2\pi\sigma^2 a}{E} \Delta a \tag{2.11}$$

The corresponding increase ΔW_s in the surface energy is

$$\Delta W_s = 4\gamma_s \Delta a \tag{2.12}$$

For the unstable extension of crack, therefore, the rate of release of strain energy per unit area of crack extension must be greater than or equal to the rate of increase of surface energy. That is,

$$\frac{\Delta W_e}{2\Delta a} \geq \frac{\Delta W_s}{2\Delta a} \quad \text{or,} \quad \frac{\sigma^2 \pi a}{E} \geq 2\gamma_s \tag{2.13}$$

Therefore, the fracture stress or critical stress in the presence of crack of size $2a$ is given by

$$\sigma_c \geq \sqrt{\frac{2E\gamma_s}{\pi a}} = \sqrt{\frac{EG_c}{\pi a}} \tag{2.14}$$

where $G_C = 2\gamma_s$ = energy required per unit area of crack extension. $\frac{\sigma^2 \pi a}{E}$ is the SERR and is represented by G. The fracture condition can also be written as

$$G \geq G_C. \tag{2.15}$$

Griffith observed experimentally with thin hollow glass tubes with longitudinal crack of different sizes under internal pressure that at the onset of crack extension, the product $\sigma\sqrt{\pi a}$, where σ is the hoop stress and acting normal to the crack direction, is constant for a wide combination of σ and a. This clearly indicates that $\sigma\sqrt{\pi a}$ is a material parameter and the onset of unstable extension is associated with a critical value of $\sigma\sqrt{\pi a}$ or the SERR G_C. Incidentally, G_C is considered to be a material property and is termed as resistance to brittle fracture. In the Griffith glass tube experiment, there was a provision for changing the axial stress, which acts parallel to the crack. He observed that the onset of crack extension was not affected by this stress.

Griffith theory cannot be easily applied to metals, which show ductile fracture and are attendant with varying degree of plastic deformation at the crack-tip prior to fracture. Furthermore, the theory poses difficulties in handling problems with complicated boundary conditions and crack at an angle with the direction of loading, where the direction of crack extension is an unknown. Specifically, for such cases, deriving an expression for strain energy release W_e presents analytical difficulties. Another shortcoming of the approach is that it is based on global energy balance and it does not attach any special importance to the stress field near the crack-tip.

2.4.1 Irwin–Orowan Modification

In order to extend the application of the Griffith theory to materials that give rise to plastic deformation around the crack-tip, Irwin (1948) and Orowan (1949, 1955) independently proposed a modification. When the crack extension takes place in the presence of plastic deformation at the crack-tip, a certain amount of energy is spent in the creation of new surfaces over and above the elastic surface energy γ_s. If the plastic zone size is assumed not to vary significantly with crack size, plastically deformed material will lie adjoining the crack edges (Fig. 2.4). The energy required for plastic deformation per unit crack extension $2\gamma_p$ can then be taken to be constant. Further, the size of the crack-tip plastic zone is so low compared with the crack dimensions or the specimen thickness that the global elastic strain energy release can still be calculated using the methods of elasticity. Consequently, the energy balance for the unstable extension is given by

$$\frac{\Delta W_e}{2\Delta a} \geq \frac{\Delta W_s}{2\Delta a} + 2\gamma_p \quad \text{or,} \quad \frac{\sigma^2 \pi a}{E} \geq 2\gamma_s + 2\gamma_p \tag{2.16}$$

Since $\gamma_p \gg \gamma_s$, the aforesaid relation can be written as

$$\sigma_c \geq \sqrt{\frac{2E\gamma_p}{\pi a}} \tag{2.17}$$

for small-scale yielding at the crack-tip. For purely elastic situation $\gamma_s \gg \gamma_p$, σ_c is given by Eq. (2.14).

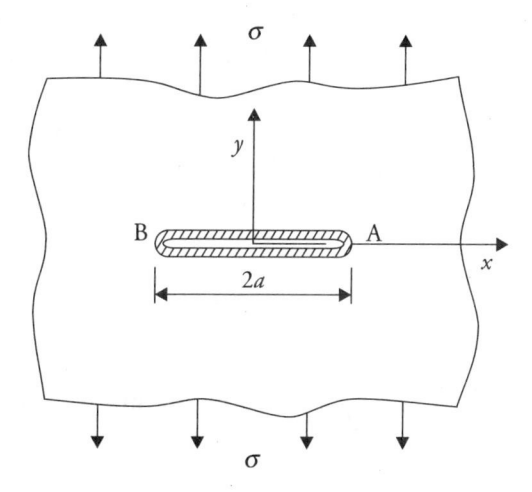

Figure 2.4 Plastically deformed material adjoining crack edges.

Thus, with the help of Griffith theory, it is possible to find out the load that a given structure will sustain when the defect size is known. Alternatively, it is possible to determine a tolerable defect dimension when the load is specified. This is the basis for damage-tolerant design. A problem with small-scale yielding can also be handled by this modified theory.

2.5 Stress Intensity Factor (SIF) Approach

In the energy balance approach, no special importance is attached to the stress–strain distribution around the crack-tip. Since the extension begins from the crack-tip, it is quite logical to expect that the crack-tip stress–strain environment influences this phenomenon. The importance of the crack-tip stress–strain field in relation to fracture was first shown by Irwin (1957a,b), and this study laid the foundation of fracture mechanics.

Irwin identified the three basic modes of crack extension (Fig. 2.5). The opening mode, Mode I, is characterized by the displacement of the crack edges/surfaces in the direction perpendicular to the plane of the crack. The sliding mode, Mode II, is characterized by the displacement of the crack edges/surfaces in the plane of the crack and perpendicular to the leading edge of the crack. In the shearing or tearing mode, Mode III, the crack surface displacement is in the plane of the crack and parallel to the leading edge of the crack. The superposition of the three modes describes the most general case. Mode I type of problems occur most often, and so this mode has been studied extensively.

14 *Fracture mechanics*

Irwin (1957a) analysed the stress-displacement field at the crack-tip and found that for each mode of crack extension, this field has the same nature irrespective of the loading and geometry. The actual magnitude of stresses and strains, however, depends on the loading and geometry. The effect of loading and geometry can be expressed through a single parameter termed as the SIF.

The stress-displacement fields in the vicinity of the crack-tip corresponding to the three basic modes are as follows (Paris and Sih 1965; Liebowitz 1968; Sih 1973b; Parton and Morozov 1978; Broek 1986; Kanninen and Popelar 1985; Anderson 2005):

Mode I (Fig. 2.5(a))

$$\sigma_x = \frac{K_I}{\sqrt{2\pi r}} \cos\frac{\theta}{2}\left(1 - \sin\frac{\theta}{2}\sin\frac{3\theta}{2}\right) + \text{terms containing } r^0, r^{1/2}, r^1, r^{3/2}, \ldots \quad (2.18a)$$

$$\sigma_y = \frac{K_I}{\sqrt{2\pi r}} \cos\frac{\theta}{2}\left(1 + \sin\frac{\theta}{2}\sin\frac{3\theta}{2}\right) + \text{terms containing } r^0, r^{1/2}, r^1, r^{3/2}, \ldots \quad (2.18b)$$

$$\tau_{xy} = \frac{K_I}{\sqrt{2\pi r}}\left(\sin\frac{\theta}{2}\cos\frac{\theta}{2}\cos\frac{3\theta}{2}\right) + \text{terms containing } r^0, r^{1/2}, r^1, r^{3/2}, \ldots \quad (2.18c)$$

$$\sigma_z = 0 \text{ for plane stress condition}$$

$$= \nu(\sigma_x + \sigma_y) \text{ for plane strain condition} \quad (2.19)$$

$$\tau_{xz} = \tau_{yz} = 0$$

$$u = \frac{K_I}{8\mu}\sqrt{\frac{2r}{\pi}}\left[(2\kappa - 1)\cos\frac{\theta}{2} - \cos\frac{3\theta}{2}\right] + \text{terms containing } r, r^{3/2}, \ldots$$

$$v = \frac{K_I}{8\mu}\sqrt{\frac{2r}{\pi}}\left[(2\kappa + 1)\sin\frac{\theta}{2} - \sin\frac{3\theta}{2}\right] + \text{terms containing } r, r^{3/2}, \ldots$$

$$w = 0 \text{ for plane strain condition}$$

$$K_I = \sigma\sqrt{\pi a} \quad (2.20)$$

$$\kappa = 3 - 4\nu \text{ for plane strain condition} \quad (2.21a)$$

$$= \frac{3 - \nu}{1 + \nu} \text{ for plane stress condition} \quad (2.21b)$$

Mode II (Fig. 2.5(b))

$$\sigma_x = \frac{-K_{II}}{\sqrt{2\pi r}} \sin\frac{\theta}{2} \left(2 + \cos\frac{\theta}{2} \cos\frac{3\theta}{2}\right) + \text{terms containing } r^0, r^{1/2}, r^1, r^{3/2}, \ldots \quad (2.22a)$$

$$\sigma_y = \frac{K_{II}}{\sqrt{2\pi r}} \sin\frac{\theta}{2} \cos\frac{\theta}{2} \cos\frac{3\theta}{2} + \text{terms containing } r^0, r^{1/2}, r^1, r^{3/2}, \ldots \quad (2.22b)$$

$$\tau_{xy} = \frac{K_{II}}{\sqrt{2\pi r}} \cos\frac{\theta}{2} \left(1 - \sin\frac{\theta}{2} \sin\frac{3\theta}{2}\right) + \text{terms containing } r^0, r^{1/2}, r^1, r^{3/2}, \ldots \quad (2.22c)$$

$\sigma_z = 0$ for plane stress condition

$\quad = \nu(\sigma_x + \sigma_y)$ for plane strain condition $\quad\quad (2.23)$

$\tau_{xz} = \tau_{yz} = 0$

$$u = \frac{K_{II}}{8\mu}\sqrt{\frac{2r}{\pi}} \left[(2\kappa + 3)\sin\frac{\theta}{2} + \sin\frac{3\theta}{2}\right] + \text{terms containing } r, r^{3/2}, \ldots \quad (2.24a)$$

$$v = -\frac{K_{II}}{8\mu}\sqrt{\frac{2r}{\pi}} \left[(2\kappa - 3)\cos\frac{\theta}{2} + \cos\frac{3\theta}{2}\right] + \text{terms containing } r, r^{3/2}, \ldots \quad (2.24b)$$

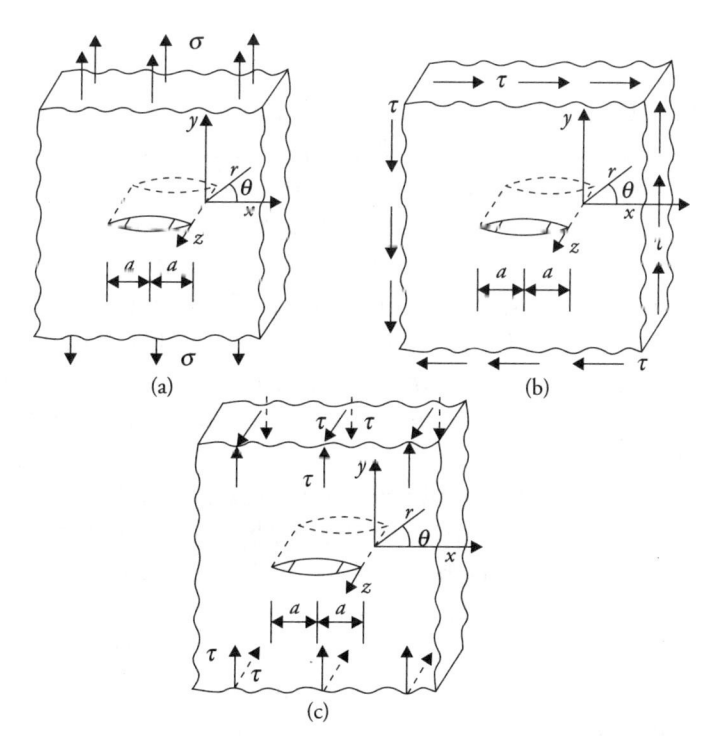

Figure 2.5 Three modes of loading of crack. (a) Mode I, (b) Mode II and (c) Mode III.

$$w = 0 \text{ for plane strain condition} \tag{2.25}$$

$$K_{II} = \tau\sqrt{\pi a} \tag{2.26}$$

$$\kappa = 3 - 4\nu \text{ for plane strain condition} \tag{2.27a}$$

$$= \frac{3-\nu}{1+\nu} \text{ for plane stress condition} \tag{2.27b}$$

Mode III (Fig. 2.5(c))

$$\tau_{xz} = -\frac{K_{III}}{\sqrt{2\pi r}} \sin\frac{\theta}{2} + \text{terms containing } r^0, r^{1/2}, r^1, r^{3/2}, \ldots \tag{2.28a}$$

$$\tau_{yz} = \frac{K_{III}}{\sqrt{2\pi r}} \cos\frac{\theta}{2} + \text{terms containing } r^0, r^{1/2}, r^1, r^{3/2}, \ldots \tag{2.28b}$$

$$\sigma_x = \sigma_y = \sigma_z = \tau_{xy} = 0$$

$$u = v = 0$$

$$w = \frac{K_{III}}{\mu}\sqrt{\frac{2r}{\pi}} \sin\frac{\theta}{2} + \text{terms containing } r, r^{3/2}, \ldots \tag{2.29}$$

$$K_{III} = \tau\sqrt{\pi a} \tag{2.30}$$

The polar coordinates (r, θ) are defined with the crack-tip as origin. The relations given above involve three quantities K_I, K_{II}, and K_{III}, which are the SIFs; the suffixes indicate the mode of crack extension or loading. These factors determine the intensity or magnitude of the local stresses and play an extremely important role in the mechanics of brittle fracture. The expressions for stresses contain higher order terms involving r^0, $r^{\frac{1}{2}}$, and so on. These terms can be considered negligible as $r \to 0$, that is, in the vicinity of the crack-tip. Therefore, the above expressions, with higher order terms in r neglected, can be regarded as good approximations, where r is small compared with other dimensions of the body and are exact as $r \to 0$.

It is, therefore, clear that the distribution of the elastic stresses and displacements in the close vicinity of the crack-tip is not affected by the geometry and loading data so long as the mode of loading is the same. Both stresses and strains have a square-root singularity at the crack-tip. The amplitude of the stress singularity is indicated by the SIFs, K_I, K_{II}, and K_{III}, respectively, for the three modes. The SIF depends on the applied load, the crack size, and the configuration of the component. The SIF affects the magnitudes of stresses and displacements but not their distribution.

2.5.1 Relationship between G and K

The release of strain energy associated with extension of a crack can be calculated from the crack-tip stress-displacement field. Thereby, it is possible to relate G and K.

Consider a crack of length $2a$ in an infinite plate (Fig. 2.6) and its extension to length $2(a + \Delta a)$ at a particular level of loading. As soon as the extension takes place, the crack edges open up. Just before the extension occurs, an element in front of the crack-tip is subjected to a normal stress σ_y, which depends on its distance ξ from the crack-tip and the SIF K_I. Immediately after the extension, crack edges open and their displacements can be obtained from the elastic

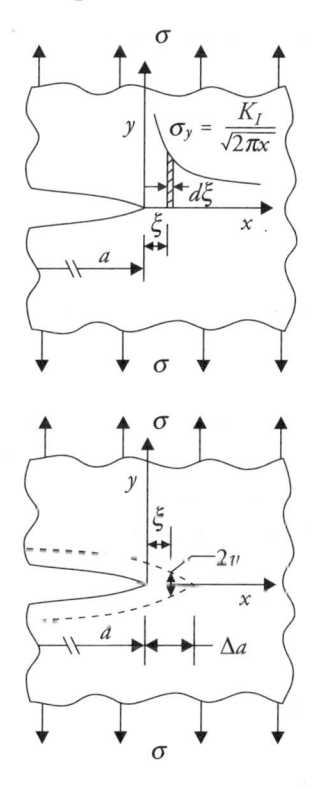

Figure 2.6 Variation of stress σ_y ahead of tip of original crack and crack opening displacements $2v$ behind the tip of extended crack.

crack-tip displacement field for the extended crack, $2(a + \Delta a)$. Therefore, at distance ξ from the crack-tip A,

$$\sigma_y = \frac{K_I}{\sqrt{2\pi\xi}} \tag{2.31}$$

Further, the crack opening ($2v$) at the same point is given by

$$2v = \frac{\sigma\sqrt{\pi(a+\Delta a)}}{4\mu} 2(\kappa+1)\left[\frac{2}{\pi}(\Delta a - \xi)\right]^{1/2} \qquad (2.32)$$

under plane stress or strain condition and with origin at the tip of extended crack $2(a+\Delta a)$. Therefore, the energy that is released because of the extension of the crack from length $2a$ to $2(a+\Delta a)$ is given by

$$W_e = \int_0^{\Delta a} \frac{1}{2}(2v)\, dF \qquad (2.33)$$

where $dF = \sigma_y B\, d\xi$ and B is specimen thickness. Irwin gave a reverse argument and suggested that the work done W_e is equal to the energy required to close the extended crack $2(a+\Delta a)$ back to its original length $2a$. If G_I is the SERR then

$$G_I B \Delta a = \frac{1}{2}\int_0^{\Delta a} \sigma_y (2v) B\, d\xi$$

$$G_I = \frac{1}{2\Delta a}\int_0^{\Delta a} \frac{K_I}{\sqrt{2\pi\xi}} \frac{\sigma\sqrt{\pi(a+\Delta a)}}{4\mu} 2(\kappa+1)\left[\frac{2}{\pi}(\Delta a - \xi)\right]^{1/2} d\xi$$

$$= \frac{K_I\, \sigma\sqrt{\pi(a+\Delta a)}}{\Delta a\, 4\pi\mu}(\kappa+1)\int_0^{\Delta a}\sqrt{\frac{\Delta a - \xi}{\xi}}\, d\xi \qquad (2.34)$$

Factor $\frac{1}{2}$ comes because the crack extension is viewed as a process taking place quasi-statically. Making a substitution $\xi = \Delta a \sin^2\theta$, the integral can be evaluated. Further, assuming $\Delta a \to 0$ and noting $k = 3 - 4\nu$ under plane strain and $E = 2\mu(1+\nu)$, the following relation is obtained.

$$G_I = \frac{K_I^2}{8\mu}(k+1) = (1-\nu^2)\frac{K_I^2}{E} \qquad (2.35)$$

Under plane stress condition, $k = \dfrac{3-\nu}{1+\nu}$ and the following simple form is obtained.

$$G_I = \frac{K_I^2}{E} \qquad (2.36)$$

Similarly, for Mode II, using the shear stress and sliding displacements

$$\tau_{xy} = \frac{K_{II}}{\sqrt{2\pi\xi}} \text{ and} \qquad (2.37a)$$

$$2u = \frac{\tau\sqrt{\pi(a+\Delta a)}}{4\mu} 2(k+1)\sqrt{\frac{2(\Delta a - \xi)}{\pi}} \qquad (2.37b)$$

and carrying out an integration of the type given by Eq. (2.33), the following results are obtained.

$$G_{II} = (1 - \nu^2)\frac{K_{II}^2}{E} \text{ for plane strain condition} \tag{2.38a}$$

$$= \frac{K_{II}^2}{E} \text{ for plane stress condition} \tag{2.38b}$$

In the similar manner, for Mode III, using stress τ_{yz} and edge sliding displacement w, the following results are obtained.

$$G_{III} = \frac{K_{III}^2}{2\mu} = (1+\nu)\frac{K_{III}^2}{E} \tag{2.39}$$

Thus, the rate of release of strain energy for a particular mode of crack extension is determined by the crack-tip SIF. So, if a crack is subjected to all the three modes of displacements simultaneously, the strain energy release rate can be obtained by superposition. That is,

$$G = G_I + G_{II} + G_{III} = \frac{K_I^2}{E} + \frac{K_{II}^2}{E} + (1+\nu)\frac{K_{III}^2}{E} \tag{2.40}$$

assuming a plane state of stress for the first two modes and an in-plane extension of crack.

The interrelation between G and K helps to propose an equivalent fracture criterion. If the onset of crack extension in a particular mode is associated with a critical SERR, it is also characterized by a critical SIF. That is, the unstable crack extension occurs when

$$G_I = G_{IC}, \quad \text{or} \quad K_I = K_{IC} \tag{2.41a}$$

$$G_{II} = G_{IIC}, \quad \text{or} \quad K_{II} = K_{IIC} \tag{2.41b}$$

$$G_{III} = G_{IIIC}, \quad \text{or} \quad K_{III} = K_{IIIC} \tag{2.41c}$$

for the three modes, respectively. Irwin termed G as the crack extension force. Consequently, the fracture is characterized by a critical crack extension force in a particular mode. Thus, for a predominantly elastic situation, the concept of a critical local crack-tip environment is entirely equivalent to the concept of a critical force driving the crack to extend, or to a critical SERR just sufficient to supply the required surface energy for the fracture process.

The importance of the SIF lies in the fact that it gives a clear picture of the stress–strain environment at the crack-tip; it has a critical value at the point of onset of unstable extension. The crack-tip stress field for a complex problem can be obtained easily through superposition. The critical value of the SIF at which crack extension begins is known as fracture toughness. The critical SIF is a material parameter, and it can be determined experimentally. Once this critical value is known, it is possible to establish what flaws are tolerable in a machine component or a structural element under a given loading. Alternatively, if the crack size in the component is known, it is possible to determine the maximum level of load the component can sustain. In either case, it is necessary to know the SIF in terms of crack size and loading corresponding to the geometry of the component.

The SIFs can be determined using analytical, numerical, and experimental techniques (Paris and Sih 1965; Liebowitz 1968; Sih 1973b). The results corresponding to a numerous geometric and loading conditions are available in the form of handbooks (Sih 1973a; Rooke and Cartwright 1975; Murakami 1987; Tada, Paris and Irwin 2000) and other sources (Hellan 1985; Broek 1986; Meguid 1989; Gdoutos 1993; Minnay 1998; Kumar 2009). In general, the SIF may be written in the form

$$K = \sigma\sqrt{\pi a}\, Y(a/w) \qquad (2.42)$$

where σ is the applied stress, a is the crack length, w is characterizing dimension of the component, may be width, and $Y(a/w)$ is a case-specific calibration function or correction factor. SIFs for a number of cases are given in Appendix 2.1.

Since the stress field equations are the same for a particular mode of loading on a crack, the SIF for a combination of loads can be obtained simply by superposition. In Fig. 2.7, in each case, the SIF is obtained by combining the SIFs due to the two constituent loadings. There are two direct loads (P and σ) in the first case; and there is one direct load (P) and a bending moment (M) in the second case.

When the loading on a component gives rise to more than one type of crack face displacements, the crack-tip stress field can be obtained through superposition. Fig. 2.8 shows two examples of mixed mode loading. In the first case (Fig. 2.8(a)), the bending moment M acting on the section AB gives rise to Mode I loading, and shear force V acting on the section gives rise to Mode II loading. Consequently, the crack-tip field is given by

$$\sigma_x = \sigma_{xI}(\text{due to } M) + \sigma_{xII}(\text{due to } V)$$

$$\sigma_y = \sigma_{yI}(\text{due to } M) + \sigma_{yII}(\text{due to } V)$$

$$\tau_{xy} = \tau_{xyI}(\text{due to } M) + \tau_{xyII}(\text{due to } V) \qquad (2.43)$$

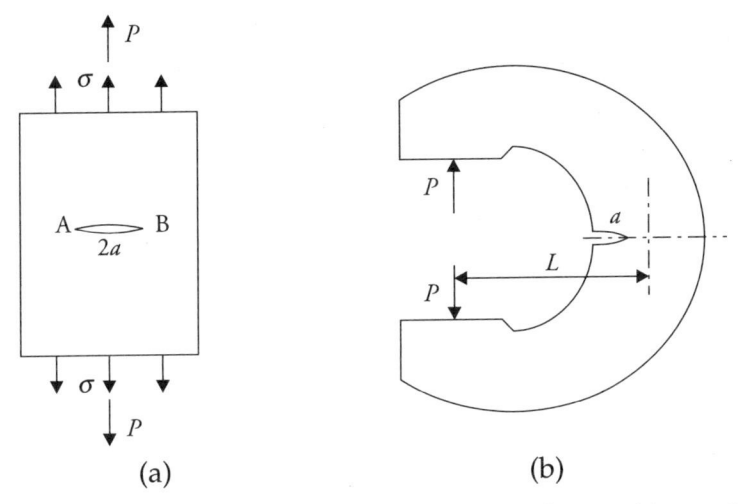

$$K_I = K_I \text{ (due to P)} + K_I \text{(due to } \sigma\text{)} \quad K_I = K_I \text{ (due to P)} + K_I \text{ (due to M)}, M = PL$$

Figure 2.7 Cracks subjected to more than one opening modes of loading.

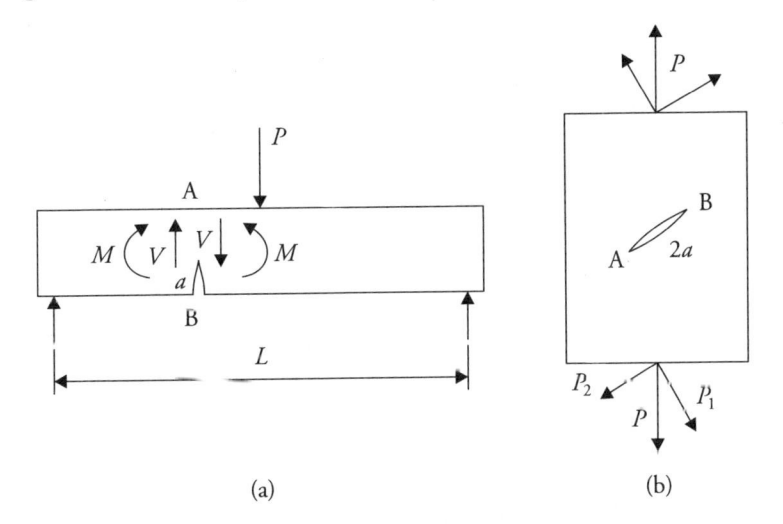

Figure 2.8 More than one mode of loading on crack. (a) Crack in beam. (b) Angled crack in plate.

In the second case (Fig. 2.8(b)), the external loading gives rise to two types of crack edge displacements: P_1 gives rise to Mode I loading and P_2 gives rise to Mode II loading. Therefore, the crack-tip stress field is given by

$$\sigma_x = \sigma_{xI}\text{(due to } P_1\text{)} + \sigma_{xII}\text{(due to } P_2\text{)}$$

$$\sigma_y = \sigma_{yI}\text{(due to } P_1\text{)} + \sigma_{yII}\text{(due to } P_2\text{)}$$

$$\tau_{xy} = \tau_{xyI}\text{(due to } P_1\text{)} + \tau_{xyII}\text{(due to } P_2\text{)} \tag{2.44}$$

Fig. 2.9 shows a circumferential crack in a shaft involving all the three modes of loading. Consider the vertical diametric section and the crack-tip located at B. For an in-plane extension, crack will propagate along x direction. The crack-tip will be subjected to opening mode due to bending moment, $M = (P_2 + P_3)a/2$, and the direct load P_1 acting in the y direction. Shear force, $V = (P_2 - P_3)a/L$, will give rise to Mode II displacements. Torque $T = P_4$ gives rise to Mode III displacements. Therefore, the crack-tip B stress field is given by

$$\sigma_x = \sigma_{xI}(\text{due to } M) + \sigma_{xII}(\text{due to } V)$$

$$\sigma_y = \sigma_{yI}(\text{due to } M) + \sigma_{yII}(\text{due to } V)$$

$$\tau_{xy} = \tau_{xyI}(\text{due to } M) + \tau_{xyII}(\text{due to } V),$$

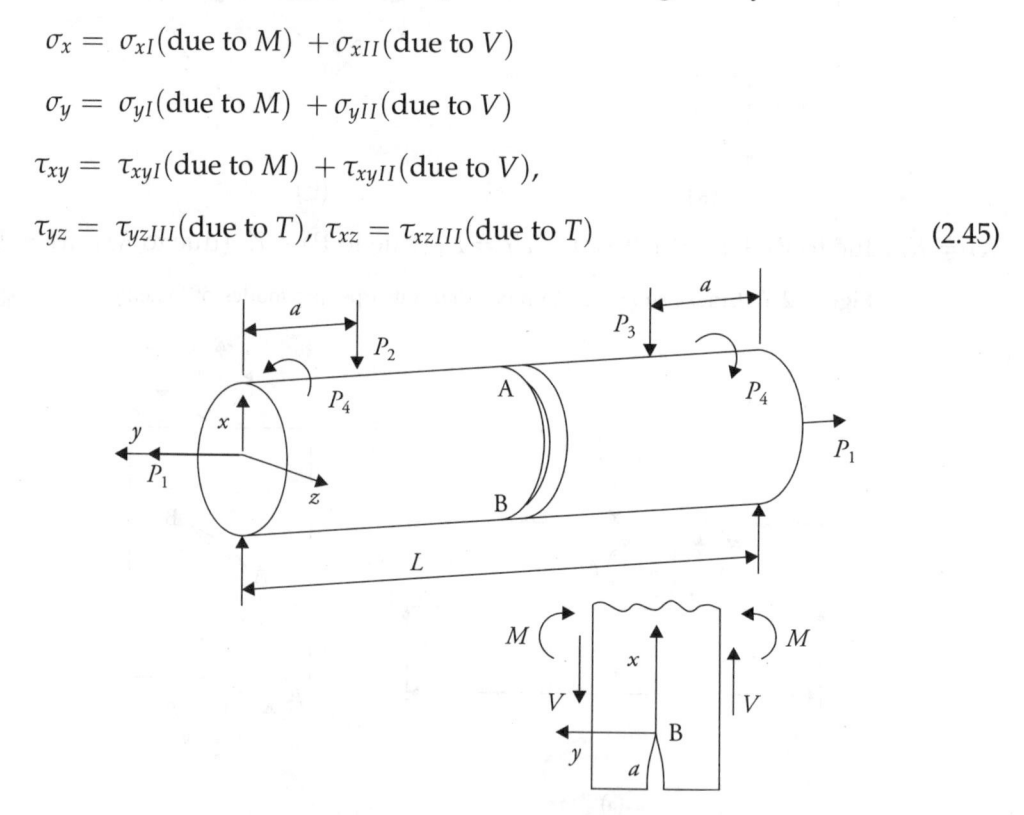

$$\tau_{yz} = \tau_{yzIII}(\text{due to } T), \quad \tau_{xz} = \tau_{xzIII}(\text{due to } T) \qquad (2.45)$$

Figure 2.9 Three modes of loading on circumferential crack in shaft.

2.6 Concepts of Strain Energy and Potential Energy Release Rates

A practical interpretation of the SERR of Griffith was provided later by Irwin and Kies (1952). They too provided a basis for experimental determination of the critical SERR. This critical value is treated as a material property.

2.6.1 Crack Extension Under Load Control (Soft Loading)

If a component with a crack (Fig. 2.10(a)) made of linear elastic material is loaded, the variation of load P with load point displacement v can be obtained (Fig. 2.11). At any level of loading, the strain energy stored U in the component is given by

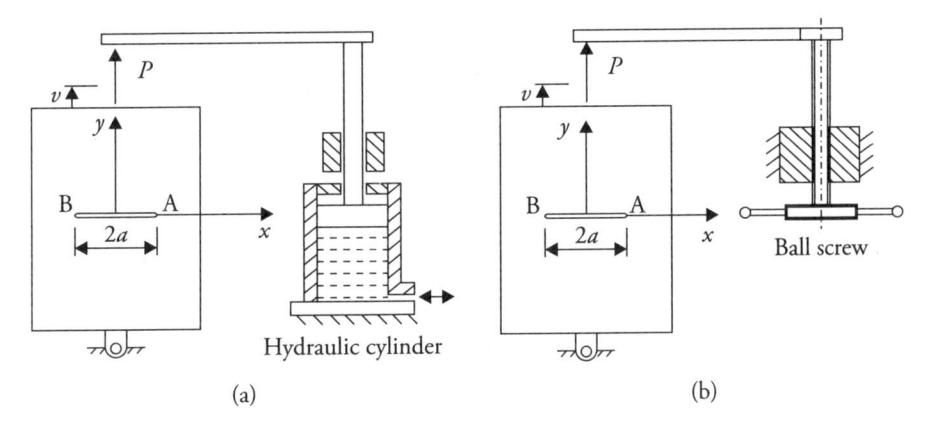

Figure 2.10 (a) Soft and (b) hard loading arrangements.

$$U = \frac{1}{2}Pv = \frac{1}{2}cP^2 \tag{2.46}$$

where c is compliance, the reciprocal of stiffness, and $v = cP$. If the crack extends from length $2a$ to $2(a + \Delta a)$ under constant load P_c, the change in strain energy $\Delta U = \frac{1}{2}P_c \Delta v$, and the corresponding change in work done $\Delta W = P_c \Delta v$. Therefore, change in potential energy $\Delta \pi = \Delta U - \Delta W = -\frac{1}{2}P_c \Delta v$. Here, $\Delta v = v_2 - v_1 = P_c \frac{\partial c}{\partial a} \Delta a$. Note that the strain energy of the system increases but the potential energy of the system decreases to the same extent. The change in strain energy of the system is equal to the hatched area OAB (Fig. 2.11). The potential energy release rate (PERR) is given by the following equation, as $\Delta a \to 0$

$$G_c = -\frac{\Delta \pi}{2B \Delta a} = \frac{1}{4B} P_c^2 \frac{\partial c}{\partial a} \tag{2.47}$$

where B is the thickness of specimen.

If the load P_c at which the crack extension occurs is known, the critical SERR (G_C) can be calculated by using Eq. (2.47) after a substitution of the proper value for $\frac{\partial c}{\partial a}$. Irwin and Kies suggested that $\frac{\partial c}{\partial a}$ as a function of crack length could be determined by measuring the compliance of specimens with various initial crack lengths.

2.6.2 Crack Extension Under Displacement Control (Hard Loading)

This situation is obtainable when loading the specimen in a screw-driven machine (Fig. 2.10(b)). As the crack extends from length $2a$ to $2(a+\Delta a)$, the change in strain

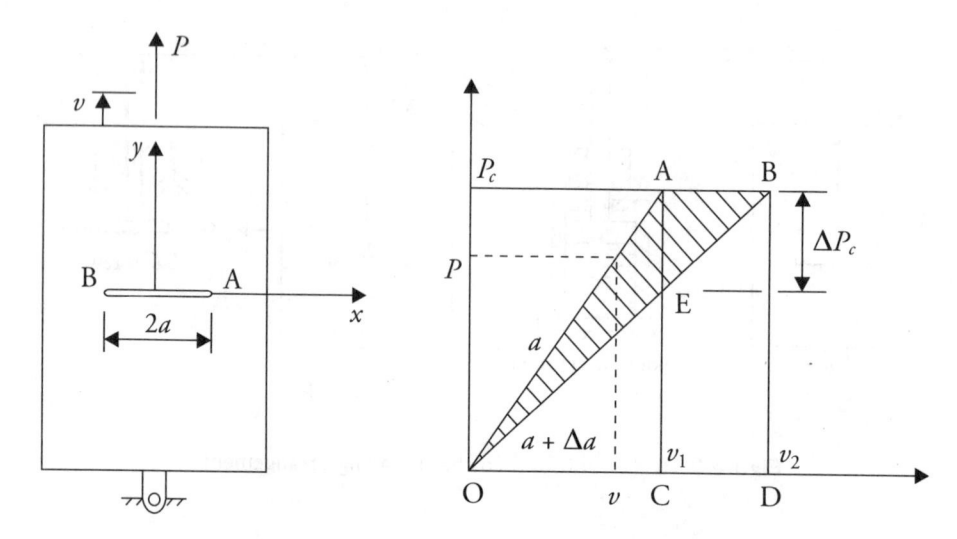

Figure 2.11 Variation of load with displacement.

energy $\Delta U = \frac{1}{2} v_1 \Delta P_c$ (Fig. 2.11) and is equal to the area OAE. In this case, the strain energy of the system reduces. The change in work done $\Delta W = 0$, since there is no displacement associated with the crack extension. The change in potential energy of the system $\Delta \pi = \Delta U - \Delta W = \frac{1}{2} v \Delta P_c$. In this case, too, the change in strain energy is equal in magnitude with the change in potential energy. Since ΔP_c is negative, the strain energy reduces. Further, since $v = cP$ and $\Delta v = 0$, $\Delta P_c = -\frac{P_c}{C} \Delta c$. The PERR is given by the following equation, as $\Delta a \to 0$

$$G_c = \frac{\Delta \pi}{2B \Delta a} = \frac{1}{4B} P_c^2 \frac{\Delta c}{\Delta a} = \frac{1}{4B} P_c^2 \frac{\partial c}{\partial a} \qquad (2.48)$$

The expression of energy release rate is the same in both the systems of loading. If the problem involves an edge crack of size a, the denominator will consist of 2 rather than 4.

It is more appropriate to consider the crack extension to be associated with a critical level of PERR rather than SERR. Griffith termed this energy release rate as the SERR; the terminology is still used by many authors.

2.7 Irwin Plastic Zone Size Correction

Although the elastic stress field at the crack-tip stipulates the existence of a very high level of stress, this does not happen in practice. Materials (especially metals) deform plastically above a certain level of stress, and this results in a stress relaxation at the crack-tip. Assuming the material to be elastic-perfectly-plastic

(Fig. 2.12) and its yield point as σ_y, and plane stress condition, the spread r_y of the plastic zone along the crack line (Irwin 1958, 1960) is given by

$$\sigma_y = \frac{K_I}{\sqrt{2\pi r_y}} = \sigma_Y \text{ or } r_y = \frac{K_I^2}{2\pi\sigma_Y^2} \qquad (2.49)$$

The corresponding stress distribution is shown by the curve ABC (Fig. 2.12(a)). Because of the stress relaxation, the load represented by area A_1, bounded by AB, y axis, and singularity variation in stress GBC, becomes extra. Since this load has to be carried through, there must be some re-adjustment of σ_y stresses for $r \geq r_y$, that is, ahead of point B. Further, since stress σ_y cannot exceed the level σ_Y over BF, the span BF is obtained by equating the area A_1 with A_2.

$$\sigma_Y \text{BF} = \int_0^{r_y} \frac{K_I}{\sqrt{2\pi r}} dr - \sigma_Y \text{AB} \qquad (2.50)$$

where specimen thickness is taken as unity. Since AB = $r_y = \frac{K_I^2}{2\pi\sigma_y^2}$, BF= r_y. Thus, the spread of plastic zone is $2r_y$ ahead of the crack-tip.

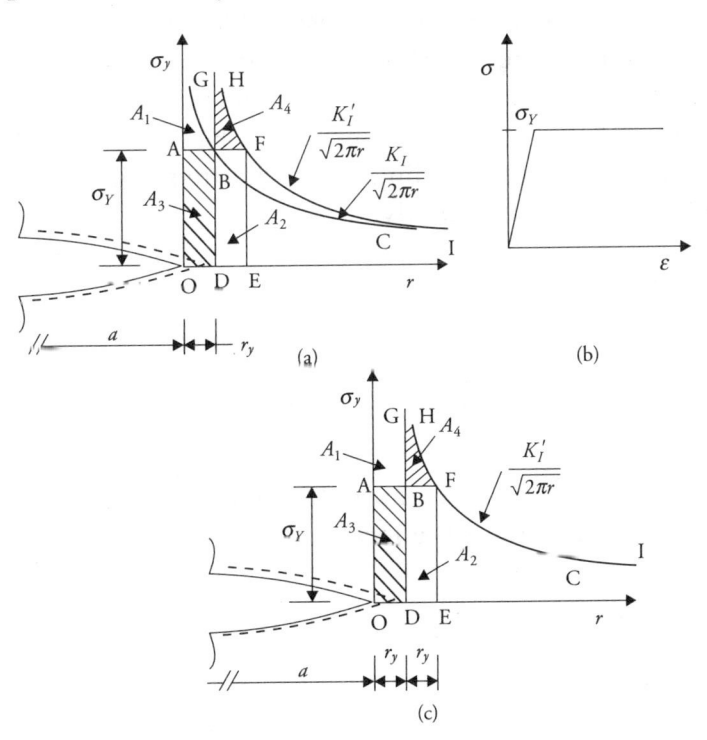

Figure 2.12 Crack-tip plastic zone and equivalent crack size. (a) Stress field ahead of given crack a. (b) Tensile stress–strain property of material considered. (c) Stress field ahead of equivalent crack $(a + r_y)$.

In the presence of plastic deformation at the crack-tip, edge displacements of a given crack are higher than the corresponding elastic crack of the same physical size. Irwin suggested that a given crack of size a can be replaced by an equivalent larger elastic crack $a + r_y$. Further, it can be considered that load A_3 transmitted by segment OD (Fig. 2.12(c)) is equivalently covered approximately by the extra load A_4 over span DE above BF. Consequently, the whole domain can be treated as elastic with a crack with tip at D and the corrected SIF, $K'_I = \sigma\sqrt{\pi(a + r_y)}$. HFI indicates approximately the new singularity variation in σ_y. r_y is known as Irwin plastic zone correction factor.

2.8 Dugdale–Barenblatt Model for Plastic Zone Size

The plastic zone size under plane stress can also be determined from a different consideration. This was done independently by Dugdale(1960) and Barenblatt(1962). Dugdale model is also known as strip-yield model. The material is again considered elastic-perfectly-plastic. The physical crack can be considered to be extended up to $a + \rho$ (Fig. 2.13) and the portion A to A_1 (or B to B_1) can be assumed to be subjected to a constant intensity of loading, which depends on the yield strength σ_Y of the material. This loading has a tendency to close the crack. The material ahead of the crack-tip B_1 can be considered as elastic. The stress σ_y at $x = a + \rho$ is finite. That is, there is no singularity. Hence, the SIF at location B_1 is zero. The SIFs due to external loading σ and closure stresses σ_Y acting over the spans BB_1 and AA_1 are given respectively by

$$K_{I\sigma} = \sigma\sqrt{\pi c}$$

$$K_{I\sigma_Y} = -\frac{2\sigma_Y \sqrt{c}}{\sqrt{\pi}} \cos^{-1}\left(\frac{a}{c}\right) \qquad (2.51)$$

where $c = a + \rho$. $K_{I\sigma_Y}$ is negative because stresses σ_Y over the spans BB_1 and AA_1 try to close the crack. Since the SIF at location B_1 is zero,

$$\sigma\sqrt{\pi c} = \frac{2\sigma_Y \sqrt{c}}{\sqrt{\pi}} \cos^{-1}\left(\frac{a}{c}\right) \quad \text{or} \quad \frac{a}{c} = \cos\frac{\pi\sigma}{2\sigma_Y} \qquad (2.52)$$

As $\sigma \to \sigma_Y$, $\frac{a}{c} \to 0$, that is, $c \to \infty$. Hence, for higher load levels, yielding spreads, as expected, over the whole ligament. For low level of loading $\left(\frac{\sigma}{\sigma_Y} \ll 1\right)$, the left side can be expanded binomially. The cosine term can be expanded in series. This gives

$$1 - \frac{\rho}{a} + \left(\frac{\rho}{a}\right)^2 - \cdots = 1 - \frac{\pi^2 \sigma^2}{8\sigma_Y^2} + \frac{1}{24}\left(\frac{\pi\sigma}{2\sigma_Y}\right)^4 - \cdots \quad (2.53)$$

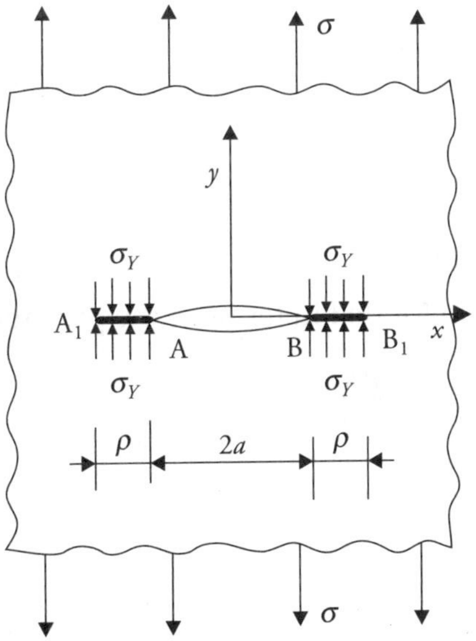

Figure 2.13 Strip-yield plastic zone.

Neglecting the third and higher order terms from both sides

$$\rho \approx \frac{\pi^2 \sigma^2 a}{8\sigma_Y^2} \approx \frac{K_I^2}{\pi \sigma_Y^2} = 2r_y \quad (2.54)$$

Thus, the results due to Dugdale–Barenblatt model and Irwin's analysis are identical. However, for higher values of σ/σ_Y, the difference increases.

2.9 Crack-Tip Plastic Zone Shape

Plastic zone develops gradually around the crack-tip. To determine the shape of the plastic zone, elastic plastic analysis is called for. Experimentally, the shape of the plastic zone can be determined by etching technique (Hahn and Rosenfield 1965) and microscopy. The zone is theoretically determined approximately by considering the material to undergo plastic deformation instantaneously. The spread of plastic deformation along a particular ray from the crack-tip can be determined (McClintock and Irwin 1965; Brown Jr. and Srawley 1965; Broek and Vlieger 1974; Broek 1986) applying either the Tresca or the Mises criterion.

According to Tresca criterion, yielding occurs when the maximum shear stress reaches the shear stress at yield point in uniaxial tension. Similarly, according to Mises criterion, yielding occurs when the distortion energy density reaches the same level at yield point in uniaxial tension. These two criteria can be expressed in the following form, respectively.

$$\sigma_1 - \sigma_3 = \sigma_Y, \text{ or } \sigma_1 - \sigma_2 = \sigma_Y, \text{ or } \sigma_1 - \sigma_2 = \sigma_Y \tag{2.55}$$

$$(\sigma_1 - \sigma_2)^2 + (\sigma_2 - \sigma_3)^2 + (\sigma_3 - \sigma_1)^2 = 2\sigma_Y^2 \tag{2.56}$$

where σ_1, σ_2, and σ_3 are the principal stresses.

2.9.1 Mode I Plastic Zone

The crack-tip stress field in Mode I in terms of the principal stresses is given by

$$\sigma_1 = \frac{K_I}{\sqrt{2\pi r}} \cos\frac{\theta}{2}\left(1 + \sin\frac{\theta}{2}\right)$$

$$\sigma_2 = \frac{K_I}{\sqrt{2\pi r}} \cos\frac{\theta}{2}\left(1 - \sin\frac{\theta}{2}\right)$$

$$\sigma_3 = \nu \frac{2K_I}{\sqrt{2\pi r}} \cos\frac{\theta}{2} \text{ for plane strain}$$

$$= 0 \text{ for plane stress} \tag{2.57}$$

The extent of plastic zone $r_y(\theta)$ along a radial direction as per $\sigma_1 - \sigma_3 = \sigma_Y$ is given by

$$r_y(\theta) = \frac{K_I^2}{2\pi\sigma_Y^2}\left[\cos\frac{\theta}{2}\left(1 + \sin\frac{\theta}{2}\right)\right]^2 \text{ for plane stress}$$

$$r_y(\theta) = \frac{K_I^2}{2\pi\sigma_Y^2}\left[\cos\frac{\theta}{2}\left(1 - 2\nu + \sin\frac{\theta}{2}\right)\right]^2 \text{ for plane strain} \tag{2.58}$$

where ν is Poisson's ratio. The shape of the zone is obtained by plotting $r_y(\theta) / \frac{K_I^2}{2\pi\sigma_Y^2}$ using the above equations. This is graphically illustrated in Fig. 2.14(a), which also includes the result by Tresca criterion, $\sigma_1 - \sigma_2 = \sigma_Y$. The spread of the plastic zone along the crack line corresponding to plane strain condition is one-ninth of the same in plane stress for $\nu = 1/3$. The shape of the plastic zone according to Mises criterion is as follows.

$$r_y(\theta) = \frac{K_I^2}{4\pi\sigma_Y^2}\left[1 + \frac{3}{2}\sin^2\theta + \cos\theta\right] \text{ for plane stress}$$

$$r_y(\theta) = \frac{K_I^2}{4\pi\sigma_Y^2}\left[\frac{3}{2}\sin^2\theta + (1-2\nu)^2(1+\cos\theta)\right] \text{ for plane strain} \quad (2.59)$$

These shapes are illustrated in Fig. 2.14(b). The size of the plastic zone as per Tresca criterion is larger than due to Mises criterion.

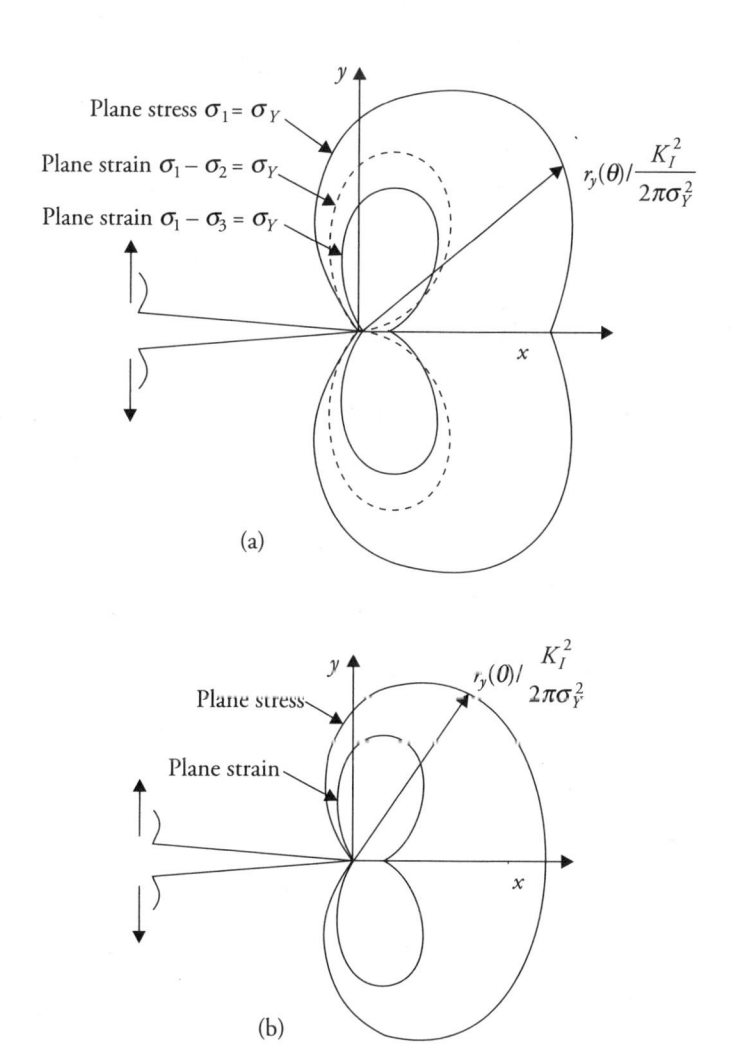

Figure 2.14 Mode I crack-tip plastic zone shape according to (a) Tresca criterion and (b) Mises criterion.

2.9.2 Plane Strain Constraint

Along the crack line ($\theta = 0$), $\sigma_1 = \sigma_2 = \alpha$, say. In the case of plane stress, the plastic zone spreads up to a distance where $\alpha = \sigma_Y$ according to Mises criterion. Under plane strain condition, assuming $\nu = 1/3$, the zone is likely to spread along the crack line up to a point [according to Eq. (2.56)], where $\sigma_1 = \sigma_2 = \alpha = 3\sigma_Y$. This indicates that the plastic zone extends over a distance from the crack-tip to a point where stress σ_1 is three times the level in the case of plane stress. Hence, a smaller plastic zone develops in the case of plane strain. This also means that there is greater stress relaxation in the case of plane stress than in the case of plane strain. In other words, there is more constraint on the growth of plastic zone in the latter case. Based on this consideration, the constraint on the plane strain plastic zone is taken as 3. Further, the plane strain plastic zone size is taken as one-third of the plane stress zone size. Therefore, Irwin plastic zone correction factor for this case is $\dfrac{K_I^2}{6\pi\sigma_Y^2}$.

2.9.3 Mode II and Mode III Plastic Zones

The shape of the plastic zones for the other two modes can be similarly obtained. These plastic zones are illustrated in Figs. 2.15(a) and 2.15(b) for Modes II and III [McClintock and Irwin 1965], respectively, according to Mises criterion.

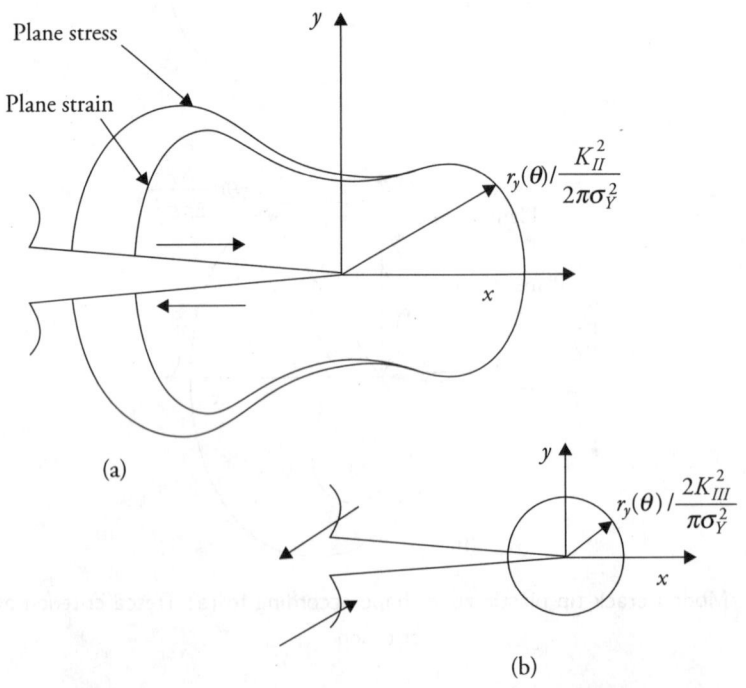

Figure 2.15 Plastic zone shape according to Mises criterion. (a) Mode II and (b) Mode III.

2.10 Triaxiality at Crack Front

When a tensile panel (Fig. 2.16) is loaded, stretching results in y direction and contraction develops in z direction. Depending on the z-coordinate at a point on the crack plane and located ahead of the crackfront, the state of stress varies (Bluhm 1961; Knott 1975).

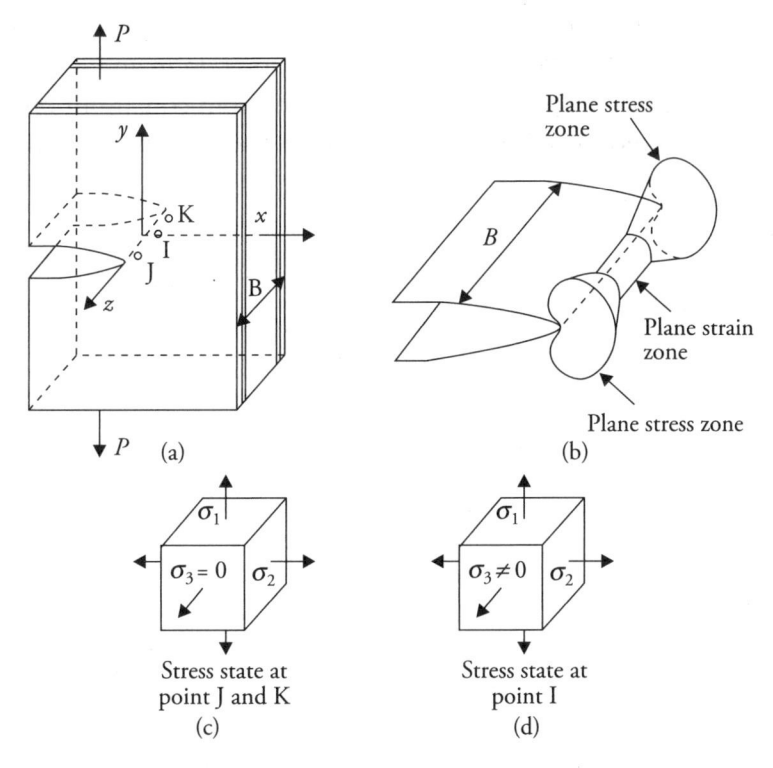

Figure 2.16 Plastic zone size variation along thickness. (a) Specimen. (b) Plastic zone shape around crack front. (c) Stress state at points J and K. (d) Stress state at point I.

For a point J or K (Fig. 2.16(a)) very close to the surface ($z = B/2$), the constraint in the z direction is absent because the outer surfaces are free of any normal stress σ_z or σ_3. On the other hand, for a point I close to the centre, tendency of contraction is restrained; some stresses σ_3 develop in the z direction because of the interaction between inner and the outer layers. This constraint gradually increases from zero at the surface to higher values at inner locations. For thicker specimens, it may develop to the full constraint of plane strain even before the mid-thickness location is reached. Hence, there is a triaxiality at the inner locations. The states of stress for the two points are illustrated in Figs. 2.16(c) and 2.16(d), respectively. Thus, there is a state of plane stress near the surfaces ($z = \pm B/2$) and there is a state of plane strain near the centre.

Since the plastic zone size is larger in plane stress than in plane strain, the size of the plastic zone varies along the thickness direction. The three-dimensional shape of the plastic zone is shown in Fig. 2.16(b).

Plastic deformation can be visualized as a shearing phenomenon on the planes of maximum shear stress. The different planes of maximum shear stress result in different patterns of failure. The shearing on planes (AC and DE) at 45° to the $x - y$ plane is a typical of plane stress condition (Fig. 2.17). The shearing on planes at 45° to the $y - z$ plane is a typical of plane strain condition (Hahn and Rosenfield 1965). The central portion of the specimen, which suffers from constraint on the growth of plastic zone, is likely to undergo flat fracture; and the surfaces undergo slant fracture.

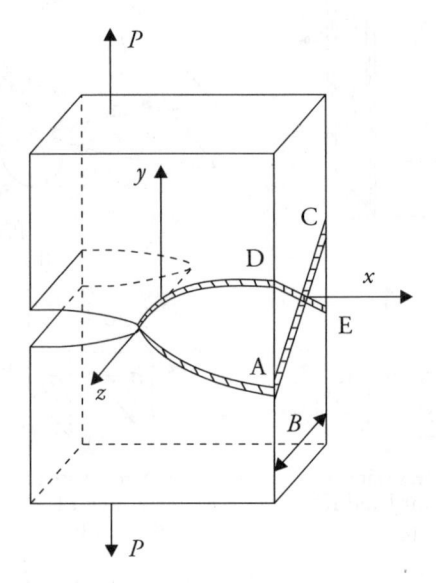

Figure 2.17 Shear planes.

The ratio of the plastic zone size to thickness is an important factor for the determination of the state of stress. If the size of the plastic zone $(2r_y)$ is of the order of the plate thickness (B), plane stress condition can develop throughout the thickness. The ratio $2r_y/B$ must be appreciably lower than unity for plane strain condition to exist over a greater part of the thickness. Experimentally, it has been observed that fracture behaviour is dominated by plane strain if $B \geq 2.5\dfrac{K_{IC}^2}{\sigma_Y^2}$, which is approximately equal to 16 times plane stress plastic zone size, where K_{IC} is plane strain fracture toughness and σ_Y is yield strength of the material.

At a given stress intensity level, the plastic zone size is proportional to K_I^2/σ_Y^2. A material with low yield stress σ_Y would give rise to a larger plastic zone and greater stress relaxation. It would require a higher stress level, hence higher SIF, for fracture. Such materials will demonstrate higher fracture resistance. This also means that a larger plate thickness will be required to maintain predominantly a state of plane strain in a material with low yield strength (σ_Y) and high toughness (K_{IC}) than in a material with high yield strength and low toughness.

2.11 Thickness Dependence of Fracture Toughness K_c

Since the state of stress near the crack front is dependent on the plate thickness, the fracture behaviour, in particular, the fracture toughness and the macroscopic nature of the fracture surface (Bluhm 1961; Isherwood and Williams 1970; Broek and Vlieger 1974), would depend on this parameter. The thickness dependence of fracture toughness and nature of the macroscopic fracture surface are illustrated in Fig. 2.18.

The thickness-dependent fracture toughness is represented by K_C and the thickness independent fracture toughness or plane strain fracture toughness in indicated by K_{IC}. There are three distinct regions. In Region I, the fracture toughness increases with thickness, and the fracture is fully 'slant'. The reason for the increase in toughness in Region I is not clearly understood. At a particular thickness, B, the toughness, attains the highest value. In Region II, the fracture toughness gradually reduces with increase in thickness and the surface appears to be a mixture of 'slant' and 'flat' zones. In Region III, the toughness is independent of thickness and mostly a state of plane strain prevails near the whole crack front, that is, over the full thickness.

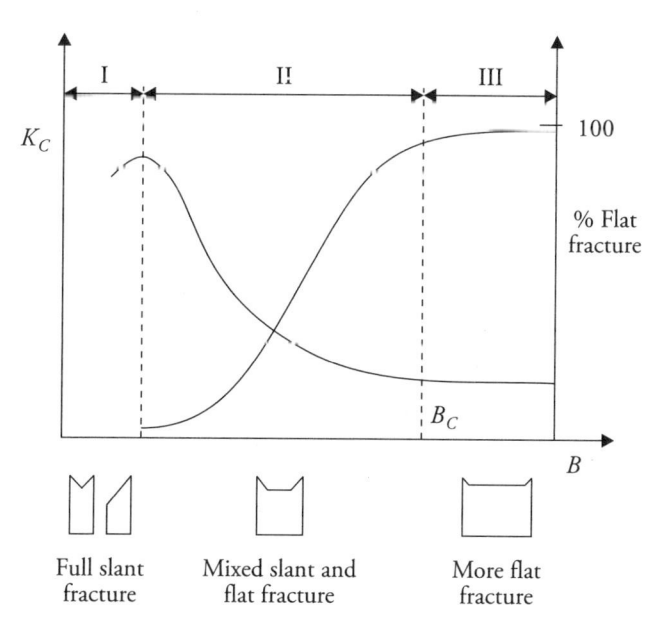

Figure 2.18 Variation of fracture toughness and percentage flat fracture with specimen thickness.

When the thickness of the specimen is small, there is a possibility that the two surface plastic zones will coalesce. This will result in a larger plastic zone all along the thickness and a larger stress relaxation. Further, this gives rise to an increase in energy requirement per unit crack extension, and hence, the fracture toughness. With the increase in thickness plastic zone size reduces, and the energy requirement for crack extension reduces, so does the fracture toughness. The percentage flat fracture too increases with the thickness up to the critical thickness B_C. Above this thickness, the fracture toughness becomes independent of thickness. Approximately, $B_C = 2.5 \dfrac{K_{IC}^2}{\sigma_Y^2}$. In view of this, the plate thickness for the measurement of plane strain fracture toughness is generally taken more than B_C.

Table 2.1 lists the fracture toughness data K_{IC} for some materials. For more data, readers may look into references by Hudson and his associates (1978, 1982, 1989, 1991).

2.12 Design Applications

Problem 2.1

For the C-frame (Fig. 2.19), calculate the safe load when $a = 5$ mm, depth h of section AB $= 40$ mm, and $L = 150$ mm. Thickness B of the section is 25 mm.

Given $K_{IC} = 59$ MPa$\sqrt{\text{m}}$, $\sigma_Y = 1500$ MPa.

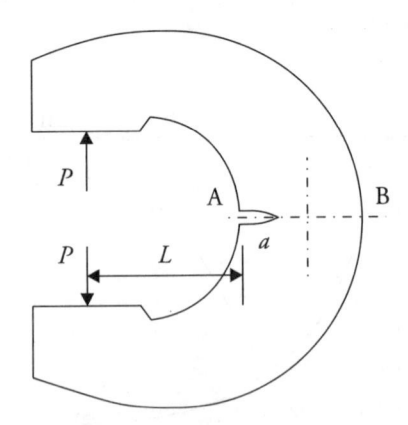

Figure 2.19 C-frame.

Solution

SIFs due to direct load P and bending moment M can be calculated using the following relations, respectively.

$$K_{IM} = \dfrac{6M}{Bh^2}\sqrt{\pi a}\, Y_M, \quad K_{IP} = \dfrac{P}{Bh}\sqrt{\pi a}\, Y_P \quad \alpha = a/h = 5/40 = 0.125$$

Collecting data from Appendix 2.1 (Figs. A2.1.2 and A2.1.3),

$$Y_M = 1.122 - 1.4\alpha + 7.33\alpha^2 - 13.08\alpha^3 + 14\alpha^4$$

$$Y_P = 1.12 - 0.23\alpha + 10.55\alpha^2 - 21.72\alpha^3 + 30.39\alpha^4.$$

Therefore, $Y_M = 1.0394$ and $Y_P = 1.221$.

$$M = P \times (L + h/2) = 0.170\, P$$

Check for plane strain thickness: $B_C = 2.5 \dfrac{K_{IC}^2}{\sigma_Y^2} = 3.87$ mm.

Since the frame thickness $B \gg B_C$, the design equation is given by $K_I = K_{IC}$. This gives

$$K_I = K_{IM} + K_{IP}$$

$$= \dfrac{6 \times 0.170 P}{0.025 \times 0.04^2} \sqrt{\pi \times 0.005} \times 1.0394 + \dfrac{P}{0.025 \times 0.04} \sqrt{\pi \times 0.005} \times 1.221 = 59 \times 10^6$$

$P = 16979$ N (Ans.).

Problem 2.2

A spherical pressure vessel (Fig. 2.20), internal diameter 3.048 m and thickness 25.4 mm, is to be made of maraging steel. The nondestructive testing technique available can detect a through the thickness crack of minimum size 2.54 mm. Three grades of maraging steels are available. Select a grade for maximum internal pressure.

Grade	Yield stress σ_Y (MPa)	Plane strain fracture toughness K_{IC} (MPa$\sqrt{m}$)
200	1510	113.25
250	1720	95.35
300	2040	66.46

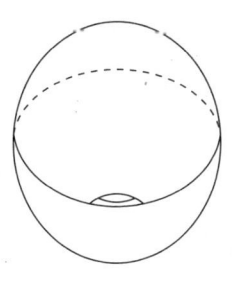

Figure 2.20 Spherical pressure vessel with crack.

Solution
If a conventional design approach based on the yield strength is considered, the design equation is given by: hoop stress $\sigma_\theta = \sigma_Y$.

Since $\sigma_\theta = \dfrac{pr}{2t}$, where p is pressure, t is vessel thickness, and r is internal radius; therefore

$$p = \frac{2t}{r}\sigma_Y; \; r = 1.524 \text{ m}, \; t = 0.0254 \text{ m}.$$

The permissible pressure capacities are:

50.33 MPa for Grade 200, 57.33 MPa for Grade 250, and 68.0 MPa for Grade 300.

If a fracture mechanics based approach to design is taken, the design equation is $K_I = K_{IC}$.

Assuming a through the thickness crack (of size $2a = 2.54$ mm) as shown, the SIF can be obtained considering the hoop stress to be the opening stress and treating the plate as an infinite plate. Hence, the correction factor is 1.0. Therefore

$$K_I = \sigma_\theta \sqrt{\pi a} = K_{IC}$$

This gives $p = \dfrac{2t}{r\sqrt{\pi a}} K_{IC}$, $r = 1.524$ m, $t = 0.0254$ m, and $a = 0.00127$ m

The permissible pressure capacities are:

59.76 MPa for Grade 200, 50.32 MPa for Grade 250, and 35.07 MPa for Grade 300.

From the point of view of classical design, Grade 300 is preferable. From the point of view of fracture mechanics based design, Grade 200 is good. From the point of view of both the approaches, either Grade 200 or Grade 250 can be used, and the operating pressure can be maintained at 50 MPa approximately.

Problem 2.3
The records of a fracture toughness test are as follows.

Crack length a (mm)	Load (kN)	Load point displacement (mm)
24.5	100	0.3050
25.5	100	0.3075

The fracture load $P_c = 157.5$ kN for $a = 25$ mm. Given $B = 25$ mm, $E = 70$ GPa and $\nu = 0.3$.

Determine G_{IC} and K_{IC}.

Solution

Compliance at $a = 24.5$ mm, $c_1 = 0.3050 \times 10^{-3}/10^5$ m/N
Compliance at $a = 25.5$ mm, $c_2 = 0.3075 \times 10^{-3}/10^5$ m/N

$\Delta c / \Delta a = (c_2 - c_1)/\Delta a = 2.5 \times 10^{-8}$ N^{-1}, noting that $\Delta a = 1$ mm.

$$G_{IC} = \frac{P_c^2}{2B} \frac{\Delta c}{\Delta a} = \frac{(157.5 \times 10^3)^2}{50 \times 10^{-3}} 2.5 \times 10^{-8} = 12403 \text{ J/m}^2$$

$G_{IC} = (1 - \nu^2)\frac{K_{IC}^2}{E}$. Using $\nu = 0.3$ and $E = 70$ GPa, $K_{IC} = 30.88$ MPa$\sqrt{\text{m}}$ (Ans.).

Problem 2.4

A plate (Fig. 2.21) with a 12 mm cut-out gave rise to a small internal crack of size 9 mm in service. Given $B = 12$ mm, $\sigma_Y = 1000$ MPa, and $K_{IC} = 65$ MPa$\sqrt{\text{m}}$, determine the safe load σ. Use a factor of safety 3.

Solution

Considering first classical design based on yield strength, the design equation is

$\sigma_{\max} = \sigma_Y/3$.

Since plate width is more than 10 times the hole diameter, the stress concentration factor at the hole edge can be taken as 3. Therefore

$\sigma_{\max} = 3\sigma = \sigma_Y/3$.

$\sigma = 1000/9 = 111$ MPa.

Considering the LEFM-based design, the design equation is

$K_I = K_{IC}/3$.

The plate width is very large compared with semi-crack size; the SIF correction factor can be taken as unity. Therefore $K_I = \sigma\sqrt{(\pi a)}$.

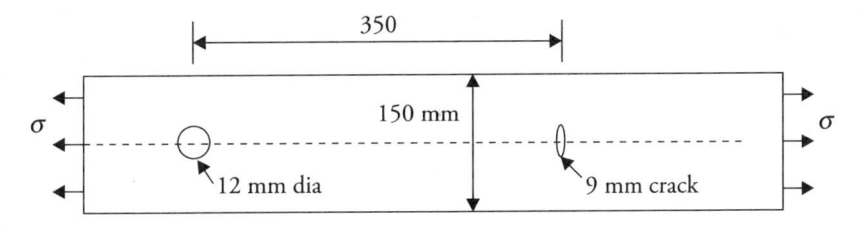

Figure 2.21 Plate with hole and crack.

Hence $K_I = \sigma\sqrt{(\pi a)} = \sigma\sqrt{(\pi \times 4.5 \times 10^{-3})} = 65/3$ MPa$\sqrt{\text{m}}$

$\sigma = 182$ MPa.

The permissible load is therefore 111 MPa (Ans.).

Problem 2.5
A proving ring (Fig. 2.22) is made of 4140 steel (tempered at 482°C) with yield point 1213 MPa and plane strain fracture toughness 75 MPa$\sqrt{\text{m}}$. The ring is subjected to a load $P = 5$ kN. The ring (radial) thickness is 40 mm and thickness in the perpendicular direction is 25 mm. Determine the level of safety with a vertical crack of depth 15 mm. Determine the maximum load that the ring can take for the same crack size.

Solution
Average ring radius $r = 250$ mm; depth $w = 0.040$ m.
Bending moment at the vertical section $M = \dfrac{Pr}{\pi} = \dfrac{5000 \times 0.25}{\pi} = 397.8$ Nm.
In the following calculation of the SIF, the effect of ring curvature is neglected.

$$\alpha = \frac{a}{w} = \frac{0.015}{0.040} = 0.375$$

The SIF correction factor Y is collected from Appendix 2.1 (Fig. A.2.1.3).

$Y = 1.122 - 1.4\alpha + 7.33\alpha^2 - 13.08\alpha^3 + 14\alpha^4 = 1.21487$

SIF $K_I = \dfrac{6M}{bw^2}\sqrt{\pi a}\ Y = \dfrac{6 \times 397.8}{0.025 \times 0.040^2}\sqrt{\pi \times 0.015} \times 1.21487 = 15.736$ MPa$\sqrt{\text{m}}$.

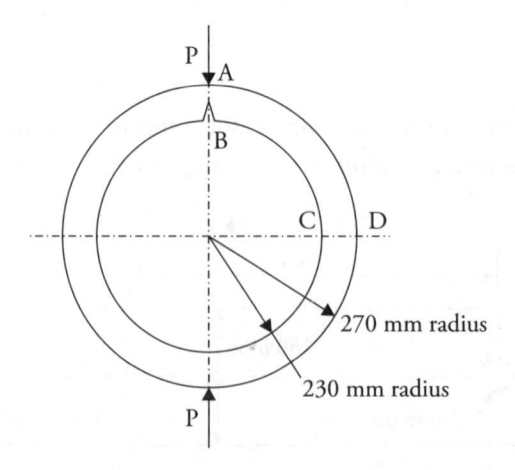

Figure 2.22 A proving ring with internal crack.

Level of safety $f = \dfrac{K_{IC}}{K_I} = \dfrac{75}{15.736} = 4.766$.

The maximum load capacity of the ring, $P_{\max} = f \times P = 4.766 \times 5 = 23.83$ kN (Ans.).

Problem 2.6

A gear of 20° involute full-depth tooth profile (Fig. 2.23) has module 8 mm and 40 teeth. It operates at 200 rpm. After some period of operation, a crack of size 8 mm was detected at the base of a tooth. The gear is made of 4340 steel with fracture toughness 57 MPa$\sqrt{\text{m}}$. Calculate the safe horse power that can be transmitted at this point. Use factor of safety 5. Face width of the gear is 80 mm.

Solution

The crack is subjected to both Mode I and Mode II loadings. Neglecting the effect of Mode II loading and Mode I effect due to the vertical component of P, and assuming width of the tooth at the addendum circle to be equal to the width of the tooth at the base circle, the following calculations are done.

Pitch circle diameter $D_p = 40 \times 8 = 320$ mm.

Base circle diameter $D_b = D_p \cos \phi = 320 \cos 20° = 300.7$ mm.

Tooth width along arc at the pitch circle, $t_p = \pi m/2$, m = module. $t_p = 12.5663$ mm.

Tooth width along arc at the base circle, $t'_b = D_b \left(\dfrac{t_p}{D_p} + \tan \phi - \phi \right)$

$= 300.7(12.5663/320 + 0.36397 - 0.34906) = 16.29$ mm.

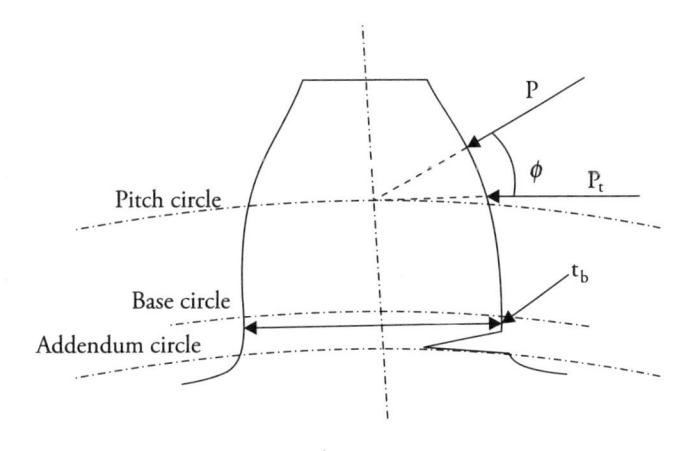

Figure 2.23 Gear tooth with crack located at base location.

Assuming chordal tooth width at the addendum circle $t_b \approx t'_b = 16.29$ mm.

The tooth section at the crack level can, therefore, be taken as a rectangular section with depth $h = t_b = 16.29$ mm and thickness $B = 80$ mm.

The relation for SIF correction factor is taken as the same as in the earlier case. The SIF is given by

$$K_I = \frac{6M}{Bh^2}\sqrt{\pi a}\, Y(\alpha), \quad Y(\alpha) = 1.122 - 1.4\alpha + 7.33\alpha^2 - 13.08\alpha^3 + 14\alpha^4, \quad \alpha = \frac{a}{h}$$

$$m = 8 \text{ mm}, \ M = P_t(1.25\,m) = 0.01\,P_t \text{ Nm}, \ \alpha = \frac{8}{16.29} = 0.491, \ Y = 1.4671.$$

Equating K_I with K_{IC} of the material

$$\frac{6 \times 0.01 \times P_t}{0.08 \times 0.01629^2}\sqrt{\pi \times 0.008}\, 1.4671 = 657.353\, P_t = 57 \times 10^6;\ P_t = 86.71 \text{ kN}.$$

The tangential load with a factor of safety 5, $P'_t = 17.34$ kN.

The horse power that can be transmitted $\text{HP} = \dfrac{2\pi n}{60}\dfrac{\left(P'_t \frac{D_p}{2}\right)}{746} = 77.9$ (Ans.).

Table 2.1 Room temperature fracture toughness of structural materials.

Material	Condition/Heat Treatment	Yield Strength (MPa)	Ultimate Limit (MPa)	Fracture Toughness K_{Ic} (MPa $\sqrt{m}$)	Fracture Toughness K_c (MPa $\sqrt{m}$)	Specimen Type, Thickness (mm)	Source
4140 steel (1.6mm forged bar)	Oil quenched from 870°C and 1h tempered at					CT, L–T	ASM Handbook (1996a)
	(i) 205°C		1448	44		15 thick	Same as above
	(ii) 395°C		1517	55		15 thick	Same as above
	(iii) 280°C		1585	55		15 thick	Same as above
4140 steel (Plate 25mm)	870°C 1h + 843°C 1h, oil quench, 1h tempered at					CT, L–T	
	(i) 482°C		1213	75		25 thick	Same as above
	(ii) 425°C		1365	44		25 thick	Same as above
4340 steel	Tempered at 260°C		1640	48.5±2		TPB 35	Same as above
4340 steel	Tempered at 425°C		1420	87±4		TPB 37	Same as above
4340 steel	Plate 16mm thick, heat treated to 51 HRC		1517	57		CT, L–T 25 thick	Same as above
4340 steel	Round bar 115mm dia.	1240	1330	117.5		CT, L–T 25 thick	Same as above
4340 steel	Plate 25mm thick, oil quenched from 843°C, tempered at					CT, L–T	
	(i) 425°C		1420		84	20 thick	Same as above
	(ii) 260°C		1640		49		Same as above
4340 steel	Bar 25mm, oil quenched, tempered at						

cont.

Table 2.1 cont.

HY 140 steel	(i) 540°C	1172	1260		110	Same as above
	(ii) 425°C	1380	1530		75	Same as above
	(iii) 205°C	1640	1950		53	Same as above
	Plate 25mm, quenched and tempered at 540°C	980	1027		≈275	Same as above
D6ac steel	Heat-treated to 46 HRC		1420		86	TPB, T–L 17.78 thick
D6ac steel	900°C ausbay quench, 525°C slack quench, 163°C, aged at 540°C		1495		49–91	CT, L–T Plate 38 thick
Mn–Cr–Mo–V (1.5%Mn) steel		372		63		Tension specimen, edge crack, 9 thick
9Ni–4Co–25 steel	Quenched and tempered plate, tempered at					
	(i) 315°C	1345	1482	82		Same as above
	(ii) 540°C	1310	1380	132		Same as above
9Ni–4Co–45 steel	Quenched and tempered plate, tempered at					
	(i) 315°C	1758	1930	49		Same as above
	(ii) 540°C	1413	1482	104		Same as above
N18 maraging steel	Aged 4h at 490°C	1695	1773	84		TPB, 19 thick
		1803	1862	117		19 thick

cont.

Table 2.1 cont.

Ni–Co–Mo maraging steel(18%Ni)	Two melts, aged 3h at 482°C		2048	136 (melt 1) 104 (melt 2)		Tension specimen, edge crack, 3.6 thick edge crack	Same as above
Ni–Co–Mo maraging steel(18%Ni)	Two melts, aged 3h at 482°C		2048	123 (melt 1) 80 (melt 2)		Tension specimen, centre crack, 3.6 thick	Same as above
Ni maraging steel(18%Ni)			2058	190			Same as above
			1920	237			Same as above
			1783	174			Same as above
			1715	204			Same as above
			1646	221			Same as above
N18 maraging steel	Vacuum melted		2058	93	217		Same as above
18Ni–Marage 200 steel	Plate aged at 480°C	1482	1550	120–154		Plate	Same as above
18Ni–Marage 250 steel	Vacuum melted sheet and plate aged at 480°C	1565 1634 1813	1634 1765 1855	75 91 77–134		Plate (T) Plate (L) Plate (T)	Same as above Same as above Same as above
18Ni–Marage 350 steel	Vacuum melted, plate aged at 480°C	2380	2427	49		Plate 25 thick	Same as above
High carbon high silicon steel	As cast	1645±49	2010±54	20.91–26.82			Putatunda(2001)
High carbon high silicon steel	Heat treated	1645±49	2010±54	48			Same as above
15-5PH (VAR) SS	H900 annealed	1280	1380	96		L–T orientation	ASM Handbook(1996b)

cont.

Table 2.1 cont.

15-5PH (VAR) SS	H900 annealed	1180	1330	81	T–L orientation	Same as above
17-4PH SS	H900 annealed	1210	1380	48	T–L orientation	Same as above
17-PH SS	H975 annealed	1160	1290	93	L–T orientation	Same as above
PH13–8MO SS	H950 austenitised at 1040°C	1360	1550	70	T–L orientation	Same as above
PH13–8MO SS	H1050 austenitised at 1040°C	1230	1320	112	T–L orientation	Same as above
15-5PH SS	Solution+precipitation heat treated	1176	1200	187–194	CT	Abdelshehid et al. (2007)
SAE 304 LN	Annealed at 1050°C +water quenched	352	687	378.8	CT, 20 thick	Ghosh et al. (2009)
A387 Cr-Mo alloy steel Class2 Grade 22	Normalized and tempered	345±57	528±33	286	Based on J_{IC} test	Suresh, Zamiski and Ritchie(1982)
Al alloy 2124-T851	Annealed, tempered, and aged	450	485	32 26 26	L–T T–L S–L 70 thick plate, % elongation 8	Metals Handbook(1990a)
Al alloy 7475-T761	Heat treated	462	524	143	1.2 thick plate, % elongation 12	Metals Handbook(1990b)
Al alloy 7175-T736	Heat treated	503	552	33 28.6	L–T T–L,and S–L, Tension test: Plate, % elongation 14	Metals Handbook(1990c)

cont.

Table 2.1 cont.

Al alloy 2024-T351	Heat treated	325		32 26 26	L–T T–L S–L Plate tested	ASM Handbook(1996c)
Al alloy 2124-T851	Heat treated	450	485	32 25 24	L–T T–L S–L orientation, Plate tested	Same as above
Al alloy 7075-T651	Heat treated	505		29 25 20	L–T T–L S–L Plate tested	Same as above
Al alloy 2048 –T8	Heat treated	420	460	37 28	Longitudinal S–T	ASM Handbook(1996d)
Al alloy 7075-T651	Heat treated	462	538	28.6 24.2 17.6	L–T T–L S–L, Plate tested	Metals Handbook(1990d)
Al alloy 7050-T73651	Heat treated	455	510	35.2 29.7 28.6	L–T T–L S–L Plate tested	Metals Handbook(1990e)
Al alloy 6061-T651		289		29.1	T–L	ASM Handbook(1996e)
Al alloy 7039-T6		381		32.3	T–L	Same as above
Al alloy 7079-T652	Hand forging	440		29 25 20	L–T T–L S–L	ASM Handbook(1996c)

cont.

Table 2.1 cont.

Al alloy 7149-T73	Die forging	460			L–T 34 T–L 24 S–L 24	Same as above
Mg alloy ZK60	Heat treated	225	266	20.6	5 thick	Somekawa and Sakai (2006)
Mg alloy AZ51–3Sn	VHN = 65	79.5	211.2	17.8	0.5 thick, %Elongation 16.5	Kim et al. (2010)
AZ51	VHN = 59	65.4	208.2	9.9	0.5 thick, %elongation 10.5	Same as above
AZ51–7Sn	VHN = 74	95.4	182.4	18.5	0.5 thick, %elongation 7.5	Same as above
Mg alloy AZ31 sample A	Annealed at 573°K	185	274	16.7 Calculated from δ_c	K_c =18.4, 0.5 thick	Sasaki et al. (2003)
Mg alloy AZ91	Heat treated	128.5	266	21.7	% elongation 19	Barbagallo and Cerri (2004)
Poly-crystalline Al$_2$O$_3$ and SiC whisker/Al$_2$O$_3$ composite (Silar SC-1)	30% by volume of whisker		641±34		8.7±0.2	Peters(1998)
SiC whisker/ Si$_3$N$_4$ composite	30% by volume of whisker		970		6.4	Same as above
	20% by volume of whisker		550		7.0	Same as above

cont.

Table 2.1 cont.

Glass–Epoxy 16 ply unidirectional laminate	$E_{11}=25.7$ GPa, $E_{22}=6.5$ GPa, $G_{12}=2.5$ GPa, $\nu_{12}=0.32$	$G_I=259$ J/m² at failure	DCB test, 5 thick	Ducept, Davies and Gamby (2000)
AS4/PEEK (APC2) unidirectional laminate (24 ply)	$E_1=E_{11}=25.7$ GPa, $E_{22}=6.5$ GPa, $G_{12}=2.5$ GPa, $\nu_{12}=0.32$	$G_{IC}=1603$ J/m² $G_{II}=0$	DCB test	Crews, Jr and Reeder (1988)

APPENDIX 2.1

Stress Intensity Factors for Various Configurations

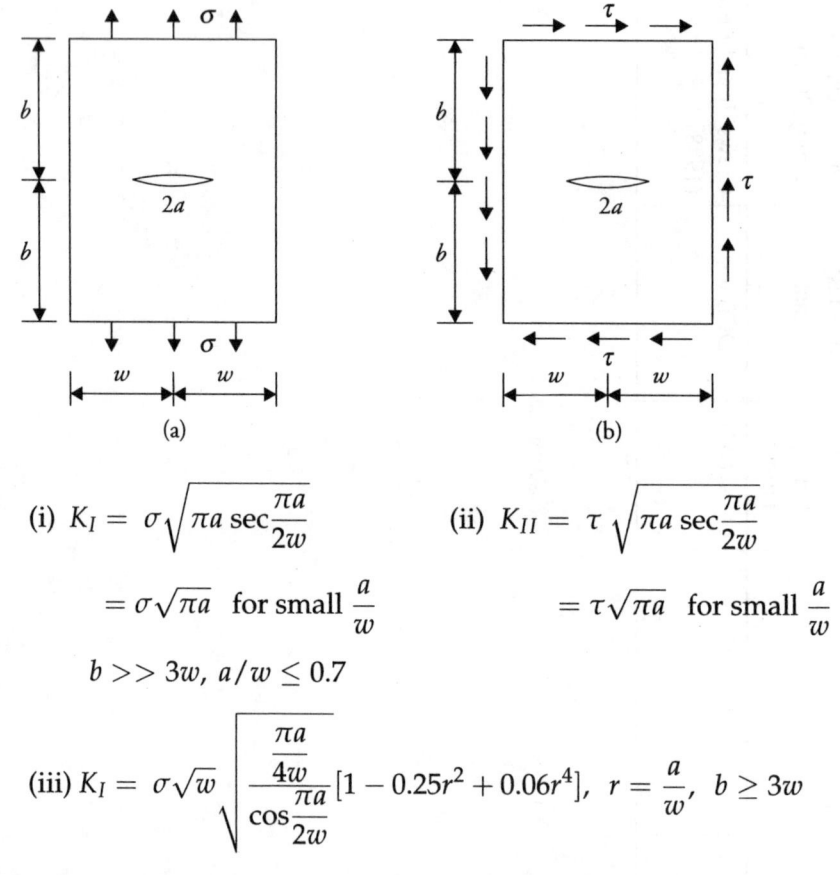

(i) $K_I = \sigma \sqrt{\pi a \sec\dfrac{\pi a}{2w}}$

   $= \sigma\sqrt{\pi a}$ for small $\dfrac{a}{w}$

   $b \gg 3w, \; a/w \leq 0.7$

(ii) $K_{II} = \tau \sqrt{\pi a \sec\dfrac{\pi a}{2w}}$

   $= \tau\sqrt{\pi a}$ for small $\dfrac{a}{w}$

(iii) $K_I = \sigma\sqrt{w} \sqrt{\dfrac{\dfrac{\pi a}{4w}}{\cos\dfrac{\pi a}{2w}}} [1 - 0.25r^2 + 0.06r^4], \; r = \dfrac{a}{w}, \; b \geq 3w$

Figure A2.1.1 Internal crack in finite plate. (a) Mode I and (b) Mode II. (i) and (ii) from [Tada, Paris and Irwin 2000, pp.40-42]. (iii) from [Miannay 1998].

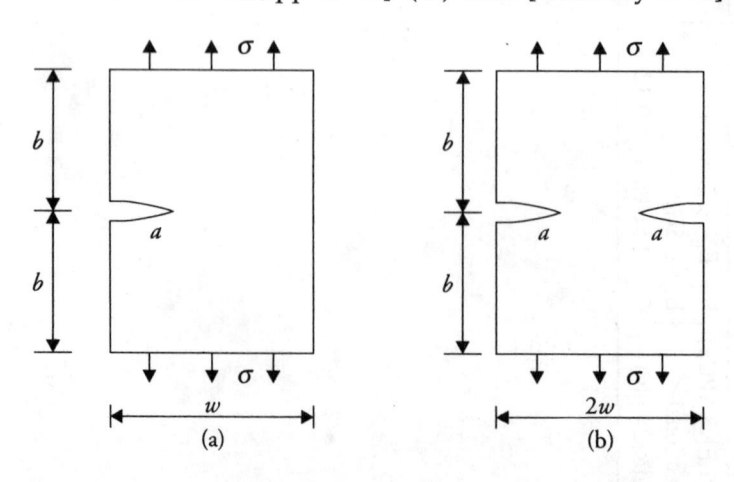

(i) $K_I = \sigma\sqrt{\pi a}\, Y(r)$, $r = \left(\dfrac{a}{w}\right)$

$r < 0.6, b/w > 1$

$Y(r) = 1.12 - 0.231r + 10.55r^2 - 21.71r^3 + 30.382r^4$

(ii) $K_I = \sigma\sqrt{\pi a}\, Y(r)$, $r = \left(\dfrac{a}{w}\right)$, $r \leq 0.7$

$Y(r) = 1.12 - 0.203r - 1.197r^2 + 1.93r^3$

(iii) $K_I = \sigma\sqrt{w}\, \dfrac{\sqrt{2\tan\dfrac{\pi a}{2w}}}{\cos\dfrac{\pi a}{2w}} Y(r)$,

$Y(r) = 0.752 + 2.02r + 0.37\left(1 - \sin\dfrac{\pi}{2}r\right)^3$

$r = \dfrac{a}{w}, b \geq 3w$

(iv) $K_I = \sigma\sqrt{w}\, \dfrac{\sqrt{\dfrac{\pi a}{4w}}}{\sqrt{1-r}} Y(r)$, $Y(r) = \Big[1.122$

$-0.561r - 0.205r^2 + 0.471r^3 - 0.19r^4\Big]$

$r = \dfrac{a}{w}, b \geq 3w$

Figure A2.1.2 (a) Single and (b) double edge crack in finite plate. (i) and (ii) from [Brown Jr. and Srawley 1965; Gdoutos 1993] for cases (a) and (b) respectively. (iii) and (iv) from [Miannay 1998] for cases (a) and (b) respectively.

$$K_I = \dfrac{6M}{Bw^2}\sqrt{\pi a}\, Y(r), \quad r = \dfrac{a}{w}, \quad r \leq 0.6$$

$$Y(r) = 1.122 - 1.4r + 7.33r^2 - 13.08r^3 + 14r^4$$

Figure A2.1.3 Beam segment with crack under bending moment. [Brown Jr. and Srawley 1965].

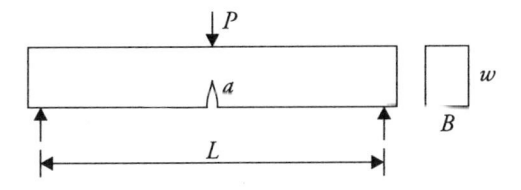

$$K_I = \dfrac{PL}{Bw^{1.5}}\dfrac{3\sqrt{r}\,[1.99 - r(1-r)(2.15 - 3.93r + 2.7r^2)]}{2(1+2r)(1-r)^{1.5}}, \quad r = \dfrac{a}{w}, \quad L = 4w$$

Figure A2.1.4 Crack in beam under three point bending load. [ASTM E399-90 2000; Srawley 1976].

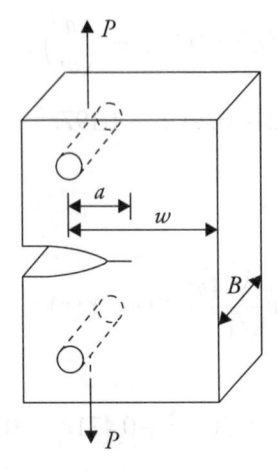

$$K_I = \frac{P}{B\sqrt{w}} f(\alpha), \quad \alpha = \frac{a}{w}$$

$$f(\alpha) = [(2+\alpha)(0.866 + 4.64\alpha - 13.32\alpha^2$$

$$+ 14.72\alpha^3 - 5.6\alpha^4)]/(1-\alpha)^{1.5}$$

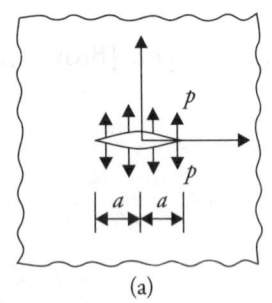

$$K_{I\,\text{max}}(\text{at B}) = 1.12 \frac{\sigma}{\Phi} \sqrt{\pi a}$$

$$K_{I\,\text{min}}(\text{at A}) = 1.12 \frac{\sigma}{\Phi} \sqrt{\pi c} \frac{a}{c}$$

$$\Phi = \int_0^{\frac{\pi}{2}} [1 - \frac{c^2 - a^2}{c^2} \sin^2 \pi] \, d\pi$$

$$\Phi \approx \frac{3\pi}{8} + \frac{\pi}{8} \frac{a^2}{c^2}$$

Figure A2.1.5 Compact tension specimen. [ASTM E399-90 2000].

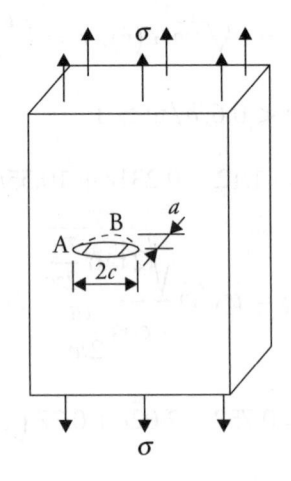

Figure A2.1.6 Thumbnail crack in plate. [Broek 1986].

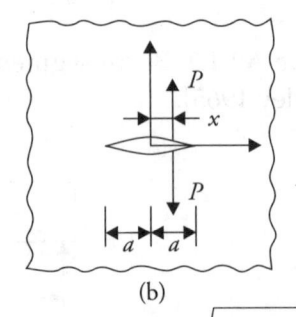

(a) (b)

$$K_I = p\sqrt{\pi a} \qquad K_I = \frac{P}{\sqrt{\pi a}} \sqrt{\left(\frac{a+x}{a-x}\right)} \text{ at the right tip, } K_I = \frac{P}{\sqrt{\pi a}} \sqrt{\left(\frac{a-x}{a+x}\right)}$$

at the left tip, P = Load per unit thickness.

Figure A2.1.7 Internal crack in infinite plate under (a) pressure load and (b) point load. [Paris and Sih 1965].

Linear elastic fracture mechanics

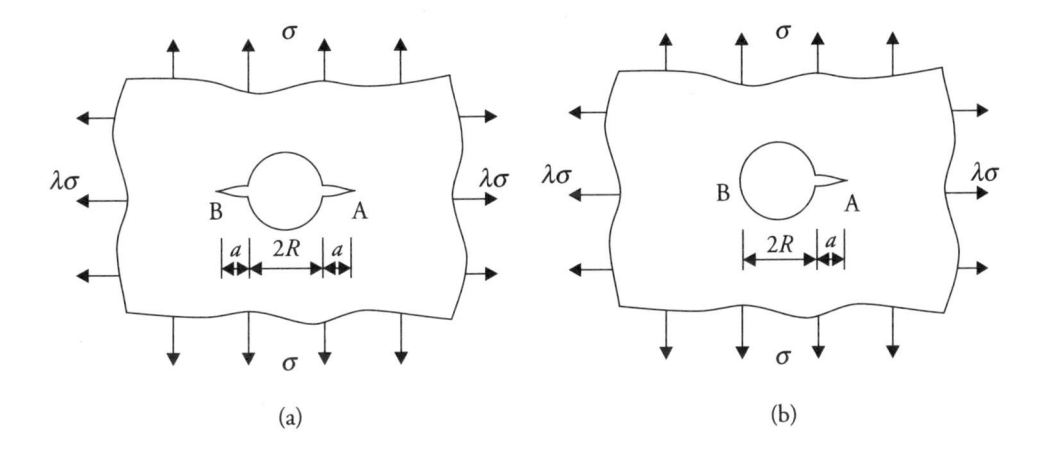

$$K_I = \sigma\sqrt{\pi a}\ f(s), \quad f(s) = (1-\lambda)f_{1(s)} + \lambda f_2(s), \quad s = \frac{a}{a+R}, \text{ for both cases (a)}$$
and (b).

$f_1(s) = 0.5(3-s)[1+1.243(1-s)^2]$ $\qquad f_1(s) = [1+0.2(1-s)+0.3(1-s)^6]f_2(s)$

$f_2(s) = 1 + [0.5 + 0.743(1-s)^3]$ $\qquad f_2(s) = 2.243 - 2.64s + 1.352s^2 - 0.248s^3$

Figure A2.1.8 Crack at hole edge in infinite plate. (a) two cracks and (b) single crack. [Tada Paris and Irwin 2000, pp.289-90]. $f_1(s)$ and $f_2(s)$ pairs given on the left and right for cases (a) and (b) repectively.

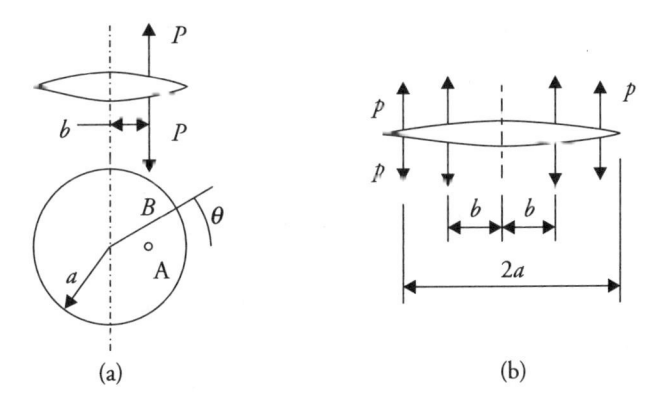

$$K_I(at\ B) = \frac{P}{\sqrt{\pi a}} \frac{\sqrt{a^2-b^2}}{(a^2+b^2-2ab\cos\theta)}, \quad K_{II} = K_{III} = 0. \quad K_I = \frac{2p}{\sqrt{\pi a}}\sqrt{a^2-b^2}$$

Figure A2.1.9 Penny-shaped crack in infinite three-dimensional body. (a) point load at A; (b) pressure loading on annular area $r = b$ to $r = a$. [Tada, Paris and Irwin 2000, p.344 and p.347].

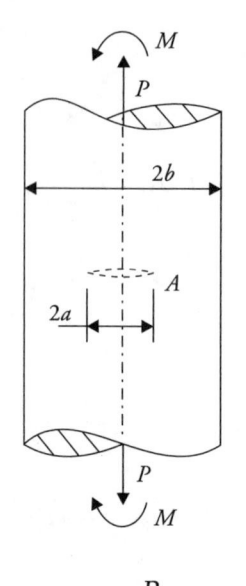

$$\sigma = \frac{P}{\pi b^2}$$

$$K_I(\text{due to } P) = \frac{2}{\pi}\sigma\sqrt{\pi a}\,f(\alpha), \quad \alpha = \frac{a}{b}, \quad f(\alpha) = \frac{1 - 0.5\alpha + 0.148\alpha^3}{\sqrt{1-\alpha}}$$

$$\sigma = \frac{4 M a}{\pi(b^4 - a^4)}, \quad \alpha = \frac{a}{b}$$

$$K_I(\text{due to M at tip } A) = \sigma\sqrt{\pi a(1-\alpha)}\,g(\alpha)$$

$$g(\alpha) = \frac{4}{3\pi}, \text{ as } \alpha \to 0$$

$$= \frac{4}{3\pi}\left[1 + 0.5\alpha + \frac{3}{8}\alpha^2 + \frac{5}{16}\alpha^3 - \frac{93}{128}\alpha^4 + 0.483\alpha^5\right]$$

Figure A2.1.10 Penny-shaped crack in shaft under tension and bending loads. [Tada, Paris and Irwin 2000, p.397 and p.399].

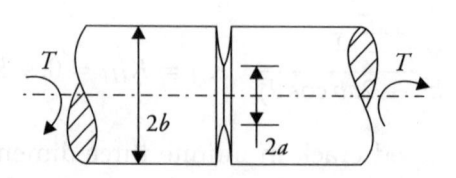

$$\tau = \frac{2T}{\pi a^3}$$

$$K_{III} = \tau\sqrt{\pi(b-a)}\, F(\alpha); \quad \alpha = \frac{a}{b}$$

$$F(\alpha) = \sqrt{\alpha}\, g(\alpha)$$

$$g(\alpha) = \frac{3}{8} \text{ as } \alpha \to 0$$

$$= \frac{3}{8}\left[1 + 0.5\alpha + \frac{3}{8}\alpha^2 + \frac{5}{16}\alpha^3 + \frac{35}{128}\alpha^4 + 0.208\alpha^5\right]$$

Figure A2.1.11 Mode III SIF for circumferential crack in shaft under torsion. [Tada, Paris and Irwin 2000, p.395].

$$K_I = \sigma\sqrt{\pi a}\, f(\alpha), \quad \alpha = \frac{D-2a}{D}$$

$$f(\alpha) = \left(\frac{1}{\alpha} + 0.5 + \frac{3}{8}\alpha - 0.361\alpha^2 + 7.33\alpha^3\right) / \sqrt{4\alpha}$$

Figure A2.1.12 Round bar with circumferential crack under tension. [Hellan 1985].

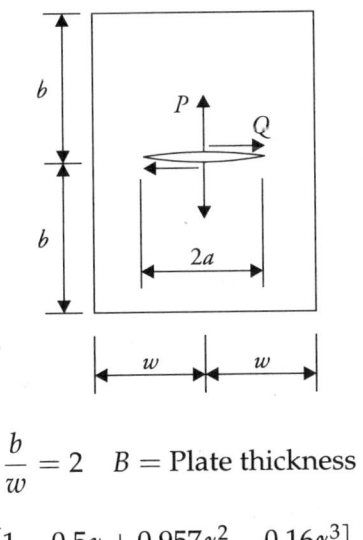

$$\frac{b}{w} = 2 \quad B = \text{Plate thickness}$$

$$K_I = \frac{P\left[1 - 0.5\alpha + 0.957\alpha^2 - 0.16\alpha^3\right]}{B\sqrt{\pi a(1-\alpha)}}, \quad \alpha = \frac{a}{w}$$

$$K_{II} = \frac{Q\left[1 - 0.5\alpha + 0.957\alpha^2 - 0.16\alpha^3\right]}{B\sqrt{\pi a(1-\alpha)}}$$

Figure A2.1.13 Crack edge loading. [Tada, Paris and Irwin 2000, p. 67].

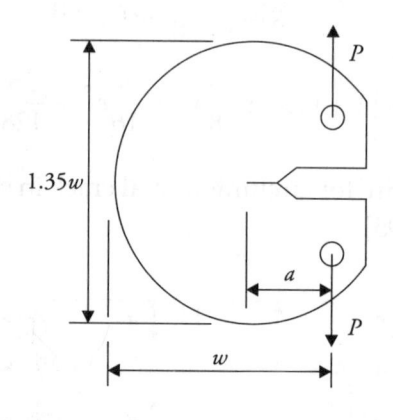

$$K_I = \frac{P}{B\sqrt{w}} f(\alpha), \quad \alpha = \frac{a}{w}$$

$$f(\alpha) = \left[(2+\alpha)(0.76 + 4.8\alpha - 11.58\alpha^2 + 11.43\alpha^3 - 4.08\alpha^4)/[(1-\alpha)^{1.5}]\right]$$

Figure A2.1.14 Disc-shaped compact tension specimen. [Tada, Paris and Irwin 2000, p. 64].

Exercise

2.1 What are the limitations of Griffith stress concentration factor based model?

2.2 Why Griffith energy balance theory could not be easily applied to practice?

2.3 In the Irwin–Orowan model, the crack-tip plastic zone size was assumed not to change with crack size. This is unrealistic. If the plastic zone size increases with crack length, what will happen to the fracture resistance?

2.4 Do you consider the SIF approach to be more convenient than the energy balance approach? Why?

2.5 Why plane strain plastic zone correction factor is taken as one-third of the plane stress correction factor?

2.6 In a thick specimen with a straight crack front under Mode I loading, where will the extension begin, at the centre or at the edges of the crack front?

2.7 For the infinite plate with a crack and loaded as shown (Fig. Q.2.7), what is the mode of loading on the crack?

Figure Q.2.7

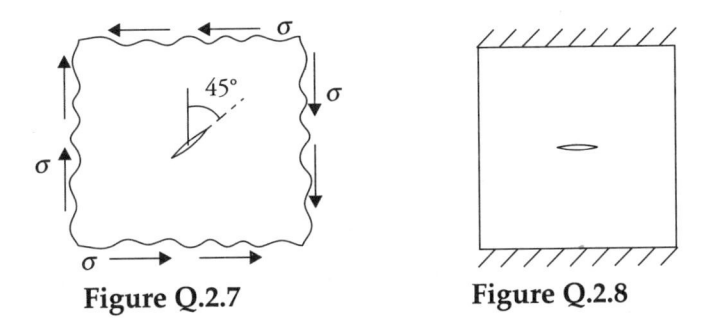

Figure Q.2.8

2.8 A uniformly thick steel plate (Fig. Q.2.8) is attached to two rigidly fixed end supports. The plate is uniformly cooled down by 20°C. What is the mode of loading the crack is subjected to?

2.9 A crack in a component is subjected to loading involving all the three modes of crack edge displacements. Would you like to give its condition of fracture in terms of SIF or energy release rate? Give reasons for your answer.

2.10 A solid shaft of a machine transmitting constant torque has an all-round circumferential crack. Obtain the energy release rate and the SIF.

2.11 Which of the two cases (Figs. Q.2.11(a) and (b)) will require more load σ to fracture?

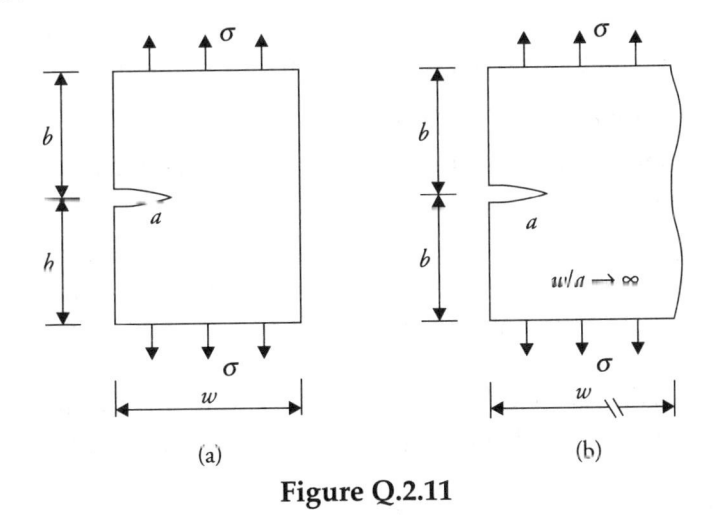

(a)

(b)

Figure Q.2.11

2.12 Why fracture toughness is more under plane stress condition than plane strain condition?

2.13 Give two practical examples of each of the three fracture modes.

2.14 A crack is opened by placing a wedge at the centre of an internal crack

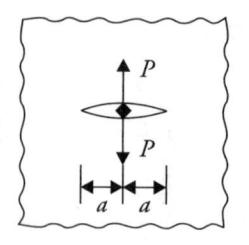

(Fig. Q.2.14). Will the crack grow stably or unstably? Explain.

2.15 18Ni-Marage 200 steel has yield point and plane strain fracture toughness 1482 MPa and 140 MPa√m, respectively. Determine the specimen thickness required for plane strain fracture toughness testing. [Ans. 22.3 mm]

2.16 What will happen to fracture load capacity if (a) there are residual compressive stresses around the crack-tip, (b) crack-tip becomes blunt, and (c) plastic zone size around the crack-tip reduces?

2.17 A cylindrical pressure vessel of radius 750 mm and thickness 30 mm is made of a material (VL-1D steel, tempered at 210°C) with $K_C = 184$ MPa√m and yield point 1372 MPa. Internal pressure is 50 MPa. Determine the required sensitivity of the inspection technique considering through the thickness longitudinal crack. [Ans. Able to detect crack size ≤13.8 mm]

2.18 Suggest some methods of increasing the fracture toughness of a material.

2.19 What is the difference between K_{IC} and K?

2.20 Solve Problem 2.5 accounting suitably for Mode I effect due to vertical component of tooth load P.

2.21 During testing of a single edge cracked specimen, the load-deflection relation was found to be of the form: $u \times P = \eta$; u in m, P in N and η is a constant. Given $K_{IC} = 73$ MPa√m, specimen thickness $B = 20$ mm, specimen width $W = 30$ mm, and yield stress $\sigma_Y = 1060$ MPa. K_{IC} was measured observing crack extension $\Delta a = 0.5$ mm. u changed from 10^{-4} m to 1.05×10^{-4} m. Determine η. Use $E = 108$ GPa and Poisson's ratio $\nu = 0.25$.

Solution

$B_C = 2.5 K_{IC}^2 / \sigma_Y^2 = 11.85$ mm, which is less than B. Hence the given toughness corresponds to the plane strain condition.

$$\text{Area AEFB} = \int_E^F P \, du = \eta \ln \frac{u_2}{u_1} = \eta \ln \frac{1.05 \times 10^{-4}}{1.00 \times 10^{-4}}$$

$$= 0.0488\eta \text{ Nm}$$

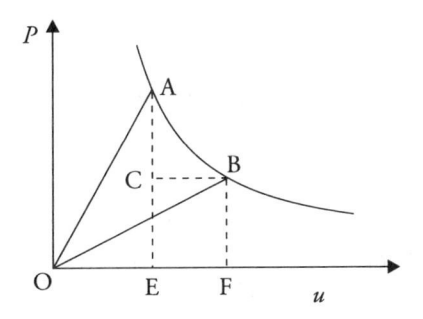

Figure Q.2.21

$P_A = 10^4 \eta$ N, $P_B = 0.952 \times 10^4 \eta$ N

Area OCA = $0.5 \times$ OE $\times$ AC = 0.024η Nm, Area OBC = 0.0238η Nm

Area ABC = $0.5 \times (P_A - P_B) \times \Delta u = 0.0012\eta$ Nm.

Note that calculation of area ABC from the two areas AEFB and CEFB gives the same answer, 0.0012η Nm.

Area OAB = work done = Δw = area (OAC + OBC + ABC) = 0.049η Nm

$G_{IC} = \Delta W/(B\Delta a) = 4.9 \times 10^3 \eta = (1-\nu^2)K_{IC}^2/E$. This gives $\eta = 9.4405$ (Ans.).

2.22 Determine the potential energy release rates associated with Mode I and Mode III loading of the double cantilever beam (DCB) specimen.

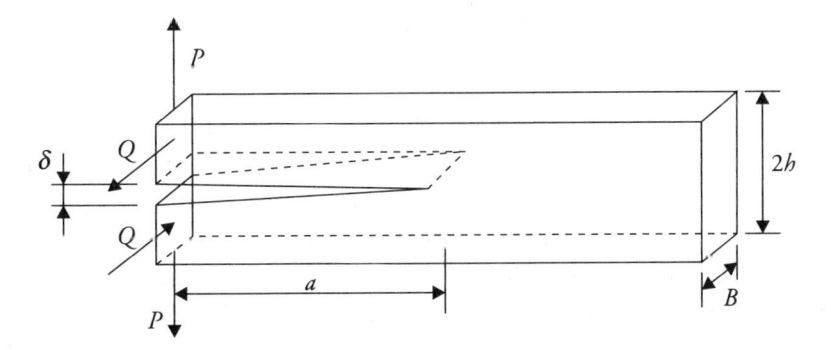

Figure Q.2.22

Hints: Calculate the crack opening δ by treating each bar segment of length a and cross-section $B \times h$ as a cantilever beam with fixed end coinciding with the crack front. Thereby, obtain compliance $c = \delta/P$. Then calculate the energy release rate using $\dfrac{P^2}{2B}\dfrac{\partial c}{\partial a}$ for Mode I. For Mode III, consider the same beam deflection parallel to the crack front.

$$\left[\text{Ans. } G_I = 12\dfrac{P^2 a^2}{EB^2 h^3},\ G_{III} = 12\dfrac{Q^2 a^2}{EB^4 h}\right]$$

2.23 Determine the safe load σ for the case shown (Fig. Q.2.23) using the particulars given. $K_{IC} = 100$ MPa$\sqrt{\text{m}}$, plate thickness $B = 25$ mm, yield point $\sigma_Y = 1200$ MPa, $E = 210$ GPa.

[Ans. 1015 MPa using SIF correction factor; 1030 MPa using no SIF correction factor].

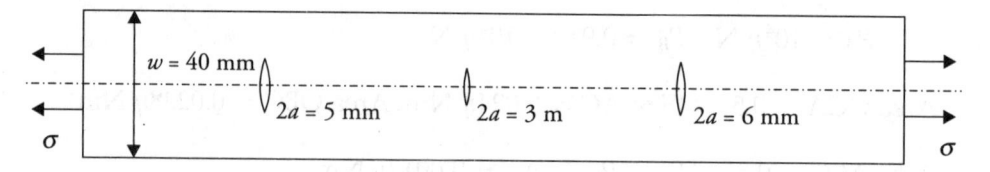

Figure Q.2.23

2.24 Determine the SIF for the case (Fig. Q.2.24). Given $L \gg 2w$ and $w \gg a$.

$$\left[\text{Ans. } \dfrac{P}{2B}\left(\dfrac{1}{2w}\sqrt{\pi a} + \dfrac{1}{\sqrt{\pi a}}\right)\right]$$

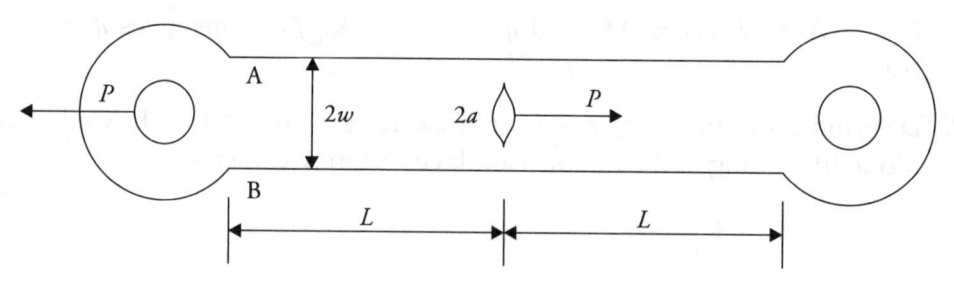

Figure Q.2.24

2.25 A beam in a mockup test (Fig. Q.2.25) developed a crack after some testing cycles. The beam central part is 15 mm thick and 40 mm deep. The beam is made of AISI 4340 steel (tempered at 400°C) with yield strength $\sigma_Y = 1400$ MPa and ultimate limit $\sigma_{ult} = 1450$ MPa, and $K_{IC} = 68$ MPa$\sqrt{\text{m}}$. The beam loading can be approximated as follows: axial load $R = 5$ kN, and $P_1 = P_2 = P$. Find the safe load the beam can carry at this stage. Comment on the loading on the crack when $P_1 > P_2$. [Ans. 3.745 kN]

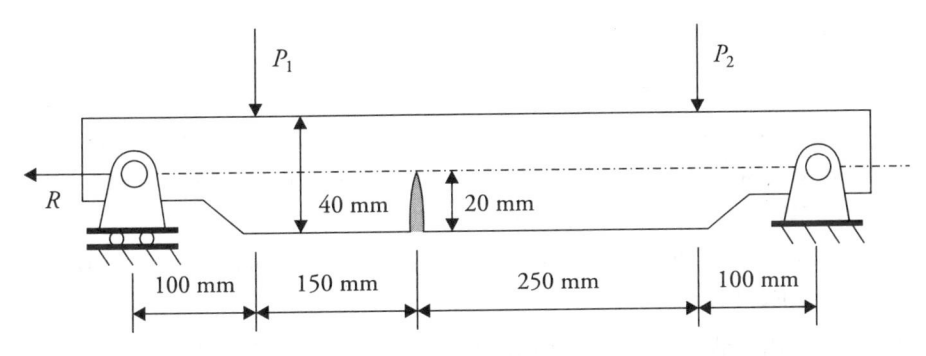

Figure Q.2.25

2.26 A long uniform rectangular bar of cross-section 40 mm × 20 mm made up of a material with fracture toughness $K_{IC} = 59$ MPa$\sqrt{m}$ and yield strength $\sigma_Y = 1500$ MPa is subjected to an axial load P. Find the load capacity P if its central section has a
   (a) a corner crack of 6 mm radius, and
   (b) an elliptical edge crack of size–semi-minor axis 5 mm and major axis 12 mm. [Ans. (a) 540 kN, (b) 445 kN]

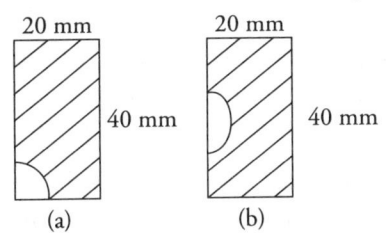

Figure Q.2.26

2.27 Suggest two minor geometric changes to reduce the SIF in a problem of given external dimensions and crack size.

2.28 What is the important limitation of the Irwin-Orowan modification of the Griffith theory of brittle fracture?

2.29 Where does the energy for creation of new surfaces come from during crack extension (i) under load control and (ii) under displacement control?

2.30 Refer Fig. 2.22. Determine the level of safety if the crack of the same size is located at the external radius at D. Account for both the bending moment and direct force on the section CD. [*Hint*: Bending moment M at the section CD $= Pr \left[\dfrac{1}{2} - \dfrac{1}{\pi} \right]$. Use appropriate correction factor for the direct compressive force]. [Ans. 10.55]

2.31 Solve Problem 2.6 accounting suitably for Mode I effect due to vertical component of tooth load P and determine % change in transmission of HP. [Ans. 20%]

References

2.1 Abdelshehid, M., K. Mohmodieh, K. Mori, L. Chen, P. Stoyanov, D. Davlantes, J. Foyos, J. Orgen, R. Clark Jr. and O.S. Es-Said. 2007. 'On the correlation between fracture toughness and precipitation hardening heat treatments in 15-5PH stainless steel.' *Engineering Failure Analysis* 14: 626–31.

2.2 Anderson, T. L. 2005. *Fracture Mechanics: Fundamentals and Applications.* Boston: CRC Press.

2.3 ASM Handbook. 1996a. 'Fatigue & Fracture', Vol. 19. Paper by S. Lapman, *Fatigue and Fracture Properties of Stainless Steels*, 621–24. Ohio: ASM International.

2.4 ———. 1996b. 'Fatigue & Fracture', Vol. 19, Paper by R.J. Bucci, G. Nordmark and E.A. Starkes Jr. *Selecting Aluminum Alloys to Resist Failure by Fracture Mechanisms*, 771–812 [refer p. 722]. Ohio: ASM International.

2.5 ———. 1996c. 'Fatigue & Fracture', Vol. 19. Paper by R.J. Bucci, G. Nordmark and E.A. Starkes Jr. *Selecting Aluminum Alloys to Resist Failure by fracture Mechanisms*, 771–812 [refer p. 776]. Ohio: ASM International.

2.6 ———. 1996d. 'Fatigue & Fracture', Vol. 19. Paper by R.J. Bucci, G. Nordmark and E.A. Starkes Jr. *Selecting Aluminum Alloys to Resist Failure by fracture Mechanisms*, 771–812 [refer p. 777]. Ohio: ASM International.

2.7 ———. 1996e. 'Fatigue & Fracture', Vol. 19. Paper by R.J. Bucci, G. Nordmark and E.A. Starkes Jr. *Selecting Aluminum Alloys to Resist Failure by fracture Mechanisms*, 771–812 [refer p. 779]. Ohio: ASM International.

2.8 ASTM E399-90. 2000 (Reapproved 1997). 'Standard Test Method for Plane-Strain Fracture Toughness of Metallic Materials.' In *Annual Book of Standards*, Section 3, Vol. 03. 01, *Metals Test Methods and Analytical Procedures*, 431–61. American Society for Testing and Materials.

2.9 Barenblatt, G.I. 1962. 'The Mathematical Theory of Equilibrium Cracks in Brittle Fracture.' In *Advances in Applied Mechanics*, Vol. VII, 55–129. New York: Academic Press.

2.10 Bluhm, J.I. 1961. 'A Model for the Effect of Thickness on Fracture Toughness'. *ASTM Proceedings* 61: 1324–31.

2.11 Barbagallo, S. and E. Cerri. 2004. 'Evaluation of K_{IC} and J_{IC} fracture parameters in a sand cast AZ91 Magnesium alloy.' *Engineering Failure Analysis* 11(1) : 127–40.

2.12 Broek, D. 1986. *Elementary Engineering Fracture Mechanics*, 4th revised edn. The Netherlands: Noordhoff.

2.13 Broek, D. and H. Vlieger. 1974. *The Thickness Effect in Plane Stress Fracture Toughness.* Amsterdam: National Aerospace Institute [Rept. TR 74032].

2.14 Brown Jr. W.F. and J.E. Srawley. 1965. 'Fracture Toughness Testing Methods'. In *Fracture Toughness Testing and its Applications*, 133–195. Philadelphia: American Society for Testing and Materials [ASTM STP 381].

2.15 ———. 1966. *Plane Strain Crack Toughness Testing of High Strength Metallic Materials.* Philadelphia: American Society for Testing and Materials [ASTM STP 410].

2.16 Crews Jr. J.H. and J.R. Reeder. 1988. *A Mixed-mode Bending Apparatus for Delamination Testing*, NASA Technical Memorandum 100662.

2.17 Ducept, F., P. Davies and D. Gamby. 2000. 'Mixed Mode Failure Criteria for a Glass/Epoxy Composite and an Adhesively Bonded Composite/Composite Joint.' *International Journal of Adhesion and Adhesives* 20: 233–44.

2.18 Dugdale, D.S. 1960. 'Yielding of Steel Sheets Containing Slits.' *Journal of Mechanics and Physics of Solids* 8: 100–04.

2.19 Gdoutos, E.E. 1993. *Fracture Mechanics – An Introduction.* Kluwer. Dordrecht/Boston/ London: Kluwer Academic Publishers.

2.20 Ghosh, S., V. Kain, A. Ray, H. Roy, S. Sivaprasad, S. Tarafdar and K.K. Ray. 2009. 'Deterioration in Fracture Toughness of 304LN Austenitic Stainless Steel due to Sensitization.' *Metallurgical and Materials Transactions A* 40(12): 2938–49.

2.21 Griffith, A.A. 1921. 'The Phenomenon of Rupture and Flow in Solid.' *Philosophical Transactions, Royal Society of London, Series A* 221: 163–69.

2.22 Hahn, G.T. and A.R. Rosenfield. 1965. 'Local Yielding and Extension of Crack Under Plane Stress.' *Acta Metallurgica* 13: 293–306.

2.23 Hellan, K. 1985. *Introduction to Fracture Mechanics.* New York: McGraw-Hill.

2.24 Hudson, C.M. and S.K. Seward. 1978. 'A Compendium of sources of Fracture Toughness and Fatigue-crack Growth Data for Metallic Alloys.' *International Journal of Fracture* 14: R151–84.

2.25 ———. 1982. 'A Compendium of sources of Fracture Toughness and Fatigue-crack Growth Data for Metallic Alloys – Part II.' *International Journal of Fracture* 20: R59–117.

2.26 ———. 1989. 'A Compendium of sources of Fracture Toughness and Fatigue-crack Growth Data for Metallic Alloys – Part III.' *International Journal of Fracture* 39: R43–63.

2.27 Hudson, C.M. and J.J. Ferrainlo. 1991. 'A Compendium of sources of Fracture Toughness and Fatigue-crack Growth Data for Metallic Alloys – Part IV.' *International Journal of Fracture* 48: R19–43.

2.28 Inglis, C.E. 1913. 'Stresses in a Plate due to the Presence of Cracks and Sharp Corners.' *Transaction of the Institute of Naval Architects* 55: 219–41.

2.29 Irwin, G.R. 1948. 'Fracture Dynamics.' In *Fracturing of Metals*, 147–66. Cleveland: ASM Publication.

2.30 ———. 1957a. 'Analysis of Stresses and Strains Near the End of a Crack Traversing a Plate.' *Journal of Applied Mechanics, Transactions of ASME* 24: 361–64.

2.31 ———. 1957b. 'Relation of stresses near a crack to the crack extension force'. *Proceedings of Ninth International Congress of Applied Mechanics*, Brussels.

2.32 ———. 1958. 'Fracture.' In *Hanbuch der Physik*, Vol. VI, ed. Flugge, S., 551–90. Berlin: Springer-Verlag.

2.33 ———. 1960. 'Plastic Zone Near a Crack and Fracture Toughness', 4–63. *Proceedings of Seventh Sagamore Conference*.

2.34 Irwin, G.R. and J.A. Kies. 1952. 'Fracture and Fracture Dynamics.' *Welding Journal Research Supplement* 31: 95S–100S.

2.35 Isherwood, D.P. and J.G. Williams. 1970. 'The Effect of Stress–Strain Properties on Notched Tensile Fracture in Plane Stress.' *Engineering Fracture Mechanics* 2: 19–22.

2.36 Kanninen, M.F. and C.H. Popelar. 1985. *Advanced Fracture Mechanics*. New York: Oxford University Press.

2.37 Kim, B., J. Do, S. Lee and I. Park. 2010. 'In Situ Fracture Observation and Fracture Toughness Analysis for Squeeze Cast AZ31-xSn Magnesium Alloys.' *Materials Science and Engineering A* 527(24–25) : 6745–757.

2.38 Knott, J.F. 1973. *Fundamentals of Fracture Mechanics*. London: Butterworths.

2.39 ———. 1975. 'The Fracture Toughness of Metals.' *Journal of Strain Analysis* 10: 201–06.

2.40 Kumar, P. 2009. *Elements of Fracture Mechanics*. New Delhi: Tata McGraw-Hill.

2.41 Liebowitz, H., ed. 1968. *Fracture – and Advanced Treatise*, Vol. II, *Mathematical Fundamentals*. New York: Academic Press.

2.42 McClintock, F.A. and G.R. Irwin. 1965. 'Plasticity Aspects of Fracture Mechanics.' In *Fracture Toughness Testing and its Applications*, 84–113. Philadelphia: American Society for Testing and Materials [ASTM STP 381].

2.43 Meguid, S.A. 1989. *Engineering Fracture Mechanics*. London: Elsevier Applied Science.

2.44 Metals Handbook. 1990a. *Properties & Selection: Nonferrous Alloys and Special-Purpose Materials*, Vol. 2. 10th edn, p. 54 and p. 77. USA: ASM International.

2.45 ———. 1990b. *Properties & Selection: Nonferrous Alloys and Special-Purpose Materials*, Vol. 2, 10th edn, 120–21. USA: ASM International.

2.46 ———. 1990c. *Properties & Selection: Nonferrous Alloys and Special-Purpose Materials*, Vol. 2, 10th edn, 117–18. USA: ASM International.

2.47 ———. 1990d. *Properties & Selection: Nonferrous Alloys and Special-Purpose Materials*, Vol. 2, 10th edn, 115–17. USA: ASM International.

2.48 ———. 1990e. *Properties & Selection: Nonferrous Alloys and Special-Purpose Materials*, Vol. 2, 10th edn, p. 112–13. USA: ASM International.

2.49 Minnay, D.P. 1998. *Fracture Mechanics*. New York: Springer-Verlag.

2.50 Murakami, Y. (Editor-in-Chief). 1987. *Stress Intensity Factors Handbook*, Vols I and II. Oxford: Pergamon Press.

2.51 Orowan, E. 1949. 'Fracture and Strength of Solids', *Report on Progress in Physics*, Vol. 12, 185–232.

2.52 ———. 1955. 'Energy Criteria of Fracture.' *Welding Journal Research Supplement* 20: 157S–160S.

2.53 Paris, P.C. and G. C. Sih. 1965. 'Stress Analysis of Cracks.' In *Fracture Toughness Testing and its Applications*, 30–81. Philadelphia: American Society for Testing and Materials [ASTM STP 381].

2.54 Parton, V.Z. and E. M. Morozov. 1978. *Elastic Plastic Fracture Mechanics*. Moscow: Mir Publishers.

2.55 Peters, S.T. , ed. 1998. *Handbook of Composites*, 2nd edn, 325–27. London: Chapman and Hall.

2.56 Putatunda, S.K. 2001. 'Fracture Toughness of a High Carbon and High Silicon Steel.' *Material Science and Engineering A* 297: 31–43.

2.57 Rooke, D.P. and D. J. Cartwright. 1975. *Compendium of Stress Intensity Factors*. Her Majesty's Stationery Office.

2.58 Sasaki, T. , H. Somekawa, A. Takara, Y. Nishikawa and K. Higashi. 2003. 'Plane-strain Fracture Toughness on Thin AZ31 Wrought Magnesium Alloy Sheets.' *Materials Transactions* 44(5): 986–90.

2.59 Sih, G.C. 1973a. *Handbook of Stress Intensity Factors*. Pennsylvania: Lehigh University.

2.60 ———, ed. 1973b. 'Methods of Analysis and Solution of Crack Problems.' In *Mechanics of Fracture*, Vol. 1. Leyden: Noordhoff International Publishing.

2.61 Somekawa, H. and T. Sakai. 2006. 'Fracture Toughness in an Extruded ZK60 Magnesium Alloy.' *Material Transactions* 47: 995–98.

2.62 Srawley, J.E. 1976. 'Wide Range Stress Intensity Factor Expressions for ASTM E399 Standard Fracture Toughness Specimens.' *International Journal of Fracture* 12: 475–76.

2.63 Suresh, S., G.F. Zamiski and R.O. Ritchie. 1982. 'Fatigue Crack Propagation Behavior of $2\frac{1}{4}$ Cr-1Mo Steels for Thick Walled Pressure Vessels.' In *Application of $2\frac{1}{4}$ Cr-1Mo Steels for Thick Walled Pressure Vessels*, eds. Sangdahl, G. S. and M. Semchyshen, 49–67. Philadelphia: American Society for Testing and Materials [ASTM STP 755].

2.64 Tada, H., P.C. Paris and G.R. Irwin. 2000. *The Stress Analysis of Cracks Handbook*, 3rd edn. New York: ASME Press.

2.65 Tetelman, A.S. and A. McEvily, Jr. 1967. *Fracture of Structural Material*. New York: John Wiley.

2.66 Timoshenko, S.P. and J.N. Goodier. 1970. *Theory of Elasticity*, International Student Edition. New York: McGraw-Hill.

3

Determination of Crack-Tip Stress Field

3.1 Introduction

In this chapter, an introduction to the Airy stress function approach and Kolosoff – Muskhelishvili potential formulation (1977) of two-dimensional elasticity is given. Westergaard stress function based solution to Mode I and Mode II crack-tip stress fields and similar solutions to all the three modes based on Williams' (1957) eigen function expansion approach are given.

3.2 Airy Stress Function Approach

Stresses at a point (Fig. 3.1) in two-dimensional elasticity is given by the three stress components σ_x, σ_y, and τ_{xy}, irrespective of whether the state of stress conforms to plane stress or plane strain. In the latter case, σ_z acting normal to the plane of the body is non-zero and it is given by $\sigma_z = \nu(\sigma_x + \sigma_y)$, where ν is Poisson's ratio. To solve for any problem of stress analysis in two dimensions, it is necessary to solve for the three stress components using the equilibrium equations.

$$\frac{\partial \sigma_x}{\partial x} + \frac{\partial \tau_{xy}}{\partial y} = 0 \tag{3.1a}$$

$$\frac{\partial \tau_{yx}}{\partial y} + \frac{\partial \sigma_y}{\partial y} = 0 \tag{3.1b}$$

$$\tau_{xy} = \tau_{yx} \tag{3.1c}$$

It is necessary to solve for the three unknown stress functions σ_x, σ_y, and τ_{xy} using the first two equilibrium equations. The problem is statically indeterminate. To overcome the indeterminacy, the compatibility condition given below is employed.

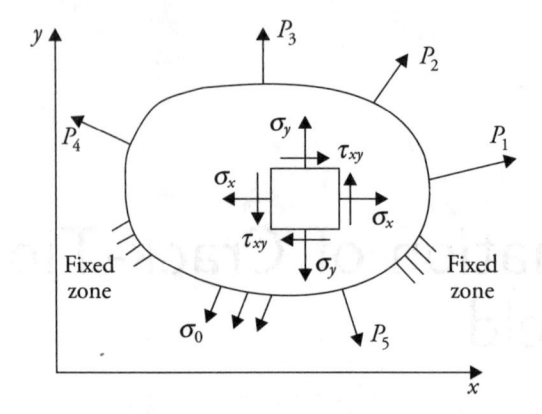

Figure 3.1 Stresses at a point.

$$\frac{\partial^2 \gamma_{xy}}{\partial x \partial y} = \frac{\partial^2 \varepsilon_x}{\partial y^2} + \frac{\partial^2 \varepsilon_y}{\partial x^2} \tag{3.2}$$

where $\varepsilon_x = \frac{\partial u}{\partial x}, \varepsilon_y = \frac{\partial v}{\partial y}$, and $\gamma_{xy} = \frac{\partial u}{\partial y} + \frac{\partial v}{\partial x}$; u and v are the displacements at a point in the two coordinate directions x and y respectively. Assuming material to be homogeneous and linear elastic, and plane state of stress, Eq. (3.2) can be written as follows.

$$2(1+\nu)\frac{\partial^2 \tau_{xy}}{\partial x \partial y} = \frac{\partial^2 (\sigma_x - \nu \sigma_y)}{\partial y^2} + \frac{\partial^2 (\sigma_y - \nu \sigma_x)}{\partial x^2} \tag{3.3}$$

Combining the equilibrium Eqs. (3.1) and the compatibility condition (3.3), it is possible to obtain the following relation.

$$\left(\frac{\partial^2}{\partial x^2} + \frac{\partial^2}{\partial x^2}\right)(\sigma_x + \sigma_y) = 0 \tag{3.4}$$

The three stresses are functions of x and y. The three functions are solved for using the three simultaneous partial differential Eqs. (3.1a), (3.1b) and (3.4). The displacements can then be obtained through integration of the strain relations.

$$\frac{\partial u}{\partial x} = \frac{1}{E}(\sigma_x - \nu \sigma_y) \tag{3.5a}$$

$$\frac{\partial v}{\partial y} = \frac{1}{E}(\sigma_y - \nu \sigma_x) \tag{3.5b}$$

where E is modulus of elasticity.

Airy expressed the stresses in terms of a single function ϕ, termed as stress function (Timoshenko and Goodier 1970; Sadd 2005), as follows.

$$\sigma_x = \frac{\partial^2 \phi}{\partial y^2}, \quad \sigma_y = \frac{\partial^2 \phi}{\partial x^2} \quad \text{and} \quad \tau_{xy} = -\frac{\partial^2 \phi}{\partial x \, \partial y}. \tag{3.6}$$

The function, irrespective of its form, satisfies the equilibrium Eqs. (3.1a) and (3.1b). To determine the function, it is necessary to ensure that it satisfies the compatibility Eq. (3.4) and the specified stress boundary conditions. After substitution of the stress function, the compatibility condition takes the form of a biharmonic equation.

$$\left(\frac{\partial^2}{\partial x^2} + \frac{\partial^2}{\partial x^2}\right)\left(\frac{\partial^2}{\partial x^2} + \frac{\partial^2}{\partial x^2}\right)\phi = 0 \quad \text{or,} \quad \nabla^2 \nabla^2 \phi = 0 \tag{3.7}$$

Many useful solutions have been obtained by writing the stress function in terms of x^n, $x^{n-1}y$, $x^{n-2}y^2$, ..., xy^{n-1}, y^n, or a combination of them involving arbitrary constants a_0, a_1, a_2, and so on. For example,

$$\phi = a_0 x^n + a_1 x^{n-1}y + a_2 x^{n-2}y^2 + \cdots + a_n y^n \tag{3.8}$$

It can be shown that the resultant forces F_x and F_y (Fig. 3.2) due to distributed tractions over the portion of boundary A to B, and the moment M_O of these tractions about the origin O are related to the stress function as follows (Timoshenko and Goodier 1970).

$$F_x = \left|\frac{\partial \phi}{\partial y}\right|_A^B, \tag{3.9a}$$

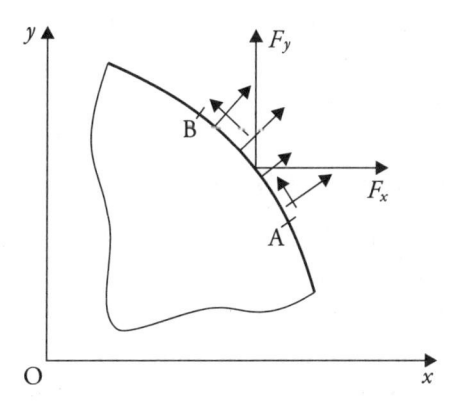

Figure 3.2 Boundary forces over span AB.

$$F_y = -\left|\frac{\partial \phi}{\partial x}\right|_A^B \tag{3.9b}$$

$$M_O = -\left|x\frac{\partial \phi}{\partial x} + y\frac{\partial \phi}{\partial y}\right|_A^B + |\phi|_A^B \tag{3.9c}$$

3.3 Kolosoff–Muskhelishvili Potential Formulation

Kolosoff and Muskhelishvili (1977) showed that the Airy stress function ϕ can be written in terms of two analytic (or regular or holomorphic) functions. These functions have real and imaginary parts, each of which satisfies the Laplace equations. Alternatively, these functions satisfy the Cauchy–Riemann conditions. They showed that

$$\phi = \operatorname{Re}\left[\bar{z}F(z) + \chi(z)\right] \tag{3.10}$$

$$\sigma_x + \sigma_y = 4\operatorname{Re}\left[F'(z)\right] \tag{3.11}$$

$$\sigma_y - \sigma_x + 2i\tau_{xy} = 2\left[\bar{z}F''(z) + \chi''(z)\right] \tag{3.12}$$

$$u + iv = \frac{3-\nu}{E}F(z) - \frac{1+\nu}{E}\left[\overline{zF'(z)} + \overline{\chi'(z)}\right] \text{ for plane stress} \tag{3.13a}$$

$$= \frac{1+\nu}{E}(3-4\nu)F(z) - \frac{1+\nu}{E}\left[\overline{zF'(z)} + \overline{\chi'(z)}\right] \text{ for plane strain} \tag{3.13b}$$

$$\sigma_\theta - \sigma_r + 2i\tau_{r\theta} = 2e^{2i\theta}\left[\bar{z}F''(z) + \chi''(z)\right] \tag{3.14}$$

$$u_\theta + u_r = e^{-i\theta}(u + iv) \tag{3.15}$$

where $F(z)$ and $\chi(z)$ are two analytic functions, $z = x + iy$, symbols (') and ('') indicate first and second derivative of the function with respect to z and the overbar sign stands for complex conjugate. σ_θ, σ_r, and $\tau_{r\theta}$ are stresses and u_θ and u_r are displacements in polar coordinates.

3.4 Examples of Analytic and Stress Functions

Consider an analytic function, $F(z) = a + ibz^2$, where a and b are real constants. Its conjugate is given by: $\overline{F(z)} = a - ib\bar{z}^2$.

$z, z^2, z^3, \ldots, z^n$ are all analytic/regular functions. For example, $F(z) = z^2 = (x + iy)^2 = x^2 - y^2 + i2xy$ is an analytic function. It can be easily verified that the real and imaginary parts, $\operatorname{Re} F(z) = (x^2 - y^2)$ and $\operatorname{Im} F(z) = 2xy$, satisfy the Laplace equation. That is, $\nabla^2 \operatorname{Re} F(z) = 0$ and $\nabla^2 \operatorname{Im} F(z) = 0$.

Consider stress functions $F'(z) = S/2$ and $\chi''(z) = Sa^2/z^2$, where S and a are both real constants. This pair of functions indicates stresses in an infinite plate with a centrally located circular cutout of radius a, with loading acting only at its outer boundary. The origin is located at the centre of the cutout and $z = re^{i\theta}$. Therefore

$$\sigma_x + \sigma_y = \sigma_\theta + \sigma_r = 4\operatorname{Re} F'(z) = 2S \tag{3.16a}$$

$$\sigma_\theta - \sigma_r + 2i\tau_{r\theta} = 2\,e^{2i\theta}\left[\bar{z}\,F''(z) + \chi''(z)\right] = 2\,e^{2i\theta}S\frac{a^2}{z^2} = 2S\frac{a^2}{r^2} \tag{3.16b}$$

$$\sigma_y - \sigma_x + 2i\tau_{xy} = 2\left[\bar{z}\,F''(z) + \chi''(z)\right] = 2S\frac{a^2}{z^2} \tag{3.16c}$$

As $r \to \infty$, $\sigma_x + \sigma_y = 2S$, $\sigma_y - \sigma_x = 0$ and $\tau_{xy} = 0$. This means that $\sigma_x = \sigma_y = S$ as $r \to \infty$. The plate is subjected to equi-biaxial normal stress S. Further at $r = a$, $\sigma_\theta + \sigma_r = 2S$, $\sigma_\theta - \sigma_r = 2S$ and $\tau_{r\theta} = 0$. Therefore, radial stress $\sigma_r = 0$ and tangential stress $\sigma_\theta = 2S$ everywhere on the hole boundary. The stress concentration factor is 2.

Similarly, for the same plate geometry with uniaxial loading, $\sigma_x = S$ acting in the x direction only the two stress functions are given by

$$F'(z) = \frac{S}{4} - \frac{S}{2}\frac{a^2}{z^2}, \quad \chi''(z) = -\frac{S}{2} + \frac{S}{2}\frac{a^2}{z^2} + 3\frac{S}{2}\frac{a^4}{z^4} \tag{3.17}$$

It can be easily shown that

$$\sigma_\theta = \frac{S}{2}\left(1 + \frac{a^2}{r^2}\right) - \frac{S}{2}\left(1 + 3\frac{a^4}{r^4}\right)\cos 2\theta \tag{3.18a}$$

$$\sigma_r = \frac{S}{2}\left(1 - \frac{a^2}{r^2}\right) + \frac{S}{2}\left(1 - 4\frac{a^2}{r^2} + 3\frac{a^4}{r^4}\right)\cos 2\theta \tag{3.18b}$$

$$\tau_{r\theta} = -\frac{S}{2}\left(1 + 2\frac{a^2}{r^2} - 3\frac{a^4}{r^4}\right)\sin 2\theta \tag{3.18c}$$

3.5 Westergaard Stress Function Approach

Westergaard (1939) examined the problems with at least one axis of symmetry and showed that stresses can be expressed in terms of only one analytic function. The steps leading to such conclusion is given as follows.

From the relations (3.11) to (3.13)

$$\sigma_y + i\tau_{xy} = 2\operatorname{Re}[F'(z)] + [\bar{z}F''(z) + \chi''(z)] \tag{3.19}$$

$$2\mu (u + iv) = \kappa F(z) - \left[\overline{zF'(z)} + \overline{\chi'(z)}\right], \quad \kappa = \frac{3-\nu}{1+\nu} \text{ for plane stress.} \quad (3.20)$$

For a problem (Fig. 3.3) with x axis as the axis of symmetry, $\tau_{xy} = 0$ along the line $y = 0$. This requires

$$\tau_{xy} = \text{Im}\,[\bar{z}F''(z) + \chi''(z)] = 0 \text{ for } y = 0. \quad (3.21)$$

This can be guaranteed by taking

$$[\bar{z}F''(z) + \chi''(z)] = A \quad (3.22)$$

where A is a real constant. There cannot be terms like $a_1 x, a_2 x^2, \ldots, b_1 y, b_2 y^2, \ldots,$ or $c_1 x^{-1}, c_2 x^{-2}, \ldots, d_1 y^{-1}, d_2 y^{-2}, \ldots,$ etc., where a_is, b_is, c_is, and d_is are real constants, on the right hand side of Eq. (3.22) because the terms with positive exponents will make stresses infinite at the outer boundary and the terms with negative exponents will make the stresses infinite on the coordinate axes. Since crack edges are traction free, $\sigma_y = 0$ and $\tau_{xy} = 0$ along the crack edges. As per Eq. (3.19), it is necessary that

$$2\,\text{Re}[F'(z)] + A = 0 \text{ for } y = 0 \text{ and } -a \geq x \leq a. \quad (3.23)$$

If we express $\chi'(z)$ as follows

$$\chi'(z) = F(z) - zF'(z) + Az \quad (3.24)$$

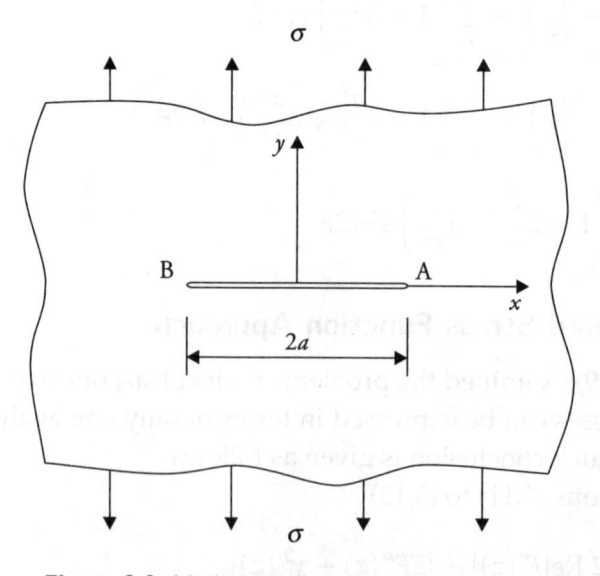

Figure 3.3 Mode I crack under remote tension.

Determination of crack-tip stress field 71

and substitute $\chi''(z)$ into the expression for τ_{xy} [Eq. (3.21)], we obtain

$$\tau_{xy} = \text{Im}[\bar{z}F''(z) + \chi''(z)] = \text{Im}[(\bar{z} - z)F''(z) + A] = -2yF''(z), \qquad (3.25)$$

which makes τ_{xy} zero for any form of $F''(z)$ along the x axis.

Introducing a new representation, $F(z) = 2F'(z)$ and $F'(z) = 2F''(z)$,

$$\sigma_x + \sigma_y = 2 \, \text{Re} \, [F(z)] \qquad (3.26a)$$

$$\sigma_y - \sigma_x + 2i\tau_{xy} = 2[(\bar{z} - z)\frac{1}{2}F'(z) + A] \qquad (3.26b)$$

This gives

$$\sigma_x = \text{Re} \, F(z) - y \, \text{Im} \, F'(z) - A \qquad (3.27a)$$

$$\sigma_y = \text{Re} \, F(z) + y \, \text{Im} \, F'(z) + A \qquad (3.27b)$$

$$\tau_{xy} = -y \, \text{Re} \, F'(z) \qquad (3.27c)$$

$F(z)$ is the Westergaard stress function for Mode I crack. The real constant A can be settled through the outer boundary conditions and Eq. (3.23) can be settled through a proper selection of the form of the Westergaard stress function. For a problem with equi-biaxial loading $A = 0$. For the stress field near the crack-tip, the constant A can be neglected.

3.5.1 Mode I Crack-Tip Field

For an internal crack in an infinite body under equi-biaxial tensile loading σ at the outer boundary (Fig. 3.4), the Westergaard stress function is given by

$$F(z) = \frac{\sigma z}{\sqrt{z^2 - a^2}} \qquad (3.28)$$

Stresses

The individual stresses are given by Eq. (3.27) with $A = 0$. It can be very easily established using Eq. (3.28) that as $z \to \infty$, $\sigma_x = \sigma_y = \sigma$ and $\tau_{xy} = 0$.

For example,

$$\sigma_x = \text{Re} \frac{\sigma z}{\sqrt{z^2 - a^2}} - y \, \text{Im} \left[\frac{\sigma}{\sqrt{z^2 - a^2}} - \frac{\sigma z^2}{(z^2 - a^2)\sqrt{z^2 - a^2}} \right]$$

$$= \text{Re} \frac{\sigma}{\sqrt{1 - a^2/z^2}} - y \, \text{Im} \left[\frac{\sigma}{z\sqrt{1 - a^2/z^2}} - \frac{\sigma}{z(1 - a^2/z^2)\sqrt{1 - a^2/z^2}} \right] \qquad (3.29)$$

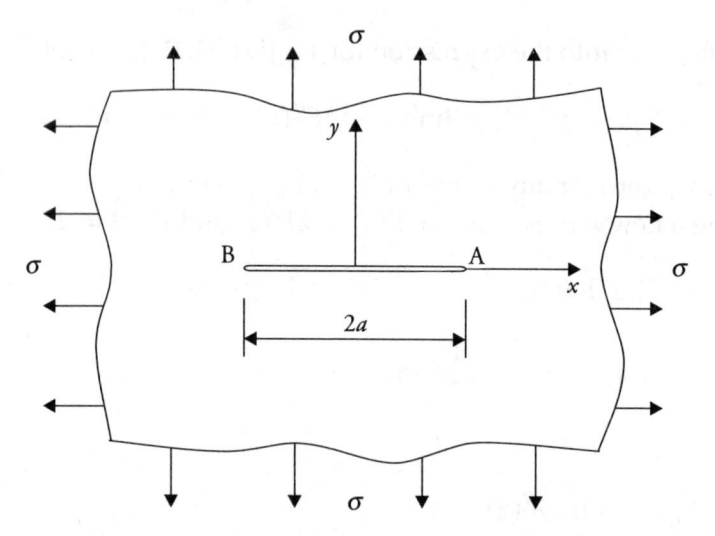

Figure 3.4 Mode I crack in infinite plate under remote equi-biaxial tension σ.

As $z \to \infty$, the term within the square bracket vanishes and denominator of the first term becomes 1. Finally, $\sigma_x = \sigma$. On the crack edges, $-a \leq x \leq a$ and $y = 0$,

$$\sigma_y = \text{Re}\frac{\sigma z}{\sqrt{z^2 - a^2}} + y \, \text{Im}\left[\frac{\sigma}{\sqrt{z^2 - a^2}} - \frac{\sigma z^2}{(z^2 - a^2)\sqrt{z^2 - a^2}}\right] = \text{Re}\frac{\sigma x}{\sqrt{x^2 - a^2}}$$

$$= \text{Re} \, i \left(\frac{\sigma x}{\sqrt{a^2 - x^2}}\right) \tag{3.30}$$

Since the quantity within the bracket is real, σ_y vanishes on the crack edges. Hence, the stress function given by Eq. (3.28) satisfies all the boundary conditions. By the uniqueness theorem of elasticity, the stress function chosen is the exact solution to the problem.

The stresses near the crack-tip (i.e., as $r \to 0$) can be obtained as follows.

$$\sigma_x = \text{Re}\frac{\sigma z}{\sqrt{z^2 - a^2}} - y \, \text{Im}\left[\frac{\sigma}{\sqrt{z^2 - a^2}} - \frac{\sigma z^2}{(z^2 - a^2)\sqrt{z^2 - a^2}}\right] \tag{3.31}$$

For $r \to 0$, the following approximation is possible: $z \approx a$, $z + a \approx 2a$. Further, selecting crack-tip as the origin, $z - a = re^{i\theta}$,

$$\sigma_x = \text{Re}\frac{\sigma a}{\sqrt{2a \, re^{i\theta}}} - r \sin\theta \, \text{Im}\left[\frac{\sigma}{\sqrt{2a \, re^{i\theta}}}\right] + r \sin\theta \, \text{Im}\left[\frac{\sigma a^2}{(2a \, re^{i\theta})\sqrt{2a \, re^{i\theta}}}\right] \tag{3.32}$$

As $r \to 0$, the first and the last terms become dominant, and the second term tends to 0. Therefore, near the crack-tip

$$\sigma_x = \frac{\sigma\sqrt{\pi a}}{\sqrt{2\pi r}}\cos\frac{\theta}{2} - \frac{\sigma\sqrt{\pi a}}{\sqrt{2\pi r}}\cos\frac{\theta}{2}\sin\frac{\theta}{2}\sin\frac{3\theta}{2} = \frac{K_I}{\sqrt{2\pi r}}\cos\frac{\theta}{2}\left(1 - \sin\frac{\theta}{2}\sin\frac{3\theta}{2}\right) \quad (3.33)$$

where $K_I = \sigma\sqrt{\pi a}$. It can be easily shown that

$$\sigma_y = \frac{K_I}{\sqrt{2\pi r}}\cos\frac{\theta}{2}\left(1 + \sin\frac{\theta}{2}\sin\frac{3\theta}{2}\right) \quad (3.34)$$

Similarly,

$$\tau_{xy} = \frac{K_I}{\sqrt{2\pi r}}\left(\sin\frac{\theta}{2}\cos\frac{\theta}{2}\cos\frac{3\theta}{2}\right) \quad (3.35)$$

Displacements

Displacements are obtained through integration of strain–stress relationship.

$$\epsilon_x = \frac{\partial u}{\partial x} = \frac{[\sigma_x - v\,\sigma_y]}{E} = \frac{1}{E}\left[(1-v)\,\mathrm{Re}\,F(z) - (1+v)\,y\,\mathrm{Im}\,F'(z)\right] \quad (3.36)$$

Upon integration

$$u = \frac{1-v}{E}\mathrm{Re}\int F(z)dx - \frac{1+v}{E}y\,\mathrm{Im}\int F'(z)dx + f(y)$$

where $f(y)$ is an arbitrary function of y. Incidentally $f(y)$ contributes to the rigid body displacements; it is neglected.

For integration with respect to x or constant y, $dz = dx + i\,dy = dx$, and noting $F(z) = \frac{\sigma z}{\sqrt{z^2 - a^2}}$,

$$u = \frac{1-v}{E}\mathrm{Re}\left[\sigma\sqrt{z^2 - a^2}\right] - \frac{1+v}{E}y\,\mathrm{Im}\left[\frac{\sigma z}{\sqrt{z^2 - a^2}}\right] \quad (3.37)$$

For the field near the crack-tip, using the approximations $z \approx a$, $z + a \approx 2a$,

$$u = \frac{1-v}{E}\mathrm{Re}\left[\sigma\sqrt{2a\,re^{i\theta}}\right] - \frac{1+v}{E}r\sin\theta\,\mathrm{Im}\left[\frac{\sigma a}{\sqrt{2a\,re^{i\theta}}}\right]$$

$$= \frac{1-v}{E}\left[\sigma\sqrt{\pi a}\sqrt{\frac{2r}{\pi}}\cos\frac{\theta}{2}\right] + \frac{1+v}{E}\left[\sigma\sqrt{\pi a}\sqrt{\frac{2r}{\pi}}\cos\frac{\theta}{2}\sin^2\frac{\theta}{2}\right]$$

$$= \frac{K_I}{E}\sqrt{\frac{2r}{\pi}}\cos\frac{\theta}{2}(1+v)\left[\frac{1-v}{1+v} + \sin^2\frac{\theta}{2}\right] \quad (3.38)$$

74 Fracture mechanics

Noting that $E = 2\mu(1+\nu)$ and $\kappa = \dfrac{3-\nu}{1+\nu}$ for plane stress

$$u = \frac{K_I}{4\mu}\sqrt{\frac{2r}{\pi}} \cos\frac{\theta}{2} \left[(\kappa - 1) + 2\sin^2\frac{\theta}{2} \right] \tag{3.39}$$

Starting from the relation for ϵ_y, y displacement is obtained.

$$\epsilon_y = \frac{\partial v}{\partial y} = \frac{[\sigma_y - \nu \sigma_x]}{E} = \frac{1}{E}[(1-\nu)\operatorname{Re} F(z) + (1+\nu)\, y\, \operatorname{Im} F'(z)] \tag{3.40}$$

$$v = \frac{1-\nu}{E} \operatorname{Re} \int F(z)\, dy + \frac{1+\nu}{E} \int y\, \operatorname{Im} F'(z)\, dy + g(x) \tag{3.41}$$

$g(x)$ is an arbitrary function of x. Since this contributes to rigid body mode, it is neglected. Further, for partial integration with respect to y or constant x, $dy = -i\, dz$, considering the second term as a product of two functions, y and $\operatorname{Im} F'(z)$,

$$v = \frac{1-\nu}{E} \operatorname{Re} \int F(z)(-i\, dz) + \frac{1+\nu}{E}\left[y \int \operatorname{Im} F'(z)\,(-i\, dz) \right.$$

$$\left. - \int \operatorname{Im}(-i) F(z)\,(-i\, dz) \right]$$

$$= \frac{1-\nu}{E} \operatorname{Re}\left[(-i)\sigma\sqrt{z^2 - a^2}\right] - \frac{1+\nu}{E}\, y\, \operatorname{Re} F(z) + \frac{1+\nu}{E} \operatorname{Im} \sigma\sqrt{z^2 - a^2}$$

$$= \frac{1-\nu}{E} \operatorname{Im}\left[\sigma\sqrt{z^2 - a^2}\right] - \frac{1+\nu}{E}\, y\, [\operatorname{Re} F(z)] + \frac{1+\nu}{E} \operatorname{Im}\left[\sigma\sqrt{z^2 - a^2}\right]$$

$$= \frac{2}{E} \operatorname{Im}\left[\sigma\sqrt{z^2 - a^2}\right] - \frac{1+\nu}{E}\, y\, \operatorname{Re}[F(z)]$$

$$= \frac{2}{E} \operatorname{Im}\left[\sigma\sqrt{z^2 - a^2}\right] - \frac{1+\nu}{E}\, y\, \operatorname{Re}\left[\frac{\sigma z}{\sqrt{z^2 - a^2}}\right]$$

$$= \frac{2}{E} \operatorname{Im}\left[\sigma\sqrt{2a\, re^{i\theta}}\right] - \frac{1+\nu}{E}\, r\sin\theta\, \operatorname{Re}\left[\frac{\sigma a}{\sqrt{2a\, re^{i\theta}}}\right]$$

$$= \frac{K_I}{E}\sqrt{\frac{2r}{\pi}} \sin\frac{\theta}{2}\left[2 - (1+\nu)\cos^2\frac{\theta}{2} \right] \tag{3.42}$$

Finally,

$$v = \frac{K_I}{4\mu}\sqrt{\frac{2r}{\pi}}\sin\frac{\theta}{2}\left[(\kappa+1)-2\cos^2\frac{\theta}{2}\right] \qquad (3.43)$$

Equations (3.39) and (3.43) are valid for plane strain when $\kappa = 3 - 4\nu$.

3.5.2 Mode II Crack-Tip Field

Irwin extended (Tada, Paris and Irwin 2000) the Westergaard stress function approach to Mode II (Fig. 3.5). The stress functions and the stresses for such a case are given by

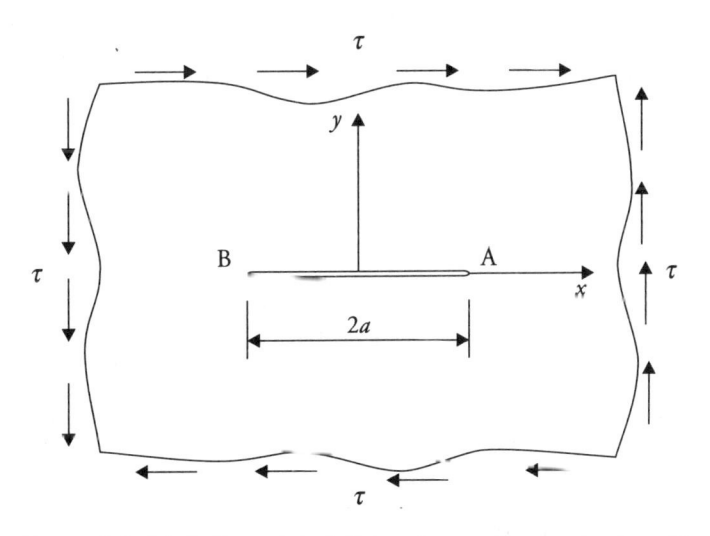

$$F(z) = \frac{-i\tau z}{\sqrt{z^2 - a^2}} \qquad (3.44)$$

$$\sigma_x = 2\,\mathrm{Re}\,F(z) - y\,\mathrm{Im}\,F'(z) \qquad (3.45a)$$

$$\sigma_y = y\,\mathrm{Im}\,F'(z) \qquad (3.45b)$$

$$\tau_{xy} = -\mathrm{Im}\,F(z) - y\,\mathrm{Re}\,F'(z) \qquad (3.45c)$$

Figure 3.5 Mode II crack in infinite plate under remote shear τ.

Starting from relation (3.45a)

$$\sigma_x = 2\,\mathrm{Re}\frac{-i\tau z}{\sqrt{z^2-a^2}} - y\,\mathrm{Im}\left[\frac{-i\tau}{\sqrt{z^2-a^2}} - \frac{-i\tau z^2}{(z^2-a^2)\sqrt{z^2-a^2}}\right]$$

$$= 2\,\text{Im}\frac{\tau z}{\sqrt{z^2-a^2}} - r\sin\theta\,\text{Re}\frac{\tau z^2}{(z^2-a^2)\sqrt{z^2-a^2}};$$

the second term is neglected as it vanishes for $r \to 0$.

Therefore, using approximations $z \approx a$, $z+a \approx 2a$, and $z-a = re^{i\theta}$,

$$\sigma_x = 2\,\text{Im}\frac{\tau a}{\sqrt{2a\,re^{i\theta}}} - r\sin\theta\,\text{Re}\frac{\tau a^2}{(2a\,re^{i\theta})\sqrt{2a\,re^{i\theta}}}$$

$$= -2\frac{\tau\sqrt{\pi a}}{\sqrt{2\pi r}}\sin\frac{\theta}{2} - \frac{\tau\sqrt{\pi a}}{\sqrt{2\pi r}}\cos\frac{\theta}{2}\sin\frac{\theta}{2}\sin\frac{3\theta}{2}$$

$$= -\frac{\tau\sqrt{\pi a}}{\sqrt{2\pi r}}\sin\frac{\theta}{2}\left[2 + \cos\frac{\theta}{2}\cos\frac{3\theta}{2}\right]$$

$$= -\frac{K_{II}}{\sqrt{2\pi r}}\sin\frac{\theta}{2}\left[2 + \cos\frac{\theta}{2}\cos\frac{3\theta}{2}\right] \quad (3.46)$$

where $K_{II} = \tau\sqrt{\pi a}$. Again,

$$\sigma_y = y\,\text{Im}\,F'(z) = y\,\text{Im}\left[\frac{-i\tau}{\sqrt{z^2-a^2}} - \frac{-i\tau z^2}{(z^2-a^2)\sqrt{z^2-a^2}}\right] \quad (3.47)$$

Based on the evaluation of the last term of σ_x it can be written that

$$\sigma_y = \frac{K_{II}}{\sqrt{2\pi r}}\sin\frac{\theta}{2}\cos\frac{\theta}{2}\cos\frac{3\theta}{2} \quad (3.48)$$

$$\tau_{xy} = -\text{Im}\,F(z) - y\,\text{Re}\,F'(z)$$

$$= -\text{Im}\frac{i\tau z}{\sqrt{z^2-a^2}} - y\,\text{Re}\left[\frac{-i\tau}{\sqrt{z^2-a^2}} - \frac{-i\tau z^2}{(z^2-a^2)\sqrt{z^2-a^2}}\right]$$

The second term tends to zero as $r \to 0$. Therefore, using earlier approximations and $z - a = re^{i\theta}$,

$$\tau_{xy} = \text{Re}\frac{\tau z}{\sqrt{z^2-a^2}} + y\,\text{Im}\frac{\tau z^2}{(z^2-a^2)\sqrt{z^2-a^2}}$$

$$= \text{Re}\frac{\tau a}{\sqrt{2a\,re^{i\theta}}} + r\sin\theta\,\text{Im}\frac{\tau a^2}{(2a\,re^{i\theta})\sqrt{2a\,re^{i\theta}}}$$

$$= \frac{K_{II}}{\sqrt{2\pi r}}\cos\frac{\theta}{2}\left[1 - \sin\frac{\theta}{2}\sin\frac{3\theta}{2}\right] \quad (3.49)$$

The displacements are again obtained starting from the expressions for strains ϵ_x and ϵ_y.

$$\epsilon_x = \frac{\partial u}{\partial x} = \frac{1}{E}[\sigma_x - \nu\, \sigma_x] = \frac{1}{E}[\,2\,\mathrm{Re}\,F(z) - (1+\nu)\,y\,\mathrm{Im}\,F'(z)] \qquad (3.50)$$

Integrating

$$u = \frac{2}{E}\mathrm{Re}\int F(z)dx - \frac{1+\nu}{E}\,y\,\mathrm{Im}\int F'(z)dx + f(y) \qquad (3.51)$$

Neglecting the arbitrary function $f(y)$ and noting the form of $F(z)$

$$u = \frac{2}{E}\mathrm{Re}(-i\,\tau\,\sqrt{z^2-a^2}) - \frac{1+\nu}{E}\,y\,\mathrm{Im}\left(\frac{-i\,\tau\,z}{\sqrt{z^2-a^2}}\right) \qquad (3.52)$$

For the crack-tip field substitution is made: $z \approx a$, $z+a \approx 2a$ and $z-a = re^{i\theta}$.

$$u = \frac{2}{E}\mathrm{Im}\left(\tau\sqrt{2a\,re^{i\theta}}\right) + \frac{1+\nu}{E}\,r\,\sin\theta\,\mathrm{Re}\left(\frac{\tau\,a}{\sqrt{2a\,re^{i\theta}}}\right)$$

$$= \tau\sqrt{\pi a}\sqrt{\frac{2r}{\pi}}\sin\frac{\theta}{2}\left[\frac{2}{E} + \frac{1+\nu}{E}\cos^2\frac{\theta}{2}\right] = \frac{K_{II}}{4\mu}\sqrt{\frac{2r}{\pi}}\sin\frac{\theta}{2}\left[(\kappa+1) + 2\cos^2\frac{\theta}{2}\right] \qquad (3.53)$$

Similarly from the expression of ϵ_y, v is obtained, neglecting the arbitrary function of x associated with integration, as follows.

$$\epsilon_y = \frac{\partial v}{\partial y} = \frac{1}{E}[\sigma_y - \nu\,\sigma_x] = \frac{1}{E}[(1+\nu)\,y\,\mathrm{Im}\,F'(z) - 2\,\nu\,\mathrm{Re}\,F(z)] \qquad (3.54)$$

$$v = \frac{1}{E}\left[-(1+\nu)\,y\,\mathrm{Im}\,\tau\sqrt{z^2-a^2} - (1-\nu)\,\mathrm{Re}\,\tau\sqrt{z^2-a^2}\right]$$

$$= -\frac{K_{II}}{4\mu}\sqrt{\frac{2r}{\pi}}\left[(\kappa-1)\cos\frac{\theta}{2} - \sin\theta\,\sin\frac{\theta}{2}\right]$$

$$= -\frac{K_{II}}{8\mu}\sqrt{\frac{2r}{\pi}}\left[(2\kappa-3)\cos\frac{\theta}{2} + \cos\frac{3\theta}{2}\right] \qquad (3.55)$$

3.6 Mode III Solution

Mode III represents a problem in three-dimension (Fig. 3.6). The solution for crack-tip field in this case cannot be obtained through the Kolosoff–Muskhelishvili complex formulation, which is suitable only for problems in two dimensions. For this problem displacements $u = v = 0$ and the displacement in z

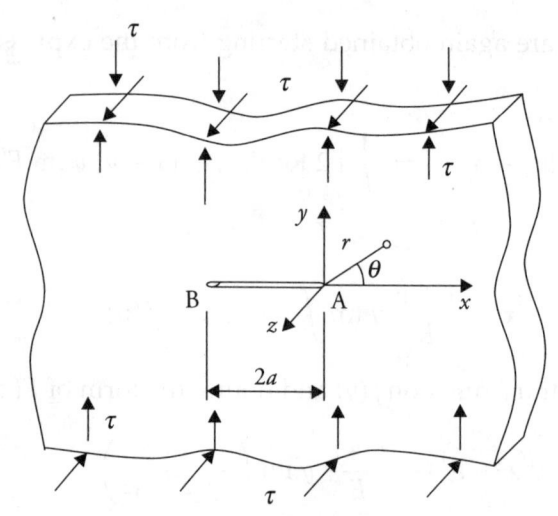

Figure 3.6 Mode III crack in infinite plate under remote out-of-plane shear τ.

coordinate direction $w = w(x, y) \neq 0$. Therefore $\epsilon_x = \epsilon_y = \epsilon_z = \gamma_{xy} = 0$, and $\gamma_{xz} = \dfrac{\partial w}{\partial x}$ and $\gamma_{yz} = \dfrac{\partial w}{\partial y}$. This also means that $\sigma_x = \sigma_y = \sigma_z = 0$ and $\tau_{xy} = \mu \gamma_{xy} = 0$, and $\tau_{xz} = \mu \dfrac{\partial w}{\partial x}$ and $\tau_{yz} = \mu \dfrac{\partial w}{\partial y}$. These stresses will satisfy identically the equilibrium equation in the x and y directions. The third equilibrium equation will be satisfied provided

$$\mu \left(\frac{\partial^2 w}{\partial x^2} + \frac{\partial^2 w}{\partial y^2} \right) = 0, \text{ or } \nabla^2 w = 0 \tag{3.56}$$

Hence, w is a harmonic function. w can be written in terms of an analytic function $F(z)$ as follows.

$$w = \frac{1}{\mu} \left[F(z) + \overline{F(z)} \right] \tag{3.57}$$

This gives

$$\tau_{xz} - i\,\tau_{yz} = \frac{\partial w}{\partial x} - i \frac{\partial w}{\partial y} = 2F'(z) \tag{3.58}$$

Assuming

$$F(z) = (A + iB)\, z^{\lambda+1} \tag{3.59}$$

where A, B and λ are real constants.

$$\tau_{xz} - i\,\tau_{yz} = 2(\lambda + 1)(A + iB)\, z^\lambda$$

$$= 2(\lambda+1) r^\lambda [A \cos\lambda\theta - B \sin\lambda\theta + i (A \sin\lambda\theta + B \cos\lambda\theta)] \quad (3.60)$$

The displacement field

$$w = \frac{1}{\mu}\left[(A+iB) z^{\lambda+1} + (A-iB) \bar{z}^{\lambda+1}\right]$$

$$= \frac{1}{\mu} r^{\lambda+1}\left[(A+iB) e^{i(\lambda+1)\theta} + (A-iB) e^{-i(\lambda+1)\theta}\right] \quad (3.61)$$

The displacement field shows a typical behaviour in the domain depending on the value of λ. For example, at the crack-tip ($r = 0$), the displacements becomes infinity if $\lambda < -1$. Therefore, for finiteness of displacements in the domain including the crack-tip, λ must be greater than -1. $\lambda = -1$ is not acceptable because it makes the stresses given by Eq. (3.60) to vanish everywhere.

Since the crack faces ($\theta = \pm\pi$) are free of any stresses, for example, τ_{yz},

$$A \sin\lambda\pi + B \cos\lambda\pi = 0 \quad (3.62a)$$

$$-A \sin\lambda\pi + B \cos\lambda\pi = 0 \quad (3.62b)$$

These two relations represent an eigenvalue problem, where A and B can be considered as modal parameters and λ as the eigenvalue. The characteristics equation from which λ can be obtained is given by

$$\begin{vmatrix} \sin\lambda\pi & \cos\lambda\pi \\ -\sin\lambda\pi & \cos\lambda\pi \end{vmatrix} = 0 \quad \text{or} \quad \sin 2\lambda\pi = 0 \quad (3.63)$$

Hence, $\lambda = \pm\frac{n}{2}$, where n is an integer. λ can have values: $-\infty, \ldots, -1, -\frac{3}{2}, -\frac{1}{2}, 0, \frac{1}{2}, 1, \frac{3}{2}, \ldots, +\infty$. In view of the restriction on λ, it can only have values: $-\frac{1}{2}, 0, \frac{1}{2}, 1, \frac{3}{2}, \ldots, +\infty$. $\lambda = -\frac{1}{2}$ gives rise to square-root stress singularity at the crack-tip. It is of interest at this point. Substituting this value of λ in Eq. (3.62), it is observed that $A = 0$ and $B \neq 0$. This gives

$$\tau_{xz} = B r^{-\frac{1}{2}} \sin\frac{\theta}{2}, \quad \tau_{yz} = -B r^{-\frac{1}{2}} \cos\frac{\theta}{2} \quad (3.64)$$

Adopting Irwin's definition of stress intensity factor (SIF)

$$K_{III} = \lim_{r \to 0} \left[\sqrt{2\pi r} \, (\tau_{yz})_{\theta=0}\right] \quad (3.65)$$

$$B = -\frac{K_{III}}{\sqrt{2\pi}} \tag{3.66}$$

Finally, the crack-tip solution is given by

$$\tau_{xz} = -\frac{K_{III}}{\sqrt{2\pi r}} \sin\frac{\theta}{2}, \quad \tau_{yz} = \frac{K_{III}}{\sqrt{2\pi r}} \cos\frac{\theta}{2} \tag{3.67}$$

$$w = \frac{K_{III}}{\mu} \sqrt{\frac{2r}{\pi}} \sin\frac{\theta}{2} \tag{3.68}$$

For other values of λ, the solutions for stresses can be obtained. Thereby, stresses can be written in the form of infinite series.

$$\tau_{xz} = -\frac{K_{III}}{\sqrt{2\pi r}} \sin\frac{\theta}{2} + A_0 + A_1 r^{\frac{1}{2}} f_1(\theta) + A_2 r^1 f_2(\theta) + \ldots\ldots \tag{3.69}$$

$$\tau_{xz} = -\frac{K_{III}}{\sqrt{2\pi r}} \cos\frac{\theta}{2} + B_1 r^{\frac{1}{2}} g_1(\theta) + B_2 r^1 g_2(\theta) + \ldots\ldots \tag{3.70}$$

where A_i and B_i are arbitrary constants to be determined from the given boundary conditions, and $f_i(\theta)$ and $g_i(\theta)$ are known functions of θ. B_0 has been excluded from Eq. (3.70) because there is no contribution to τ_{xz} corresponding to $\lambda = 0$.

3.7 Williams' Eigenfunction Expansion for Mode I

Williams too solved the Mode I and Mode II problems through the Kolosoff–Muskhelishvili potential formulation. Noting the possibility of existence of stress singularity at the crack-tip and symmetric stress-displacement field about the crack plane in Mode I, he assumed the two stress functions involved in Eqs. (3.11) to (3.13) in the following forms.

$$F(z) = A z^{\lambda+1}, \; \chi'(z) = B z^{\lambda+1} \tag{3.71}$$

where $A = C + iD$ and $B = E + iF$. C, D, E, F, and λ are all real constants. He considered the origin to be located at the crack-tip and x axis collinear with the crack. Since in this case, stresses σ_x and σ_y are symmetric about x axis, and τ_{xy} is anti-symmetric about the same axis, it can be easily established that $D = 0$ and $F = 0$. From Eq. (3.11)

$$\sigma_y + \sigma_x = 4\mathrm{Re}\left[F'(z)\right] = 4(\lambda+1) r^\lambda \left[C \cos\lambda\theta - D \sin\lambda\theta\right] \tag{3.72}$$

For symmetric distribution of $\sigma_y + \sigma_x$ about x axis, the sum must be only a cosine function of $(\lambda\theta)$. That means $D = 0$. In a similar manner, starting from Eq. (3.12), the expression for $\tau_{xy} = (\lambda + 1) r^\lambda [F \cos \lambda\theta + E \sin \lambda\theta]$ is obtained. Since it must be only a sine function of $(\lambda\theta)$, it is obvious that $F = 0$.

From Eq. (3.13), it is noted that u and v is proportional to $r^{\lambda+1}$. For finiteness of displacements in the whole domain including the crack-tip ($r = 0$), λ must be greater than -1. Using $F(z) = C z^{\lambda+1}$ and $\chi'(z) = E z^{\lambda+1}$ using the Eqs. (3.11) and (3.12),

$$\sigma_y = 2C(\lambda + 1) r^\lambda \cos \lambda\theta + C\lambda(\lambda + 1) r^\lambda \cos(\lambda - 2)\theta$$
$$+ E(\lambda + 1) r^\lambda \cos \lambda\theta \tag{3.73}$$

$$\tau_{xy} = C\lambda(\lambda + 1) r^\lambda \sin(\lambda - 2)\theta + E(\lambda + 1) r^\lambda \sin \lambda\theta \tag{3.74}$$

In order to ensure $\sigma_y = 0$ and $\tau_{xy} = 0$ for $\theta = \pi$, the following two equations are obtained.

$$C(\lambda + 2) \cos \lambda\pi + E \cos \lambda\pi = 0 \tag{3.75a}$$

$$C\lambda \sin \lambda\pi + E \sin \lambda\pi = 0 \tag{3.75b}$$

These equations represent an eigenvalue problem. The characteristic equation to solve for λ gives $\lambda = \pm \dfrac{n}{2}$, where n is an integer varying from $-\infty$ to ∞. Therefore, λ can have values $-\infty, \ldots, -1, -\dfrac{1}{2}, 0, \dfrac{1}{2}, \dfrac{3}{2}, \ldots, +\infty$. In view of the restriction on λ, it can have values: $-\dfrac{1}{2}, 0, \dfrac{1}{2}, 1, \dfrac{3}{2}, \ldots +\infty$. $\lambda = -\dfrac{1}{2}$ gives rise to square-root stress singularity at the crack-tip. It is again of interest at this point. Substituting this value of λ in Eq. (3.75), it is observed that $E = C/2$. This gives

$$\sigma_y = C r^{-\frac{1}{2}} \left[\frac{5}{4} \cos \frac{\theta}{2} - \frac{1}{4} \cos \frac{5\theta}{2} \right] \tag{3.76a}$$

$$\tau_{xy} = C r^{-\frac{1}{2}} \left[\frac{1}{4} \sin \frac{5\theta}{2} - \frac{1}{4} \sin \frac{\theta}{2} \right] \tag{3.76b}$$

Using Irwin's definition of SIF

$$K_I = \lim_{r \to 0} \left[\sqrt{2\pi r} \, (\sigma_y)_{\theta=0} \right], \quad C = \frac{K_I}{\sqrt{2\pi}} \tag{3.77}$$

Finally, therefore

$$\sigma_y = \frac{K_I}{\sqrt{2\pi r}} \left[\frac{5}{4} \cos \frac{\theta}{2} - \frac{1}{4} \cos \frac{5\theta}{2} \right] = \frac{K_I}{\sqrt{2\pi r}} \cos \frac{\theta}{2} \left[1 + \sin \frac{\theta}{2} \sin \frac{3\theta}{2} \right] \tag{3.78a}$$

$$\tau_{xy} = \frac{K_I}{\sqrt{2\pi r}} \left[\frac{1}{4} \sin \frac{5\theta}{2} - \frac{1}{4} \sin \frac{\theta}{2} \right] = \frac{K_I}{\sqrt{2\pi r}} \sin \frac{\theta}{2} \cos \frac{\theta}{2} \cos \frac{3\theta}{2} \quad (3.78b)$$

From Eq. (3.72)

$$\sigma_y + \sigma_x = 2C \cos \frac{\theta}{2} = 2 \frac{K_I}{\sqrt{2\pi r}} \cos \frac{\theta}{2}$$

$$\sigma_x = \frac{K_I}{\sqrt{2\pi r}} \cos \frac{\theta}{2} \left[1 - \sin \frac{\theta}{2} \sin \frac{3\theta}{2} \right] \quad (3.79)$$

The displacements are given by

$$u + iv = \frac{3-\nu}{E} F(z) - \frac{1+\nu}{E} \left[\overline{zF'(z) + \chi'(z)} \right]$$

$$= \frac{3-\nu}{E} \frac{K_I}{\sqrt{2\pi}} z^{\frac{1}{2}} - \frac{1+\nu}{E} \left[z \frac{K_I}{\sqrt{2\pi}} \frac{1}{2} (\bar{z})^{-\frac{1}{2}} + \frac{1}{2} \frac{K_I}{\sqrt{2\pi}} (\bar{z})^{\frac{1}{2}} \right]$$

$$= \frac{K_I}{2E} \sqrt{\frac{2r}{\pi}} \left[(3-\nu) \left(\cos \frac{\theta}{2} + i \sin \frac{\theta}{2} \right) \right.$$

$$\left. - \frac{1+\nu}{2} \left\{ \left(\cos \frac{3\theta}{2} + i \sin \frac{3\theta}{2} \right) + \left(\cos \frac{\theta}{2} - i \sin \frac{\theta}{2} \right) \right\} \right] \quad (3.80)$$

Hence

$$u = \frac{K_I}{4\mu} \sqrt{\frac{2r}{\pi}} \cos \frac{\theta}{2} \left[(\kappa - 1) + 2 \sin^2 \frac{\theta}{2} \right], \quad v = \frac{K_I}{4\mu} \sqrt{\frac{2r}{\pi}} \sin \frac{\theta}{2} \left[(\kappa + 1) - 2 \cos^2 \frac{\theta}{2} \right] \quad (3.81)$$

The stresses corresponding to the other values of λ (0, $\frac{1}{2}$, 1, $\frac{3}{2}$, etc.) can be similarly obtained. The general form of the stress function can be written by summing up all the eigen solutions as follows.

$$F(z) = \sum_{m=1}^{\infty} A_m z^{\lambda_m + 1}, \quad (3.82a)$$

$$\chi'(z) = \sum_{m=1}^{\infty} B_m z^{\lambda_m + 1} \quad (3.82b)$$

where $m = 1, 2, 3, 4, \ldots, \infty$ \quad (3.83a)

$$\lambda_m = -\frac{1}{2}, 0, \frac{1}{2}, 1, \frac{3}{2}, \ldots \quad (3.83b)$$

It may be noted here that for $\lambda_m = 0$, σ_x is constant and is independent of θ, and $\sigma_y = \tau_{xy} = 0$.

3.8 Williams' Eigenfunction Expansion for Mode II and Mixed Mode

Proceeding in the same manner as in the case of Mode I, it is possible to obtain the solution for this case. Since it is anti-symmetric problem about the x axis, τ_{xy} is a symmetric function of θ, and σ_x and σ_y are anti-symmetric functions of θ. This will lead to conclusion that the constants C and D involved in the stress functions

$$F(z) = C z^{\lambda+1}, \qquad (3.84a)$$

$$\chi'(z) = D z^{\lambda+1} \qquad (3.84b)$$

are purely imaginary. The general form of the stress function can be written by summing up all the eigen solutions as follows.

$$F(z) = \sum_{m=1}^{\infty} -i\, C_m z^{\lambda_m+1}, \qquad (3.85a)$$

$$\chi'(z) = \sum_{m=1}^{\infty} -i\, D_m z^{\lambda_m+1} \qquad (3.85b)$$

where $m = 1, 2, 3, 4, \ldots, \infty$, C_m and D_m are all real, and

$$\lambda_m = -\frac{1}{2},\, 0,\, \frac{1}{2},\, 1,\, \frac{3}{2}, \ldots \qquad (3.86)$$

Combining the solutions for the two modes, general stress function for any two dimensional problem is obtained in the following form.

$$F(z) = \sum_{m=1}^{\infty} (A_m - i\, C_m) z^{\lambda_m+1}, \qquad (3.87a)$$

$$\chi'(z) = \sum_{m=1}^{\infty} (B_m - i\, D_m) z^{\lambda_m+1} \qquad (3.87b)$$

From the definition of stress intensity factors, K_I and K_{II} are obtained in terms of the arbitrary constants A_m and B_m.

$$K_I - iK_{II} = \lim_{z \to 0} 2\sqrt{2\pi z}\, F'(z) \qquad (3.88)$$

That is, $K_I = \sqrt{2\pi}\, A_1$ and $K_{II} = \sqrt{2\pi}\, C_1$. Therefore for an assessment of the SIFs for a problem, it is necessary to know only the coefficient of the first term of $F(z)$.

For a mixed crack in a finite plate (Fig. 3.7), stresses can be calculated using a truncated series for the stress functions (Eq. (3.87)). Boundary collocation

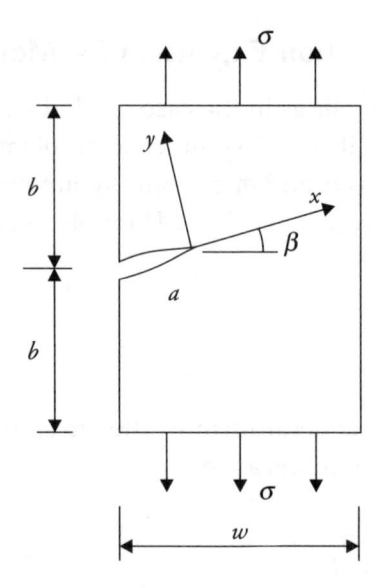

Figure 3.7 Mixed mode crack.

technique (e.g., Bowie 1973) can be employed to solve for the arbitrary constants A_m, B_m, C_m, and D_m using the specified stress boundary conditions. The SIFs are obtained by just knowing the constants A_1 and C_1. While applying the boundary collocation technique care must be exercised to ensure that the coordinate axis x is aligned with the crack (Fig. 3.7) and originates from the crack-tip.

Exercise

3.1 Determine the Mode II crack-tip stress and displacement field by the eigenfunction approach.

3.2 Determine the stress field for a rectangular domain, length $= 2l$ and depth $= 2h$, given by the Airy stress function $\phi = c\,xy$, where c is a real constant and origin is at the centre of the domain. Axis y is parallel to depth direction.
[Ans. $\tau_{xy} = -c$]

3.3 Determine the Airy stress function for the above domain if it is subjected to a constant tensile stress $\sigma_x = \sigma_0$ at its edges parallel to the y axis.
[Ans. $\phi = \tfrac{1}{2}\sigma_0 y^2$]

References

3.1 Bowie, O.L. 1973. 'Solution of Crack Problems by Mapping Technique.' In *Methods of Analysis and Solution of Crack Problems, Mechanics of Fracture*, Vol. 1, ed. Sih, G.C. Leyden: Noordhoff International Publishing.

3.2 Muskhelishvili, N.I. 1977. *Some Basic Problems of the Mathematical Theory of Elasticity*, 4th edn. Leyden: Noordhoff International Publishing.

3.3 Sadd, M.H. 2005. *Elasticity*. New Delhi: Academic Press.

3.4 Tada, H., P.C. Paris and G.R. Irwin. 2000. *The Stress Analysis of Cracks Handbook*, 3rd edn. New York: ASME Press.

3.5 Timoshenko, S.P. and J.N. Goodier. 1970. *Theory of Elasticity*, International Student Edition. New York: McGraw-Hill.

3.6 Westergaard, H.M. 1939. 'Bearing Pressure and Cracks.' *Journal of Applied Mechanics, Transactions of ASME* 6: 49–53.

3.7 Williams, M.L. 1957. 'On Stress Distributions at the Base of a Stationary Crack.' *Journal of Applied Mechanics, Transactions of ASME* 24: 109–14.

4

Crack Opening Displacement, J Integral, and Resistance Curve

4.1 Introduction

The fracture mechanics based on stress intensity factor (SIF) helped to characterize fracture in terms of the critical SIF, or fracture toughness. The application of linear elastic fracture mechanics (LEFM) became very limited for metals, in which plastic deformation preceded any crack extension. Wells (1961) experimented with variety of metals. He observed that before the onset of extension, the crack-tip blunts and there is a definite opening at the original crack-tip location. The extent of the opening is dependent on the fracture resistance of the material. The opening increases as the resistance of the material to fracture increases. He estimated the crack opening displacement (COD) at the original crack-tip location and presented the condition of fracture in terms of this parameter. This forms the basis of COD criterion of fracture mechanics. For small-scale plastic deformation at the tip, this condition is equivalent to the fracture condition in terms of Griffith potential energy release rate G_I for Mode I, that is, $G_I = G_{IC}$, where G_{IC} is the fracture resistance of the material.

In the presence of linear or nonlinear elastic deformation at the crack-tip, the deformation field is conservative. Stress–strain relation for a material showing plastic deformation at the crack-tip, under monotonically increasing loading, is very similar to that of a nonlinear elastic material. Provided there is no unloading or crack extension, the material obeys the deformation theory of plasticity, that is, the total strain at any stage is related to the total stress, and the relationship is path independent. Rice (1968) showed that under such an elastic (linear or non-linear) deformation of a component with a crack, there exists an integral, called J integral, which is path independent when calculated joining any two points on the opposite crack flanks. Further, this integral indicates the potential energy release rate associated with the crack extension. It can characterize the onset of crack growth in the same fashion as the SIF, but it is valid even beyond the linear elastic limit. The path independence of this parameter along with its energy release rate character is shown in this chapter. Further its graphical interpretation is also given.

It has been shown in the Chapter 2 that the fracture resistance of a material is constant for a purely elastic material. It is also constant for high strength and low toughness material under plane strain condition, that is, for specimen thickness greater than the critical thickness $B_C = 2.5 \frac{K_{IC}^2}{\sigma_Y^2}$, where K_{IC} is the plane strain fracture toughness of the material and σ_Y is its yield strength. For situations under plane stress, the fracture resistance increases with crack extension. It may happen under plane strain conditions too for some materials (Anderson 2005). This increased resistance is due to the fact that the energy required for plastic deformation at the crack-tip increases as the crack length increases. During such crack extension, the plastic zone size continually increases. Through experiments with different materials, it has been observed that the variation of fracture resistance expressed in terms of G_R with crack extension Δa is independent of the starting crack length. This type of curve is known as the resistance curve. This curve is useful for finding out the point of instability, provided the driving force G variation against crack size is known for a given load level. This is illustrated in this chapter.

These three approaches for specifying the onset of unstable fracture in the presence of small-scale plastic deformation are discussed in this chapter. COD and J can be useful even if there is large-scale plastic deformation at the crack-tip. Their experimental determination is presented in Chapter 9.

4.2 Crack Opening Displacement

According to the Westergaard stress function approach, the crack edge displacement v is given [refer Eq. (3.42) in Chapter 3] by

$$v = \frac{2}{E} \text{Im} \left[\sigma \sqrt{z^2 - a^2} \right] - \frac{1+\nu}{E} y \, \text{Re} \left[\frac{\sigma z}{\sqrt{z^2 - a^2}} \right] \quad (4.1)$$

For $y = 0$ (Fig. 4.1)

$$v = \frac{2\sigma}{E} \text{Im} \left[\sigma \sqrt{x^2 - a^2} \right] = \frac{2\sigma}{E} \sqrt{a^2 - x^2}, \text{ since } x \leq a \quad (4.2)$$

Hence

$$\text{COD} = \delta = \frac{4\sigma}{E} \sqrt{a^2 - x^2} \quad (4.3)$$

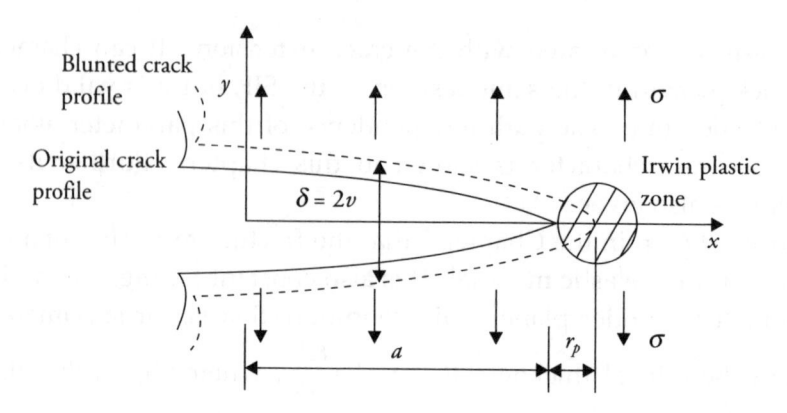

Figure 4.1 Crack-tip blunting and crack opening displacement δ for centre crack under remote Mode I loading.

If there is plastic deformation at the crack-tip and the virtual crack size is $a+r_p$, then the crack opening is given by

$$\delta = \frac{4\sigma}{E}\sqrt{(a+r_p)^2 - x^2} \tag{4.4}$$

Therefore, the opening at the actual crack-tip ($x = a$) location is approximated by

$$\delta = \frac{4\sigma}{E}\sqrt{2ar_p} \tag{4.5}$$

Since Irwin's plastic zone correction factor under plane stress condition is $r_p = \frac{K_I^2}{2\pi\sigma_Y^2}$, and $K_I = \sigma\sqrt{\pi a}$ for an infinite plate with a central crack,

$$\delta = \frac{\pi\sigma^2 a}{E} = \frac{\pi}{4}\sigma_Y\delta = \frac{4\sigma^2 a}{E\sigma_Y} \tag{4.6a}$$

or $\quad G = \dfrac{\pi\sigma^2 a}{E} = \dfrac{\pi}{4}\sigma_Y\delta \approx \sigma_Y\delta \tag{4.6b}$

Therefore, if crack extension is associated with a critical potential energy release rate G, fracture is also associated with a critical value of COD δ.

According to Dugdale strip yield model (Dugdale 1960), or Barenblatt cohesive zone based calculations (Barenblatt 1962), COD δ is given by

$$\delta = \frac{8\sigma_Y a}{\pi E}\ln\left[\sec\frac{\pi\sigma}{2\sigma_Y}\right]$$

$$= \frac{8\sigma_Y a}{\pi E}\left[\frac{1}{2}\left(\frac{\pi\sigma}{2\sigma_Y}\right)^2 + \frac{1}{12}\left(\frac{\pi\sigma}{2\sigma_Y}\right)^4 + \cdots\cdots\cdots\right] \tag{4.7}$$

For fracture occurring at small load levels, $\dfrac{\sigma}{\sigma_Y} \ll 1$,

$$\delta = \frac{8\sigma_Y a}{\pi E} \frac{1}{2} \left(\frac{\pi \sigma}{2\sigma_Y}\right)^2 = \frac{\pi \sigma^2 a}{E \sigma_Y} \tag{4.8a}$$

or $\quad G = \sigma_Y \delta \tag{4.8b}$

In general, $G = \lambda \sigma_Y \delta$, λ varies from $\dfrac{\pi}{4}$ to 2.2. Experimental data on δ can be collected through metallographic sectioning.

In the case of single edge crack specimen with bending dominated loading at the crack-tip as in the case of compact tension (CT) specimen, or three point bend specimen (TPB), the plastic deformation gives rise to the formation of a plastic hinge at a point ahead of the crack-tip. The calculation of COD is then done differently. This is discussed later in Chapters 8 and 9.

4.3 Special Integrals

In a gravitational field, if a body of weight W is moved along a closed contour ABA (Fig. 4.2),

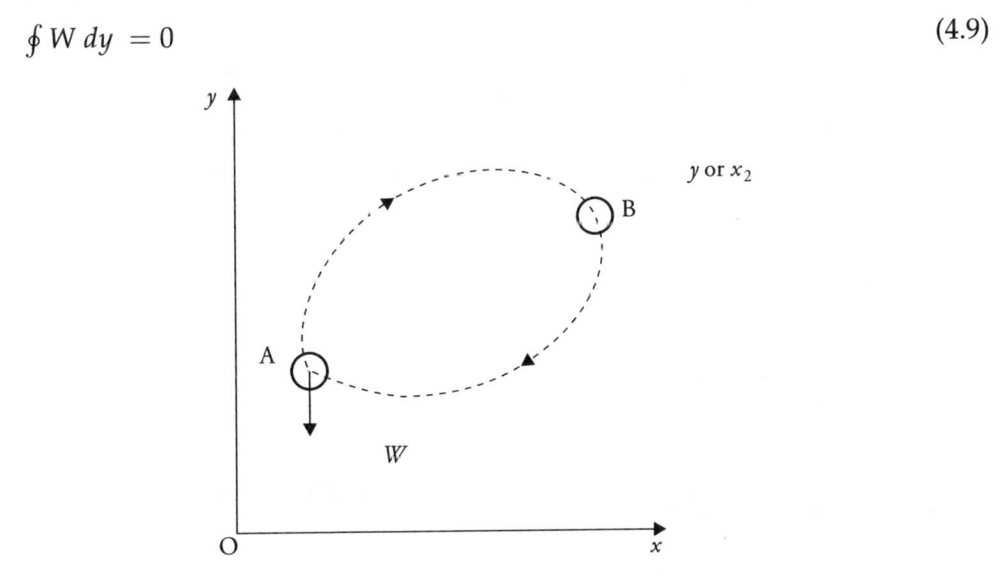

$$\oint W\, dy = 0 \tag{4.9}$$

Figure 4.2 Contour for movement of W.

Similarly, in a conservative stress–strain filed irrespective of whether the material is elastic or non-linear elastic, there are certain integrals, which when evaluated along a close contour, become zero. For example,

$$I = \oint \left[W\, dy - T_i \frac{\partial u_i}{\partial x} dS \right] = \oint \left[W\, dy - \left(T_1 \frac{\partial u_1}{\partial x} + T_2 \frac{\partial u_2}{\partial x} \right) dS \right]. \quad (4.10)$$

When I is evaluated along a close contour S (Fig. 4.3), it becomes zero, where $W =$ strain energy density $= \int \sigma_{ij}\, d\epsilon_{ij}$, $T_i = l_j \sigma_{ij} =$ traction components at a point on S, $i = 1$ and 2, and $j = 1$ and 2, and

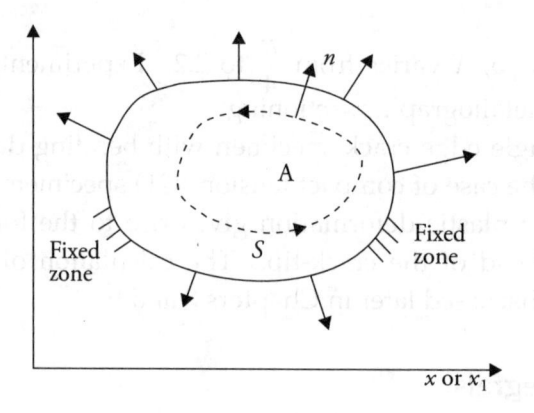

Figure 4.3 Close contour for line integral in stress field.

$$T_1 = l_1 \sigma_{11} + l_2 \sigma_{12} = l_1 \sigma_x + l_2 \tau_{xy}, \quad (4.11a)$$

$$T_2 = l_1 \sigma_{21} + l_2 \sigma_{22} = l_1 \tau_{xy} + l_2 \sigma_y, \quad (4.11b)$$

$l_j =$ direction cosines of local normal n at a point to the contour S, $A =$ area bounded by S and $x = x_1$ and $y = x_2$. It can be established that I is zero as follows.

$$I = \oint \left[W\, dx_2 - T_i \frac{\partial u_i}{\partial x_1} dS \right] = \int \frac{\partial W}{\partial x_1} dA - \oint T_i \frac{\partial u_i}{\partial x_1} dS$$

$$= \int \frac{\partial W}{\partial x_1} dA - \oint \sigma_{ij} l_j \frac{\partial u_i}{\partial x_1} dS$$

$$= \int \frac{\partial W}{\partial x_1} dA - \int \frac{\partial}{\partial x_j} \left(\sigma_{ij} \frac{\partial u_i}{\partial x_1} \right) dA, \text{ by Green's formula,}$$

$$= \int \frac{\partial W}{\partial x_1} dA - \int \frac{\partial \sigma_{ij}}{\partial x_j} \frac{\partial u_i}{\partial x_1} dA - \int \sigma_{ij} \frac{\partial}{\partial x_j} \left(\frac{\partial u_i}{\partial x_1} \right) dA$$

$$= \int \frac{\partial W}{\partial x_1} dA - \int \sigma_{ij} \frac{\partial}{\partial x_j} \left(\frac{\partial u_i}{\partial x_1} \right) dA, \text{ since } \frac{\partial \sigma_{ij}}{\partial x_j} = 0 \text{ because of equilibrium conditions,}$$

$$= \int \frac{\partial W}{\partial x_1} dA - \int \frac{\partial W}{\partial \epsilon_{ij}} \frac{\partial}{\partial x_1}\left(\frac{\partial u_i}{\partial x_j}\right) dA$$

$$= \int \frac{\partial W}{\partial x_1} dA - \int \frac{\partial W}{\partial \epsilon_{ij}} \frac{\partial \epsilon_{ij}}{\partial x_1} dA, \quad \text{since } \epsilon_{ij} = \frac{\partial u_i}{\partial x_j},$$

Finally, therefore

$$I = \int \frac{\partial W}{\partial x_1} dA - \int \frac{\partial W}{\partial x_1} dA = 0. \tag{4.12}$$

4.4 Rice's Path-Independent Integral J

Consider a crack aligned with x axis. Taking I integral over a close contour MNOPQ RM, which encloses an area A, we get

$$I = \int_{\text{MNOPQRM}} \left(W\, dy - T_i \frac{\partial u_i}{\partial x} dS\right)$$

$$= \int_{\text{MNO}} \left(W\, dy - T_i \frac{\partial u_i}{\partial x} dS\right) + \int_{\text{OP}} \left(W\, dy - T_i \frac{\partial u_i}{\partial x} dS\right)$$

$$+ \int_{\text{PQR}} \left(W\, dy - T_i \frac{\partial u_i}{\partial x} dS\right) + \int_{\text{RM}} \left(W\, dy - T_i \frac{\partial u_i}{\partial x} dS\right) \tag{4.13}$$

Since dy is zero all along O to P and R to M, the first part of the second and fourth integrals is zero. Further, since these segments are stress-free, and the tractions components on these two segments are zero. Therefore, the second part of the second and fourth integrals is also zero. Hence,

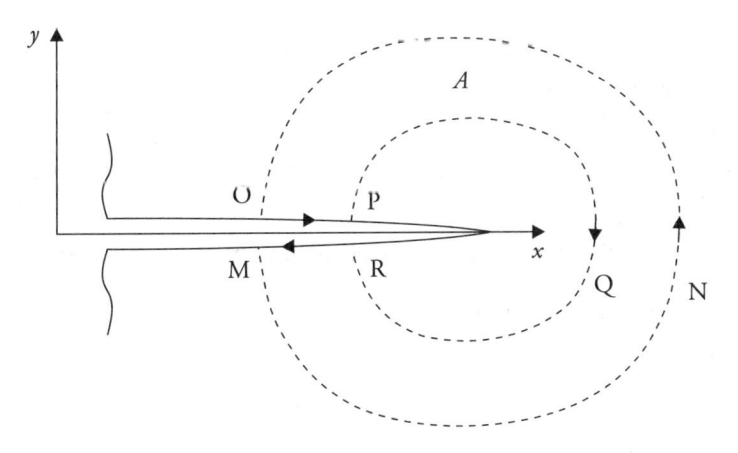

Figure 4.4 Contour for J integral.

$$\int_{MNO} \left(W\,dy - T_i \frac{\partial u_i}{\partial x} dS \right) + \int_{PQR} \left(W\,dy - T_i \frac{\partial u_i}{\partial x} dS \right) = 0$$

This means that

$$\int_{MNO} \left(W\,dy - T_i \frac{\partial u_i}{\partial x} dS \right) - \int_{RQP} \left(W\,dy - T_i \frac{\partial u_i}{\partial x} dS \right) = 0$$

$$\int_{MNO} \left(W\,dy - T_i \frac{\partial u_i}{\partial x} dS \right) = \int_{RQP} \left(W\,dy - T_i \frac{\partial u_i}{\partial x} dS \right) = \text{constant}.$$

Hence, the integral taken on contour S, for example, MNO or RQP, joining two points on the opposite crack flanks, or taken along the outer boundary of an edge crack problem, is constant. This integral has been introduced by Rice (1968). It is well known as Rice's J integral. Finally

$$J = \int_S \left(W\,dy - T_i \frac{\partial u_i}{\partial x} dS \right) \tag{4.14}$$

where S represents the contour. Rice showed that for a linear elastic material, J is the same as G, the potential energy release rate. The proof is given in the next section.

4.5 J As Potential Energy Release Rate

Consider a rectangular plate with boundary loading (Fig. 4.5) and a crack along x axis. $+x$ axis is directed along the crack extension direction. Potential energy π of the system is given by

$$\pi = \int_A W\,dA - \int_{PQRSTU} T_i u_i \, dS \tag{4.15}$$

where A is the area of the plate. Therefore, the potential energy release rate under constant boundary loading or T_i is given by

$$-\frac{d\pi}{da} = \frac{d}{da}\left[\int_{PQRSTU} T_i u_i \, dS - \int_A W\,dA \right] = \left[\int_{PQRSTU} T_i \frac{du_i}{da} dS - \int_A \frac{dW}{da} dA \right] \tag{4.16}$$

Shifting the origin to the crack-tip, and noting that $x = x_1 - a$ and $y = y_1$

$$\frac{du_i}{da} = \frac{\partial u_i}{\partial a} + \frac{\partial u_i}{\partial x}\frac{\partial x}{\partial a} + \frac{\partial u_i}{\partial y}\frac{\partial y}{\partial a} = \frac{\partial u_i}{\partial a} - \frac{\partial u_i}{\partial x}$$

$$\frac{dW}{da} = \frac{\partial W}{\partial a} + \frac{\partial W}{\partial x}\frac{\partial x}{\partial a} + \frac{\partial W}{\partial y}\frac{\partial y}{\partial a} = \frac{\partial W}{\partial a} - \frac{\partial W}{\partial x} \tag{4.17}$$

$$-\frac{d\pi}{da} = \left[\int_{PQRSTU} T_i \frac{\partial u_i}{\partial a} dS - \int_A \frac{\partial W}{\partial a} dA\right] - \left[\int_{PQRSTU} T_i \frac{\partial u_i}{\partial x} dS - \int_A \frac{\partial W}{\partial x} dA\right]$$

$$= \left[\int_A \frac{\partial W}{\partial x} dA - \int_{PQRSTU} T_i \frac{\partial u_i}{\partial x} dS\right] - \left[\int_A \frac{\partial W}{\partial a} dA - \int_{PQRSTU} T_i \frac{\partial u_i}{\partial a} dS\right]$$

$$= \left[\int_{PQRSTU} W\, dy - \int_{PQRSTU} T_i \frac{\partial u_i}{\partial x} dS\right] - \left[\int_A \frac{\partial W}{\partial a} dA - \int_{PQRSTU} T_i \frac{\partial u_i}{\partial a} dS\right]$$

$$= J - \int_A \frac{\partial W}{\partial a} dA + \int_{PQRSTU} l_j\, \sigma_{ij} \frac{\partial u_i}{\partial a} dS$$

$$= J - \int_A \frac{\partial W}{\partial a} dA + \int_A \frac{\partial}{\partial x_j}\left(\sigma_{ij} \frac{\partial u_i}{\partial a}\right) dA$$

$$= J - \int_A \frac{\partial W}{\partial a} dA + \int_A \sigma_{ij} \frac{\partial}{\partial x_j}\left(\frac{\partial u_i}{\partial a}\right) dA, \text{ since } \frac{\partial \sigma_{ij}}{\partial x_j} = 0 \text{ because of}$$

equilibrium conditions,

$$= J - \int_A \frac{\partial W}{\partial a} dA + \int_A \sigma_{ij} \frac{\partial}{\partial a}\left(\frac{\partial u_i}{\partial x_j}\right) dA = J - \int_A \frac{\partial W}{\partial a} dA + \int_a \frac{\partial W}{\partial \epsilon_{ij}} \frac{\partial \epsilon_{ij}}{\partial a} dA,$$

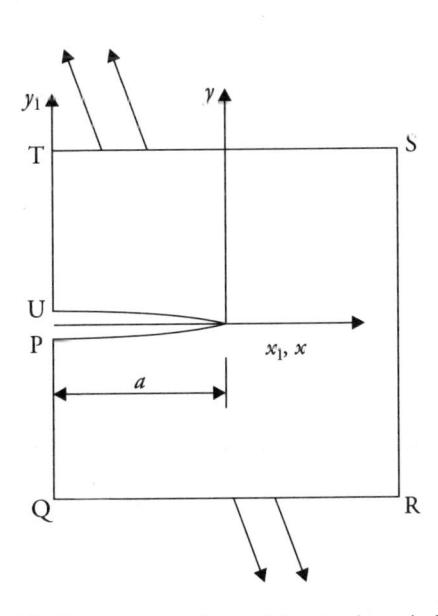

as $\epsilon_{ij} = \dfrac{\partial u_i}{\partial x_j}$. Finally

Figure 4.5 Rectangular plate with mixed mode loading.

94 Fracture mechanics

$$-\frac{d\pi}{da} = J - \int_A \frac{\partial W}{\partial a} dA + \int_a \frac{\partial W}{\partial a} dA = J \qquad (4.18)$$

Therefore J indicates the potential energy release rate.

4.6 Graphical Representation of J for Non-linear Elastic Case

Consider a plate with a crack a. The load deflection diagram for a nonlinear elastic material is given by OA. The crack begins to extend at load P^* at constant displacement u^* (Fig. 4.6(a)). The potential energy at A

$$\pi = \int_0^{u^*} P du - P^* u^* \qquad (4.19)$$

Therefore

$$J = -\frac{d\pi}{da} = -\frac{d}{da} \int_0^{u^*} P du = -\sum \left(\frac{\Delta P}{\Delta a} \Delta u\right)_{u=\text{constant}}$$

$$= \sum \frac{\Delta P}{\Delta a} \Delta u = \frac{\text{Area OAC}}{\Delta a} \qquad (4.20)$$

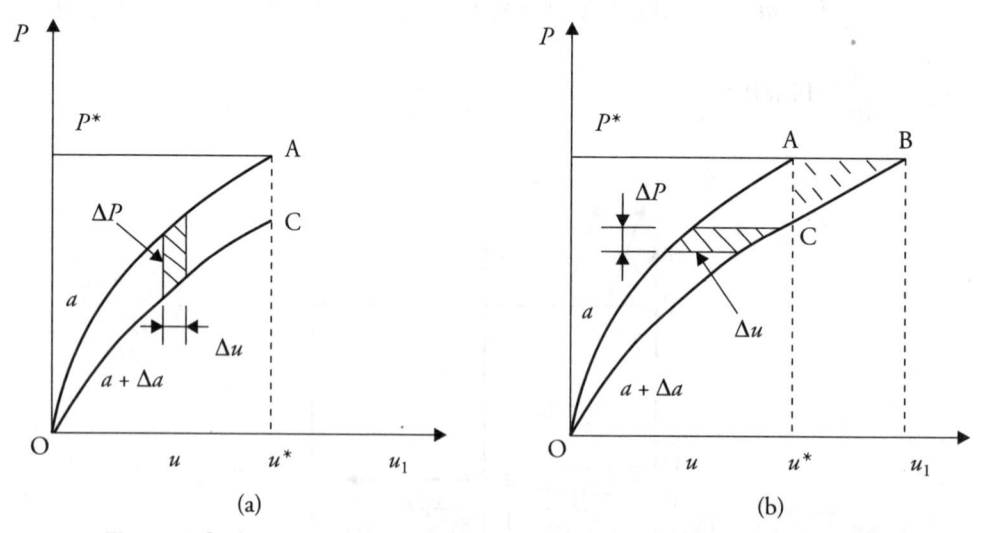

Figure 4.6 Crack extension under constant (a) displacement and (b) load.

When the crack extension occurs at constant load P^* (Fig. 4.6(b)), strain energy $U = P^* u^* - \int_0^{P^*} u\, dP$. Therefore, potential energy at A, $\pi = U - P^* u^* = -\int_0^{P^*} u\, dP$. This gives

$$J = -\frac{d\pi}{da} = \frac{d}{da} \int_0^{P^*} u\, dP = \sum \left(\frac{\Delta u}{\Delta a} \Delta P\right)_{P=\text{constant}} = \sum \frac{\Delta P}{\Delta a} \Delta u$$

$$= \frac{\text{Area OAB}}{\Delta a} \qquad (4.21)$$

As $\Delta a \to 0$, difference ABC between the two areas OAC and OAB reduces to zero. J can be calculated from the area OAC in the two cases.

4.7 Resistance Curve

For an elastic material, the resistance to fracture in terms of energy required for creation of unit surface area G_R remains constant. Therefore, variation of G_R with crack extension Δa will be given by line DE as shown in Fig. 4.7. In general, variation of crack driving force G with crack size for an internal crack in an infinite body depends on load level, $G = \frac{\pi \sigma^2 a}{E}$, where E is material modulus of elasticity and $a = a_0 + \Delta a$. Typical variations of G with a or Δa for a Griffith crack is as shown in Fig. 4.7 by solid lines. The variation can be nonlinear (e.g., AHI in Fig. 4.7) for an internal crack in a finite plate like geometry. AE correspond to the variation of G with a for load level σ_1, AF corresponds to the variation of G with a for load level σ_2, and so on. For load level σ_1, the crack does not grow if its length is a_0. If the load level is increased to σ_3, the instability sets in at D, because

$$G = G_R, \quad \frac{\partial G}{\partial a} \geq \frac{\partial G_R}{\partial a}. \qquad (4.22)$$

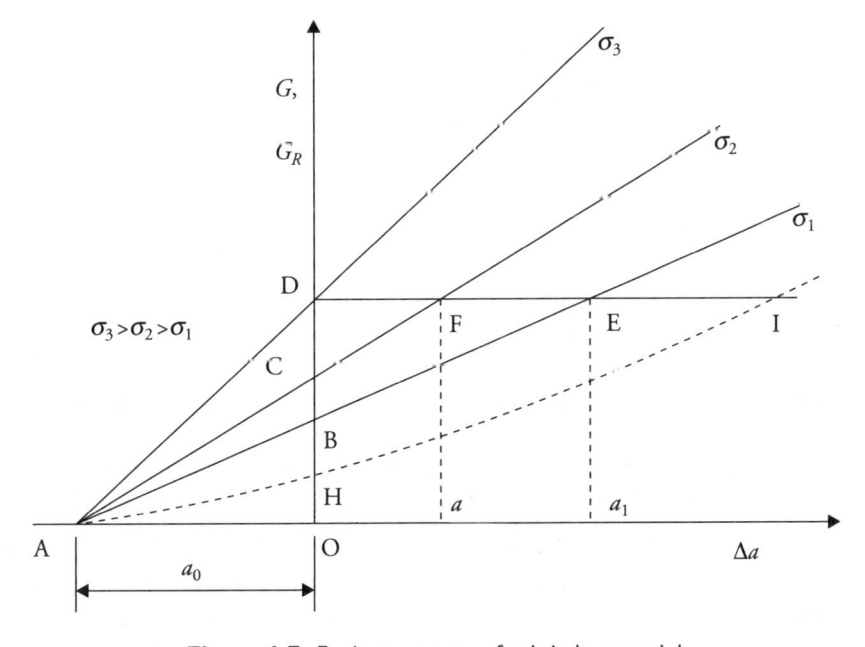

Figure 4.7 Resistance curve for brittle material.

The second condition indicates that at point D, the rise of crack driving force G for a further small crack extension is more than the rise of crack growth resistance G_R corresponding to the same crack extension. At load level σ_1, instability can occur at E if the starter crack size is a_1. This type of plot G_R versus Δa is known as the resistance curve. The same curve can be plotted considering crack resistance in terms of K_R (or J_R).

For high strength and low toughness metals, this type of behaviour may be seen under plane strain conditions (Anderson 2005). But for low and intermediate strength metals under plane stress and plane strain conditions, the resistance curve OBCFED (Fig. 4.8) shows gradually increasing resistance with crack extensions. This is due to the fact that as the crack size increases, the plastic zone size increases. This, in turn, increases the energy required per unit area of further crack extensions. The crack driving force in terms of G or K shows a curve with gradually increasing slope under a fixed load level σ_1, or σ_2 (Fig. 4.8), and so on.

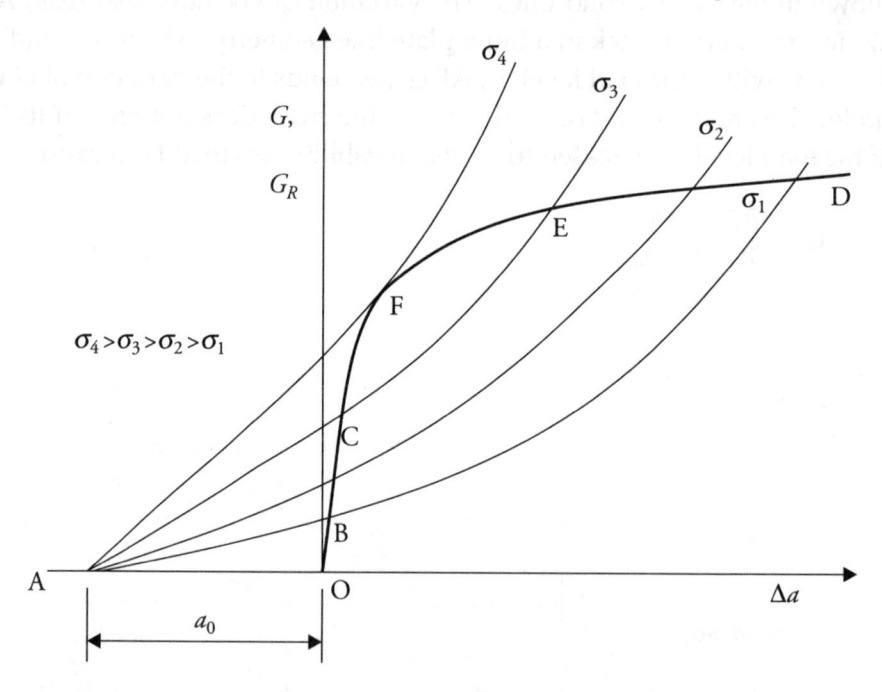

Figure 4.8 Resistance curve for materials with small-scale plastic deformation at crack-tip.

If the initial crack size is slightly greater than a_0 and corresponds to the span AB on the curve of constant load level σ_1, the crack grows a little, but it does not lead to instability because the loading curve ABD has a slope at B much lower than the resistance curve OBCFED. As the load level increases from σ_1 to σ_3, again the crack grows stably. When load level reaches level σ_4, the instability sets in. This occurs at the point F, because at this point Eq. (4.22) becomes satisfied.

That is,

$$G = G_R, \quad \frac{\partial G}{\partial a} \geq \frac{\partial G_R}{\partial a}.$$

Noting that $G = \frac{K^2}{E}$ under plane stress condition, the same instability condition can be written in terms of K and K_R as follows.

$$K = K_R, \quad \frac{\partial K}{\partial a} \geq \frac{\partial K_R}{\partial a} \qquad (4.23)$$

σ_4 corresponds to the unstable fracture load. In general, the fracture resistance is supposed to be independent of the specimen geometry, but it depends on the specimen dimensions like thickness, the three-dimensional constraint at the crack-tip, and nearness of the crack front to the free boundary.

By testing a specimen of particular thickness with a starter crack and recording variation of fracture load with physical crack size during the stable crack growth, the K-resistance curve can be generated. The procedure for the calculation of K_R corresponding to a physical crack size is illustrated in one of the following examples.

4.8 Stability of Crack Growth

The stability of crack growth, which is discussed in the previous section, refers to a situation under load control. If the same test is done under displacement control, the driving force curves will be different. These are identified as u_1, u_2, and u_3 in Fig. 4.9. With an increase in crack length, for example at F, the crack driving force G reduces. Hence, further crack extension can occur with an increase in displacement. However, under load control, instability can be observed at load level σ_4.

The stability and instability of crack extension are not only influenced by the stiffness of the specimen, but are also affected by the stiffness of machine or loading system. This has been shown by Hutchinson(1979) and Hutchinson and Paris (1979). It was shown in Chapter 2 that under load control (or soft loading), the energy release rate for a unit crack extension is higher than the energy release rate associated with displacement controlled loading (or hard loading). This difference is due to the area ABE shown in Fig. 2.11. If the resistance curve of a specimen shows increasing resistance with crack extension as shown in Fig. 4.9, a hard loading system, as mentioned earlier, will not show instability at point F. Such a loading system leads to stable crack extension.

98 *Fracture mechanics*

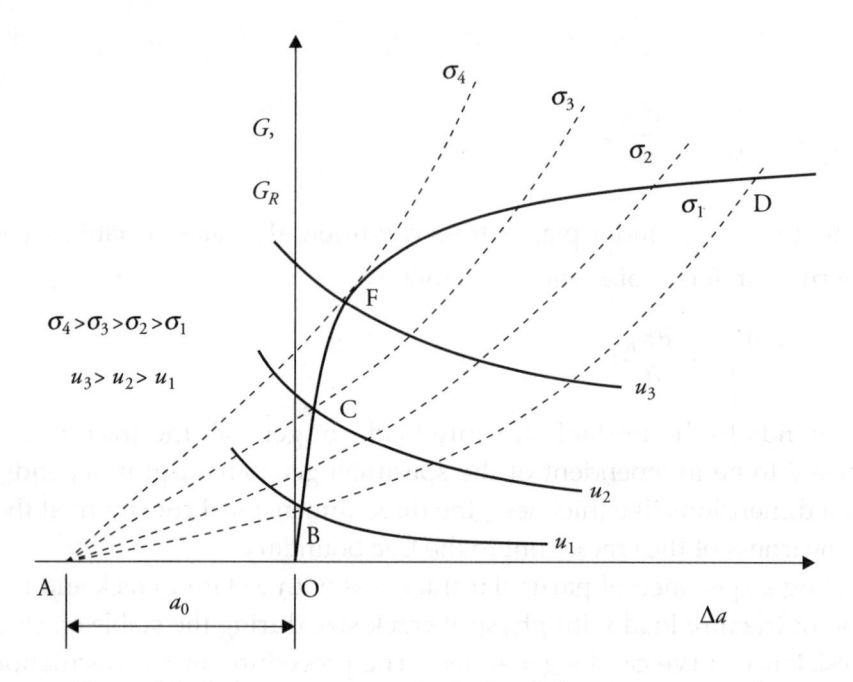

Figure 4.9 Resistance curves under load and displacement controls.

Problem 4.1
During a J test, load P was observed to vary with deflection u as follows: $= bu + c\sqrt{u}$, where the crack length is a. b and c are two specimen geometry-dependent constants. As the crack length increases by 5%, the constants b and c decrease by 2% and 1%, respectively. The specimen thickness is B. Determine J corresponding to the given crack size a_0 and displacement u_0.

Solution
Area under P versus u diagram up to final displacement u_0 corresponding to crack size a_0,

$$U_1 = \int_0^{u_0} P\,du = \frac{b}{2}u_0^2 + c\,\frac{2}{3}u_0^{1.5}.$$

Area U_2 under the P versus u diagram up to final displacement u_0 corresponding to the crack size $1.05a_0$ is given by

$$U_2 = \frac{0.98\,b}{2}u_0^2 + 0.99\,c\,\frac{2}{3}u_0^{1.5}.$$

Therefore

$$J \approx \frac{\Delta U}{B\, \Delta a} = \frac{U_1 - U_2}{B\, 0.05\, a_0} = \frac{0.01 b\, u_0^2 + \frac{0.02}{3} c\, u_0^{1.5}}{B\, 0.05\, a_0} \quad \text{(Ans.)}.$$

Problem 4.2

In Problem 4.1, if the constants b and c reduce by 4.5% and 2%, respectively, when the crack size increases from the original size a_0 by 10%, calculate the value of J corresponding to the range $1.05 a_0$ to $1.10 a_0$.

Solution

Area U_3 under the P versus u diagram corresponding to the crack size $1.10 a_0$ and the same final displacement level u_0 is given by

$$U_3 = \frac{0.955\, b}{2} u_0^2 + 0.98\, c\, \frac{2}{3} u_0^{1.5}.$$

Therefore, using U_2 from the previous problem,

$$J \approx \frac{\Delta U}{B\, \Delta a} = \frac{U_2 - U_3}{B\, 0.05 a_0} = \frac{0.0125 b\, u_0^2 + \frac{0.02}{3} c\, u_0^{1.5}}{B\, 0.05 a_0} \quad \text{(Ans.)}.$$

Problem 4.3

One tensile panel 15mm thick × 55 mm wide of 4340 steel with an edge crack of size 25 mm was tested for K_R measurement. The load obtained at the onset of crack extension is 172 kN. Determine the value of K_R. Given material properties are: $\sigma_Y = 1240$ MPa and $K_{IC} = 117.5$ MPa$\sqrt{\text{m}}$.

Solution

$$r = \frac{a_0}{w} = \frac{25}{55} = 0.4545$$

The SIF correction factor obtained from Gdoutos (1993) is as follows.

$$Y(r) = 1.12 - 0.23 r + 10.55 r^2 - 21.72 r^3 + 30.39 r^4 = 2.4526$$

Load intensity $\sigma = \dfrac{P}{B\, w} = \dfrac{72000}{0.015 \times 0.055} = 208.48$ MPa

Therefore, approximate $K_R = \sigma \sqrt{\pi a_0}\, Y(r) = 208.48\, \sqrt{\pi \times 0.025} \times 2.4526$

$$= 143.30\ \text{MPa}\sqrt{\text{m}}.$$

For the material, the critical thickness for K_{IC} measurement is 22.44 mm. Therefore, plane stress condition is prevailing in the specimen. The effective crack size a with plastic zone correction is as follows.

$$a = a_0 + r_p = 25 + \frac{K_R^2}{2\pi\sigma_Y^2} = 25 + \frac{(143.30 \times 10^6)^2}{2\pi(1240 \times 10^6)^2} \times 1000$$

$$= 25 + 2.126 = 27.126 \text{ mm}.$$

Corrected crack size ratio $r = \frac{27.126}{55} = 0.4932$, and the SIF correction factor $Y(r) = 1.12 - 0.23r + 10.55r^2 - 21.72r^3 + 30.39r^4 = 2.765$.

The corrected K-resistance $K_R = \sigma\sqrt{\pi a_0}\,Y(r) = 208.48\sqrt{\pi \times 0.027126} \times 2.765$

$$= 168.30 \text{ MPa}\sqrt{m}.$$

This can be quoted as the K-resistance corresponding to a crack extension of 2.126 mm.

With another step of iteration, $a = a_0 + r_p = 25 + \frac{(168.30 \times 10^6)^2}{2\pi(1240 \times 10^6)^2} \times 1000$

$$= 25 + 2.932 = 27.932 \text{ mm}.$$

The new crack size ratio $r = 0.578$, $Y(r) = 2.901$ and $K_R = 179.10$ MPa$\sqrt{m}$, which corresponds to a crack growth of 2.932 mm. After the next eight iterations, the convergence is observed. Thereby, $K_R = 191.8$ MPa$\sqrt{m}$ and crack extension $\Delta a = 3.80$ mm is obtained.

Exercise

4.1 Draw the resistance curve K_R versus Δa and K versus Δa for a Mode I edge crack in a finite plate of high strength low fracture resistance material.

4.2 Explain why J defined by Eq. (4.18) is valid for both non-linear and linear elastic materials?

4.3 For Problem 4.3, the following fracture loads and the corresponding instantaneous crack sizes were recorded. Draw the variation of K_R with crack extension over the span of crack growth. The specimen is under plane stress condition.

Instantaneous physical crack size (mm)	Load corresponding to onset of crack growth (kN)
22	160
23	165
24	168
24.5	170

[Ans. Values of K_C and Δa = 119.2 MPa$\sqrt{m}$, 1.47 mm; 136.8 MPa$\sqrt{m}$, 1.938 mm; 157.5 MPa$\sqrt{m}$, 2.567 mm; 172.1 MPa$\sqrt{m}$, 3.065 mm].

References

4.1 Anderson, T.L. 2005. *Fracture Mechanics: Fundamentals and Applications*. Boston: CRC Press.

4.2 Barenblatt, G.I. 1962. 'The Mathematical Theory of Equilibrium Cracks in Brittle Fracture.' In *Advances in Applied Mechanics*, eds. Dryden, H.L., Th von Karman, F.H. van den Dungen and L. Howarth, Vol. VII, 55–129. New York: Academic Press.

4.3 Dugdale, D.S. 1960. 'Yielding of Steel Sheets Containing Slits.' *Journal of Mechanics and Physics of Solids* 8: 100–04.

4.4 Gdoutos, E.E. 1993. *Fracture Mechanics–An Introduction*, Kluwer. Dordreaht/Boston/London: Kluwer Academic Publishers.

4.5 Hutchinson, J.W. 1979. *A Course on Nonlinear Fracture Mechanics*. Department of Solid Mechanics, Technical University of Denmark.

4.6 Hutchinson, J.W. and P.C. Paris. 1979. 'Stability of J-controlled crack growth', in *Elastic Plastic Fracture*, eds. Landes, J.D., J.A. Begley and G.A. Clarke, 37–64. Philadelphia: American Society for Testing and Materials [ASTM STP 668].

4.7 Rice, J.R. 1968. 'A Path Independent Integrals and the Approximate Analysis of Strain Concentration by Notches and Cracks.' *Journal of Applied Mechanics, Transactions of ASME* 35: 379–86

4.8 Wells, A.A. 1961. 'Unstable crack propagation in metals: cleavage and fast fracture', Vol. 1, 210–30. *Proceedings of the Crack Propagation Symposium*, College of Aeronautics, Cranfield.

5

Determination of Stress Intensity Factors

5.1 Introduction

The stress intensity factor (SIF) plays the most pivotal role in the application of linear elastic fracture mechanics (LEFM) principles to practice. It is useful in the assessment of safety or reliability of a machine or structural component with a crack. It enables the calculation of crack growth rate through a component under fatigue loading, stress corrosion, etc. For the safety assessment, two things are needed: the SIF corresponding to the loading on component and the fracture toughness of its material. The latter is a material data obtained through experiment. The former is obtainable in some situations from handbooks of SIFs (Sih 1973a; Rooke and Cartwright 1976; Murakami et al. 1987; Tada, Paris and Irwin 2000), while in others it has to be determined using either an analytical method, or a numerical method, or an experimental technique. The analytical techniques include complex stress function based approaches, boundary collocation method, integral transform technique (Sneddon and Lowengrub 1969; Sneddon 1973), Green's function method, weight function method, etc. The numerical methods have been very widely employed for their versatility and capability for handling complex geometry easily. The three important numerical techniques are: finite element method (Wilson 1973; Atluri 1986), boundary element method (Aliabadi, Rooke and Cartwright 1987; Cruse 1996; Mukhopadhyay, Maiti and Kakodkar 2000; Rabczuk 2013), and meshless method (Belytschko et al. 1996; Atluri and Zhu 1998). The experimental techniques include strain gauge based method (Dally and Sanford 1987), photoelasticity (Kobayashi 1975; Dally and Riley 1991; Ramesh 2000), and method of caustics (Theocaris and Gdoutos 1976; Theocaris 1981; Rosakis and Zehnder 1985). A very good account of the analytical and finite element based methods is given in compilations by Sih (1973b) and Atluri (1986).

In this chapter, some important analytical methods, numerical technique based on finite element method, strain gauge based technique, and photoelasticity are only discussed.

5.2 Analytical Methods

In the analytical methods, the SIFs are calculated using the following relations for Mode I, II, and III, respectively, provided the crack-tip stress field is given in terms of r and θ.

$$K_I = \lim_{r \to 0} \sqrt{2\pi r}\, (\sigma_y)_{\theta=0} \tag{5.1}$$

$$K_{II} = \lim_{r \to 0} \sqrt{2\pi r}\, (\tau_{xy})_{\theta=0} \tag{5.2}$$

$$K_{III} = \lim_{r \to 0} \sqrt{2\pi r}\, (\tau_{yz})_{\theta=0} \tag{5.3}$$

In all these cases, origin is at the crack-tip. x axis ($\theta = 0$) is aligned with the crack plane and it points towards the direction of crack extension (Fig. 5.1).

If the Westergaard stress function $F_i(z)$, $i =$ I or II for a problem is known, the SIF can be obtained from

$$K_i = \lim_{z \to a} \sqrt{2\pi(z-a)}\, F_i(z), \quad i = I \text{ or } II \tag{5.4}$$

The Westergaard stress functions are generally defined with centre of the crack (size $2a$) as the origin, and the crack-tip is located at $z = a$.

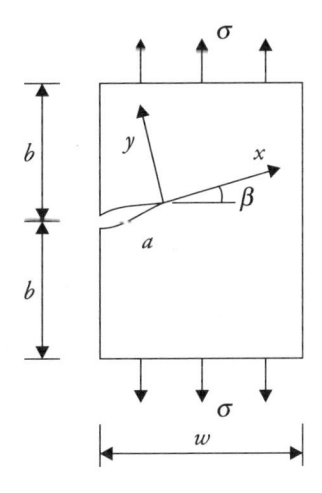

Figure 5.1 Plate with angled crack and crack-tip coordinates.

If the stresses are given in terms of two Williams stress functions $F(z)$ and $\chi(z)$, which are generally defined with the crack-tip as origin, Modes I and II SIFs are given by

$$K_I - iK_{II} = \lim_{z \to 0} 2\sqrt{2\pi z} \ F'(z) \tag{5.5}$$

If the stresses are given in terms of two analytic functions, with crack centre as the origin and the crack-tip is located at $z = z_1$, the SIFs are given (Bowie 1973) by

$$K_I - iK_{II} = \lim_{z \to z_1} 2\sqrt{2\pi(z - z_1)} \ F'(z) \tag{5.6}$$

If conformal mapping, $z = \omega(\zeta)$, is used to map the given problem to a convenient geometry like a circle in the mapping plane ζ (zeta) and the crack-tip is located at $\zeta = \zeta_1$, the SIFs for mixed mode problem in two dimensions are given by

$$K_I - iK_{II} = \lim_{\zeta \to \zeta_1} 2\sqrt{2\pi[\omega(\zeta) - \omega(\zeta_1)]} \ \frac{F'(\zeta)}{\omega'(\zeta)}, \quad F'(\zeta) = \frac{dF}{d\zeta} \tag{5.7}$$

Problem 5.1

Determine K_I and K_{II} for the case when stresses along the x axis (Fig. 5.2) are given (Maiti and Smith 1983) by

$$\sigma_y + i\tau_{xy} = \frac{P + iQ}{\pi (x - b)} \sqrt{\frac{a^2 - b^2}{x^2 - a^2}} \tag{5.8}$$

where P and Q are specified in N/m.

Solution
Using Eq. (5.1)

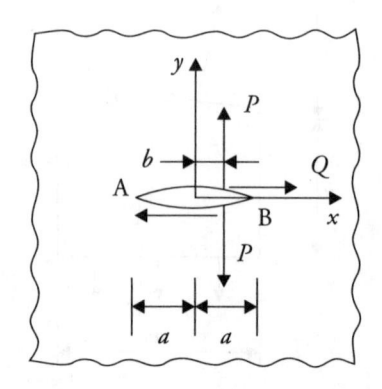

Figure 5.2 Point loading on crack edges.

$$K_{IB} = \lim_{r \to 0} \sqrt{2\pi r} \, (\sigma_y)_{\theta=0}$$

$$= \lim_{(x-a) \to 0} \sqrt{2\pi(x-a)} \, (\sigma_y)_{\theta=0}$$

$$= \lim_{x \to a} \sqrt{2\pi(x-a)} \, \frac{P}{\pi(x-b)} \sqrt{\frac{a^2 - b^2}{x^2 - a^2}}$$

$$= \frac{P}{\pi} \sqrt{2\pi(a^2 - b^2)} \lim_{x \to a} \frac{1}{x-b} \sqrt{\frac{1}{x+a}} = \frac{P}{\sqrt{\pi a}} \sqrt{\frac{a+b}{a-b}}$$

Therefore, $K_{IB} = \dfrac{P}{\sqrt{\pi a}} \sqrt{\dfrac{a+b}{a-b}}$ \hfill (5.9a)

Similarly, $K_{IA} = \dfrac{P}{\sqrt{\pi a}} \sqrt{\dfrac{a-b}{a+b}}$ (Ans.). \hfill (5.9b)

Further,

$$K_{IIB} = \lim_{r \to 0} \sqrt{2\pi r} \, (\tau_{xy})_{\theta=0} = \lim_{(x-a) \to 0} \sqrt{2\pi(x-a)} \, (\tau_{xy})_{\theta=0}$$

$$= \lim_{x \to a} \sqrt{2\pi(x-a)} \, \frac{Q}{\pi(x-b)} \sqrt{\frac{a^2 - b^2}{x^2 - a^2}} = \frac{Q}{\sqrt{\pi a}} \sqrt{\frac{a+b}{a-b}} \qquad (5.10a)$$

It can be shown that

$$K_{IIA} = \frac{Q}{\sqrt{\pi a}} \sqrt{\frac{a-b}{a+b}} \text{ (Ans.).} \qquad (5.10b)$$

Problem 5.2
Westergaard stress function for uniform pressure loading on a crack of size $2a$ in an infinite plate is $F(z) = p \left[\dfrac{z}{\sqrt{z^2 - a^2}} - 1 \right]$, where p is pressure. Determine the SIF.

Solution
From Eq. (5.4)

$$K_I = \lim_{z \to a} \sqrt{2\pi(z-a)} \, F(z) = \lim_{z \to a} \sqrt{2\pi(z-a)} \, p \left[\frac{z}{\sqrt{z^2 - a^2}} - 1 \right] = p\sqrt{2\pi} \, \frac{a}{\sqrt{2a}}$$

$$= p\sqrt{\pi a} \text{ (Ans.).}$$

Problem 5.3
Given the stress functions for the case shown (Fig. 5.3) in $x - y$ cordinates (Maiti and Smith 1983):

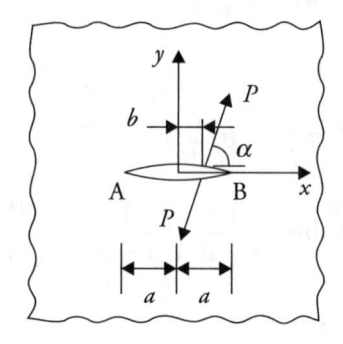

Figure 5.3 Crack edge loading.

$$F'_1(z) = \psi'_1(z) = \frac{P(\sin\alpha - i\cos\alpha)}{2\pi(z-b)}\sqrt{\frac{a^2 - b^2}{z^2 - a^2}},$$

where $\sigma_y + \sigma_x = 2\left[F'_1(z) + \overline{F'_1(z)}\right]$

$$\sigma_y - \sigma_x + 2i\tau_{xy} = 2\left[(\bar{z} - z)F''_1(z) - F'_1(z) + \overline{\psi_1'(z)}\right].$$

Determine the SIFs.

Solution
Along the crack line, that is, $y = 0$ or $\theta = 0$, $\sigma_y + i\tau_{xy} = F'_1(z) + \overline{\psi_1'(z)}$; $\bar{z} = z$.

Therefore, $\sigma_y = \dfrac{P\sin\alpha}{\pi(x-b)}\sqrt{\dfrac{a^2 - b^2}{x^2 - a^2}}$, and $\tau_{xy} = \dfrac{P\cos\alpha}{\pi(x-b)}\sqrt{\dfrac{a^2 - b^2}{x^2 - a^2}}$

$$K_{I\ B} = \lim_{(x-a)\to 0}\sqrt{2\pi(x-a)}\,(\sigma_y)_{\theta=0} = \frac{P\sin\alpha}{\sqrt{\pi a}}\sqrt{\frac{a+b}{a-b}}\ \text{(Ans.).} \quad (5.11)$$

Similarly, $K_{II\ B} = \lim_{(x-a)\to 0}\sqrt{2\pi(x-a)}\,(\tau_{xy})_{\theta=0} = \dfrac{P\cos\alpha}{\sqrt{\pi a}}\sqrt{\dfrac{a+b}{a-b}}$ (Ans.). (5.12)

Problem 5.4
Airy stress function for the infinite plate (Fig. 5.4) in the mapping plane ζ, where the mapping function, $z = \omega(\zeta) = \dfrac{a}{2}\left(\zeta + \dfrac{1}{\zeta}\right)$, is given by

$$F(\zeta) = -\sigma a \, \frac{1 - e^{2i\alpha}}{4\zeta}.$$

Determine the SIFs.

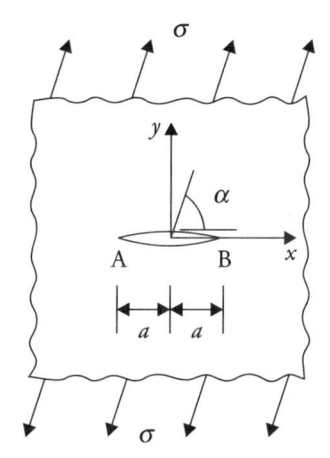

Figure 5.4 Loading on edges of infinite plate.

Solution

Since the crack-tip is located at $z = \pm a$, through the mapping function the corresponding location in the ζ-plane is $\zeta_1 = \pm 1$. Further, $\omega'(\zeta_1) = 0$ at the crack-tip.

Using Eq. (5.7)

$$K_I - iK_{II} = \lim_{\zeta \to \zeta_1} 2\sqrt{2\pi [\omega(\zeta) - \omega(\zeta_1)]} \, \frac{F'(\zeta)}{\omega'(\zeta)}$$

$$= \lim_{h \to 0} 2\sqrt{2\pi [\omega(\zeta_1 + h) - \omega(\zeta_1)]} \, \frac{F'(\zeta_1 + h)}{\omega'(\zeta_1 + h)}, \zeta = \zeta_1 + h.$$

Expanding $\omega(\zeta_1 + h)$, $\omega'(\zeta_1 + h)$, and $F'(\zeta_1 + h)$ about ζ_1 by Taylor's series, it can be seen that

$$\lim_{h \to 0} \sqrt{\omega(\zeta) - \omega(\zeta_1)} \, \frac{F'(\zeta)}{\omega'(\zeta)}$$

$$= \lim_{h \to 0} \left[\sqrt{h\omega'(\zeta_1) + \frac{h^2}{2}\omega''(\zeta_1) + \ldots} \, \frac{F'(\zeta_1) + hF''(\zeta_1) + \frac{h^2}{2}F'''(\zeta_1) + \ldots}{\omega'(\zeta_1) + h\omega''(\zeta_1) + \frac{h^2}{2}\omega'''(\zeta_1) + \ldots} \right]$$

$$= \frac{F'(\zeta_1)}{\sqrt{2\omega''(\zeta_1)}}, \text{ noting that } \omega'(\zeta_1) = 0$$

$$K_I - iK_{II} = 2\sqrt{\pi}\,\frac{F'(\zeta_1)}{\sqrt{\omega''(\zeta_1)}} \qquad (5.13)$$

Substituting the values

$$K_I - iK_{II} = 2\sqrt{\pi}\left[\frac{\sigma a}{4}\frac{(1-e^{2i\alpha})\zeta^{-2}}{\sqrt{a\,\zeta^{-3}}}\right]_{\zeta=\zeta_1} = \frac{\sigma\sqrt{\pi a}}{2}[(1-\cos 2\alpha) - i\sin 2\alpha]$$

Therefore

$$K_I = \sigma\sqrt{\pi a}\,\sin^2 2\alpha, \qquad (5.14a)$$

$$K_{II} = \sigma\sqrt{\pi a}\,\cos^2 2\alpha \text{ (Ans.).} \qquad (5.14b)$$

5.2.1 Boundary Collocation Method

This method has been mentioned in Chapter 3 earlier. In many cases, the stresses can be expressed in terms of two analytic functions, which can be in the form of finite series. Particularly when Williams' eigenfunction expansions are used, each term satisfies the stress-free conditions at the crack edges (Fig. 5.1). To get the coefficients of the finite series, it is just necessary to fit or collocate the stress boundary conditions at a few selected number of stations over the remaining part of the boundary. This is why the method is known as the collocation method. The method does not guarantee any convergence with an increase in number of terms of the finite series. Nevertheless, it has been exploited to get some solutions (Hartranft and Sih 1973).

5.2.2 Green's Function Approach

In this method, known solutions for concentrated load on the crack edges are employed to get the solution for distributed loading on the crack edges.

For a crack loaded as shown in Fig. 5.5, the SIF can be easily determined using the solution for the standard case shown in Fig. 5.2 as the Green function. The Green functions [see Eqs. (5.9) and (5.10)] for the two crack-tips corresponding to Mode I (load P) and Mode II (Q) are as follows.

$$K_{IA} = \frac{P}{\sqrt{\pi a}}\sqrt{\frac{a-b}{a+b}}, \quad K_{IB} = \frac{P}{\sqrt{\pi a}}\sqrt{\frac{a+b}{a-b}}$$

$$K_{IIA} = \frac{Q}{\sqrt{\pi a}}\sqrt{\frac{a-b}{a+b}}, \quad K_{IIB} = \frac{Q}{\sqrt{\pi a}}\sqrt{\frac{a+b}{a-b}}$$

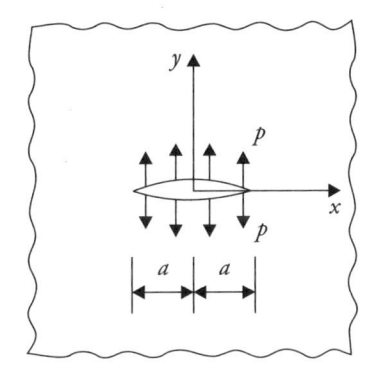

Figure 5.5 Uniform pressure loading on crack edges.

For the case (Fig. 5.5), which is symmetric, the SIFs at the two tips are the same and is given by

$$K_I = \int_{-a}^{a} \frac{1}{\sqrt{\pi a}} \sqrt{\frac{a-x}{a+x}} \, p \, dx = p\sqrt{\pi a}$$

The above integral can be easily integrated by making a substitution $x = a \sin^2 \theta$.

In case the loading is specified as outer boundary loading rather than explicit crack edge loading, it is possible to obtain first the loading on the crack-line in the corresponding crack-free configuration. The required SIF is then given by

$$K_{IA} = -\int_{-a}^{a} \frac{1}{\sqrt{\pi a}} \sqrt{\frac{a-x}{a+x}} f(x) \, dx \tag{5.15}$$

$$K_{IIA} = -\int_{-a}^{a} \frac{1}{\sqrt{\pi a}} \sqrt{\frac{a-x}{a+x}} g(x) \, dx \tag{5.16}$$

where $f(x)$ is the crack-line normal stress distribution, and $g(x)$ is the crack-line shear stress distribution.

5.2.3 Method of Superposition

If we have to find the SIF for the case (Fig. 5.6(b)), we can consider a plate (Fig. 5.6(a)) without any crack. In this case, the SIF at A or B is zero. The two cases (Figs. 5.6(b) and (c)) together are equivalent to the case (Fig. 5.6(a)). This means, cases (Figs. 5.6(b) and (c)) are complementary. Therefore,

$$K_I^{(a)} = 0 = K_I^{(b)} + K_I^{(c)} = K_I^{(b)} + p\sqrt{\pi a}, \text{ that is, } K_I^{(b)} = -p\sqrt{\pi a}$$

In case the loading on crack edges in the case (Fig. 5.6(b)) is in the opening mode, $K_I^{(b)} = p\sqrt{\pi a}$.

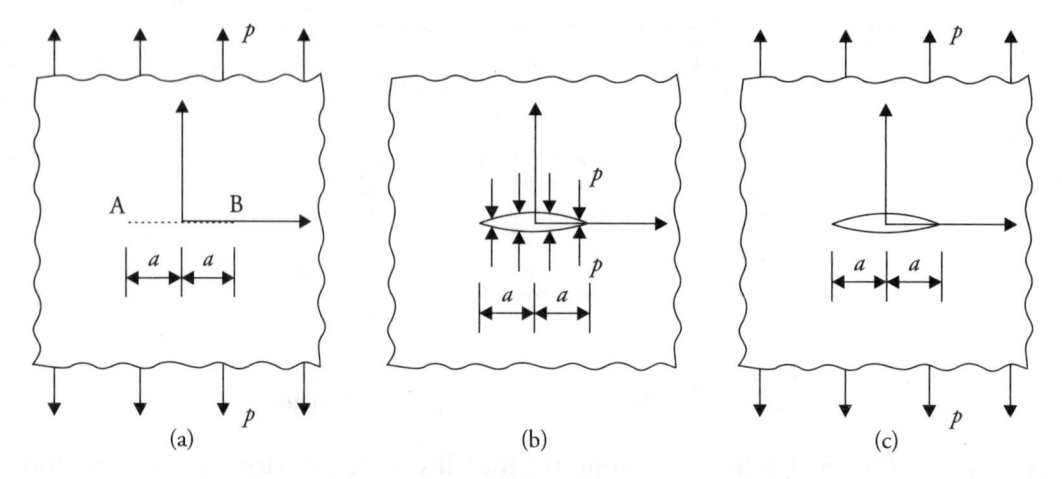

Figure 5.6 (a) Crack-free plate. (b) and (c) Two cases of complementary crack loadings.

5.2.4 Weight Function Method

This method makes use of the existence of a common crack edge profile in the case of a family of crack problems (Wu and Carlsson 1991). The family may consist of symmetrically located internal crack, or edge crack. A function of this type can be used to calculate the SIF for an unknown problem by knowing the stresses on the crack-line in the corresponding crack-free geometry. They were introduced by Bueckner and were later shown to be independent of loading on the outer boundary by Rice (1972). They can be derived as follows. The discussion is limited to Mode I loading only.

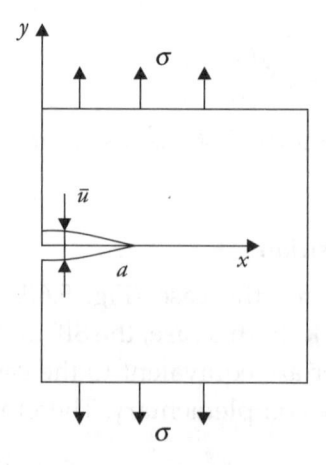

Figure 5.7 Edge crack under opening load.

For a plate of unit thickness, as the edge crack extends from size 0 to a (Fig. 5.7), the work done to extend the crack is given by

$$W = \frac{1}{2} \int_0^a p(x,0) \bar{u}\, dx \qquad (5.17)$$

where $p(x,0)$ is crack-line normal stress distribution in the corresponding crack-free body and $\bar{u}$ is the total crack opening at x from the origin, that is, $\bar{u} = 2v$. The energy release rate G as the crack opens up is given by

$$G = \frac{dW}{da} = \frac{1}{2} \int_0^a p(x,0) \frac{d\bar{u}}{da} dx$$

Under external load σ on the boundary parallel to x axis, $p(x,0) = \sigma$ in the corresponding crack-free body. Noting that $G = K_I^2/E$,

$$K_I = \int_0^a p(x,0) \left[\frac{E}{2K_I} \frac{d\bar{u}}{da} \right] dx \qquad (5.18)$$

For a class of problems (Rice 1972), with the same type of symmetry, the quantity within the square brackets in Eq. (5.18) is a constant. This is known as weight function. Hence, the SIF K_I^* for the case of a different geometry but with the same type of symmetry under Mode I loading can be determined, provided the crack-line opening stress $p^*(x,0)$ in the corresponding crack-free geometry is known. That is,

$$K_I^* = \int_0^a p^*(x, 0) \left[\frac{E}{2K_I} \frac{d\bar{u}}{da} \right] dx = \int_0^a p^*(x,0)\, m(x,a)\, dx$$

where $m(x, a)$ is the weight function. An illustrative example is given subsequently (Fig. 5.8). The opening mode displacement for the case (Fig. 5.8(b)) is given by $v = \frac{2\sigma}{E} \sqrt{a^2 - x^2}$. Therefore,

$$\bar{u} = \frac{4\sigma}{E} \sqrt{a^2 - x^2}$$

Hence, $m(x, a) = \dfrac{E}{2K_I} \dfrac{d\bar{u}}{da} = \dfrac{2a}{\sqrt{\pi a}} \dfrac{1}{\sqrt{a^2 - x^2}}$, using $K_I = \sigma\sqrt{\pi a}$ for the case (Fig. 5.8(b)).

Since the two cases shown in Fig. 5.8 have similar symmetry, the SIF for the case (Fig. 5.8(a)) is given by

$$K_I^* = \int_0^a p(x,0)\, m(x,a)\, dx = \int_0^a p\, \frac{2a}{\sqrt{\pi a}} \frac{1}{\sqrt{a^2 - x^2}}\, dx = p\sqrt{\pi a} \text{ (Ans.)}.$$

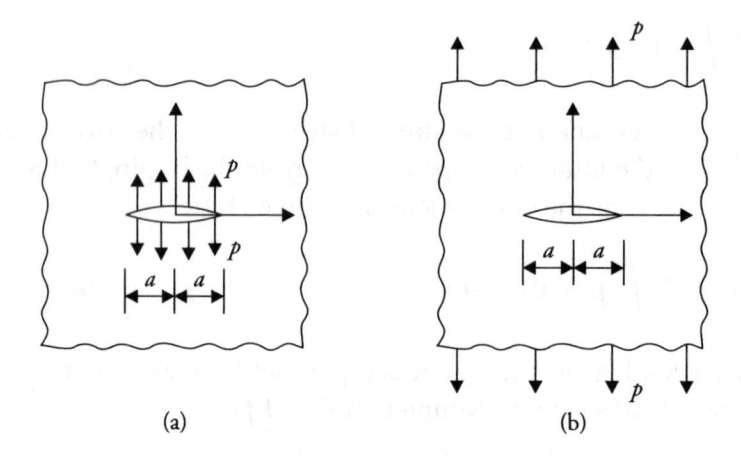

Figure 5.8 Internal crack with two different loadings but similar symmetry. (a) Uniform pressure on crack edges. (b) Uniform tension on plate edges.

Problem 5.5

The weight function for an edge crack in finite plate of width w (Fig. 5.9) is given by (Parker 1981)

$$m(x,a) = \frac{2}{\sqrt{2\pi(a-x)}} \left[1 + m_1 \frac{a-x}{a} + m_2 \left(\frac{a-x}{a}\right)^2\right],$$

$$m_1 = A_1 + B_1 r^2 + C_1 r^6, \quad m_2 = A_2 + B_2 r^2 + C_2 r^6,$$

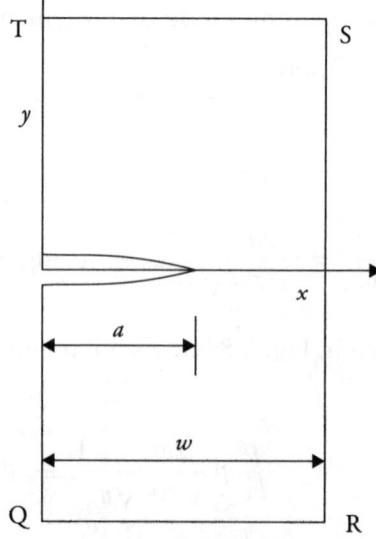

Figure 5.9 Edge crack geometry.

for $0 \leq r \leq 1$, where $r = \dfrac{a}{w}$.

$A_1 = 0.6147 \quad B_1 = 17.1884 \quad C_1 = 8.7822$

$A_2 = 0.2502 \quad B_2 = 3.2899 \quad C_2 = 70.0444$

(a) Calculate K_I for remote end loading σ and $r = 0.45$.
(b) Calculate K_I for remote bending load M and $r = 0.5$.

Assume thickness as unity.

Solution
(a) For $r = 0.45$, $m_1 = 4.1682$, and $m_2 = 1.4980$.

$$K_I = \int_0^a \sigma \frac{2}{\sqrt{2\pi(a-x)}} \left[1 + m_1 \frac{a-x}{a} + m_2 \left(\frac{a-x}{a}\right)^2\right] dx, \text{ since } p(x, 0) = \sigma.$$

Substituting $a - x = z$, the integral can be easily integrated to give $K_I = 2.4206 \sigma \sqrt{\pi a}$ (Ans.).

(b) For this case, $p(x, 0) = \dfrac{12M}{w^3}(a-x)$, assuming the beam thickness as unity.
For $r = 0.5$, $m_1 = 5.0490$, and $m_2 = 2.1671$.

$$K_I = \int_0^a \frac{12M}{w^3}(a-x) \frac{2}{\sqrt{2\pi(a-x)}} \left[1 + m_1 \frac{a-x}{a} + m_2 \left(\frac{a-x}{a}\right)^2\right] dx$$

Upon integration, $K_I = 1.4879 \sigma_{max} \sqrt{\pi a}$, where $\sigma_{max} = \dfrac{6M}{w^2}$ (Ans.).

5.3 Numerical Technique: Finite Element Method

The numerical technique like the finite element method (FEM) has been very widely applied for the determination of the SIFs. It has been so because of its applicability to a wide range of problems, irrespective of complexity of loading and geometry of components.

For finite element analysis of a given domain, it is discretized into a convenient number of elements of finite dimensions (Zienkiewicz and Taylor 2000; Cook et al. 2002). The elements are interconnected at their nodal points. Field variables (e.g., displacements in the case of displacement finite element formulation) are associated with each of these nodes. A distribution, which may be linear, quadratic, cubic, and so on, of the variables is assumed within each of the elements and a functional such as potential energy functional is then derived in terms of these nodal variables. When the restrictions are imposed to make the derivative of the functional zero with respect to each of the variables, a set of simultaneous equations is obtained. These are nothing but the global equilibrium equations. The solution of this set gives the displacements at the nodal points corresponding to the specified boundary and loading conditions.

Through gradual refinements of the discretization, convergence to the exact solution can be guaranteed, provided the assumed displacement field meets the convergence criteria. In particular, the assumed displacement field should satisfy the rigid body mode, the constant strain condition, and the compatibility at the inter-element boundary.

The application of FEM to different areas involve routine steps: idealization of the geometry, discretization, element stiffness calculation and assembly, insertion of boundary conditions, solution of simultaneous equations, and calculation of output data, which may include element stresses, strains, total strain energy, etc.

A large number of element choices, for example, triangular, quadrilateral, triangle with curved boundaries, quadrilateral with curved boundaries, etc., are available for the discretization. The choices of elements to a large extent depend on the particular application. Generally, the geometric configurations, local high stress gradient, if any, in the domain, accuracy required, and available computational facilities are some of the issues, which influence the selection. In the early stages of application of FEM to fracture mechanics, the conventional elements like the constant strain triangles, linear strain triangles, 4-noded quadrilaterals, 8-noded quadrilaterals, and their analogues in the three dimensions, were widely utilized to discretize the body, and special techniques were adopted to obtain the SIFs. Later, it was realized that for any problem involving stress singularity, the convergence rate is dominated by the singular nature of the solution (Tong and Pian 1973). Therefore, the convergence can be enhanced by employing elements that can approximate the singular field properly. Now there are a large number of special or singularity elements available, which can be constructed at the crack-tip to facilitate a faster convergence rate (Tracey 1971; Byskov 1970; Henshell and Shaw 1975; Barsoum 1976; Tracey and Cook 1977; Stern 1979; Dutta, Maiti and Kakodkar 1990; Maiti, 1992 a, b and c. etc.). These help to eliminate the need for a very fine crack-tip discretization, and some formulations even permit direct and accurate determination of SIFs (Rao, Raju and Krishna Murthy 1971; Tong and Pian 1973; Atluri, Kobayashi and Nakagaki 1975; Atluri 1986).

Consider a problem with eight elements and nine nodes (Fig. 5.10). The external load is acting at 8 and the plate does not move in the y direction at the

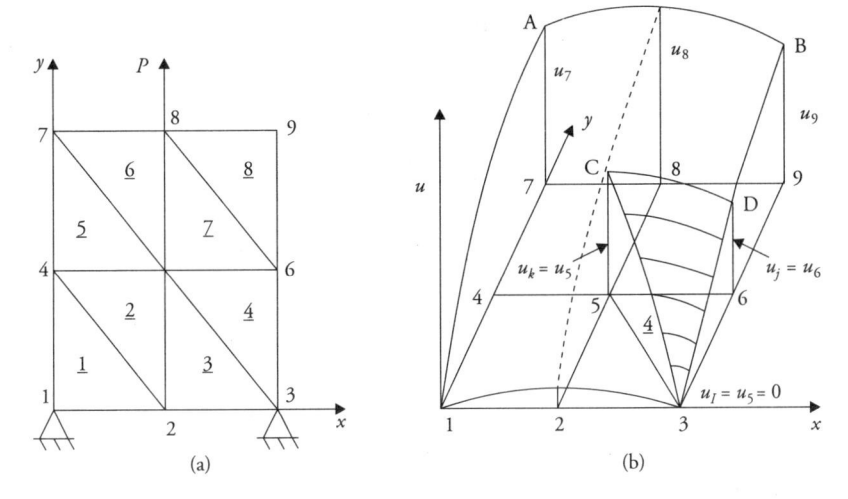

Figure 5.10 (a) Typical FE discretization and (b) u-displacement surface.

nodes 1 and 3. Under the action of the load, all the nodes move in both x and y directions due to deformations except the ones under constraint. If we plot the displacement u in the vertical direction, we get a surface 13AB as shown (Fig. 5.10(b)). This is u-surface. Similarly, we can get v-surface by plotting v displacements. The portion of the surface above a typical element, say 4, is a 'triangular' surface. It can be approximated by a plane. As the size of the element reduces, the accuracy of approximation as a plane increases. Similarly, the displacement surface v above element 4 can also be approximated by a plane. These two surfaces can be written as follows.

$$u = \alpha_1 + \alpha_2 x + \alpha_3 y, \tag{5.19a}$$

$$u = \alpha_4 + \alpha_5 x + \alpha_6 y \tag{5.19b}$$

where α_1 to α_6 are known as generalized coordinates. These coordinates can be eliminated by noting that these surfaces pass through the nodal points. That is,

$$u_i = \alpha_1 + \alpha_2 x_i + \alpha_3 y_i + \alpha_4 0 + \alpha_5 0 + \alpha_6 0$$

$$v_i = \alpha_1 0 + \alpha_2 0 + \alpha_3 0 + \alpha_4 + \alpha_5 x_i + \alpha_6 y_i$$

$$u_j = \alpha_1 + \alpha_2 x_j + \alpha_3 y_j + \alpha_4 0 + \alpha_5 0 + \alpha_6 0$$

$$v_j = \alpha_1 0 + \alpha_2 0 + \alpha_3 0 + \alpha_4 + \alpha_5 x_j + \alpha_6 y_j$$

$$u_k = \alpha_1 + \alpha_2 x_k + \alpha_3 y_k + \alpha_4 0 + \alpha_5 0 + \alpha_6 0$$

$$v_k = \alpha_1 0 + \alpha_2 0 + \alpha_3 0 + \alpha_4 + \alpha_5 x_k + \alpha_6 y_k$$

where $i = 3$, $j = 6$, and $k = 5$ for element $n = 4$. Alternatively,

$$\begin{Bmatrix} u_i \\ v_i \\ u_j \\ v_j \\ u_k \\ v_k \end{Bmatrix} = \begin{bmatrix} 1 & x_i & y_i & 0 & 0 & 0 \\ 0 & 0 & 0 & 1 & x_i & y_i \\ 1 & x_j & y_j & 0 & 0 & 0 \\ 0 & 0 & 0 & 1 & x_j & y_j \\ 1 & x_k & y_k & 0 & 0 & 0 \\ 0 & 0 & 0 & 1 & x_k & y_k \end{bmatrix} \begin{Bmatrix} \alpha_1 \\ \alpha_2 \\ \alpha_3 \\ \alpha_4 \\ \alpha_5 \\ \alpha_6 \end{Bmatrix}, \quad \{u\}_n = [A]_n \{\alpha\}_n \quad \text{(5.20 a and b)}$$

Therefore, $\{\alpha\}_n = [A]_n^{-1} \{u\}_n$

Substituting $\{\alpha\}_n$ in Eq. (5.19)

$$\begin{Bmatrix} u \\ v \end{Bmatrix}_n = \begin{bmatrix} N_i & 0 & N_j & 0 & N_k & 0 \\ 0 & N_i & 0 & N_j & 0 & N_k \end{bmatrix} \begin{Bmatrix} u_i \\ v_i \\ u_j \\ v_j \\ u_k \\ v_k \end{Bmatrix} = [N]\{u\}_n \quad (5.21)$$

N_i, N_j, and N_k are the shape functions or interpolation functions and are given as follows.

$$N_i = \frac{a_i + b_i x + c_i y}{2\Delta}, \quad N_j = \frac{a_j + b_j x + c_j y}{2\Delta}, \quad N_k = \frac{a_k + b_k x + c_k y}{2\Delta} \quad (5.22)$$

Δ = area of triangle n with vertices i, j, and k or 3, 6, and 5.

$$a_i = x_j y_k - x_k y_j \quad b_i = y_j - y_k, \quad c_i = x_k - x_j$$
$$a_j = x_k y_i - x_i y_k \quad b_j = y_k - y_i, \quad c_j = x_i - x_k$$
$$a_k = x_i y_j - x_j y_i \quad b_k = y_i - y_j, \quad c_k = x_j - x_i \quad (5.23)$$

Thus, it is possible to express an element displacement field in terms of displacements of its nodes. The strain field within the element n is given by

$$\varepsilon_x = \frac{\partial u}{\partial x} = \frac{1}{2\Delta}[b_i u_i + b_j u_j + b_k u_k], \quad \varepsilon_y = \frac{\partial v}{\partial y} = \frac{1}{2\Delta}[c_i v_i + c_j v_j + c_k v_k]$$

$$\gamma_{xy} = \frac{\partial u}{\partial y} + \frac{\partial v}{\partial x} = \frac{1}{2\Delta}[c_i u_i + b_i v_i + c_j u_j + b_j v_j + c_k u_k + b_k v_k],$$

$$\begin{Bmatrix} \varepsilon_x \\ \varepsilon_y \\ \gamma_{xy} \end{Bmatrix} = \frac{1}{2\Delta} \begin{bmatrix} b_i & 0 & b_j & 0 & b_k & 0 \\ 0 & c_i & 0 & c_j & 0 & c_k \\ c_i & b_i & c_j & b_j & c_k & b_k \end{bmatrix} \{u\}_n, \quad \{\varepsilon\}_n = [B]_n \{u\}_n \quad \text{(5.24 a and b)}$$

Since b_i, b_j, b_k, c_i, c_j, and c_k are constants for an element, the strains within the element remain constant. That is why this element is known as constant stain triangular (CST) element.

The stresses and strains in two dimensions are related by the following relationship for an isotropic elastic material.

$$\begin{Bmatrix} \sigma_x \\ \sigma_y \\ \tau_{xy} \end{Bmatrix} = \frac{E}{1-\nu^2} \begin{bmatrix} 1 & \nu & 0 \\ \nu & 1 & 0 \\ 0 & 0 & \frac{1-\nu}{2} \end{bmatrix} \begin{Bmatrix} \varepsilon_x \\ \varepsilon_y \\ \gamma_{xy} \end{Bmatrix} \text{ for plane stress condition,}$$

$$= \frac{E}{(1+\nu)(1-2\nu)} \begin{bmatrix} 1-\nu & \nu & 0 \\ \nu & 1-\nu & 0 \\ 0 & 0 & \frac{1-2\nu}{2} \end{bmatrix} \begin{Bmatrix} \varepsilon_x \\ \varepsilon_y \\ \gamma_{xy} \end{Bmatrix} \text{ for plane strain condition.} \quad (5.25)$$

In short, $\quad \{\sigma\}_n = [D]_n \{\varepsilon\}_n \quad\quad\quad\quad\quad\quad\quad\quad\quad\quad\quad\quad\quad\quad\quad\quad (5.26)$

where $[D]_n$ is material property matrix, and $\{\sigma\}_n$ is stress matrix. Substituting the value of strain matrix in terms of element nodal displacements $\{u\}_n$,

$$\{\sigma\}_n = [D]_n [B]_n \{u\}_n \quad\quad\quad\quad\quad\quad\quad (5.27)$$

Since elements of the $[D]_n$ matrix are all constants, the stresses in the element are all constants. In general, therefore, for an element of finite dimensions, the field given by the above relations indicates a constant stress–strain field. The strain energy stored U_n in an element is given by

$$U_n = \int_{A_n} \frac{1}{2} \{\varepsilon\}_n^T \{\sigma\}_n h \, dA = \frac{1}{2} \{\varepsilon\}_n^T \{\sigma\}_n h A_n \{u\}_n$$

$$= \frac{1}{2} \{u\}_n^T [B]_n^T [D]_n [B]_n h A_n \{u\}_n = \frac{1}{2} \{u\}_n^T [k]_n \{u\}_n \quad (5.28)$$

where h is plate thickness, A_n is area of element n and $[k]_n$ is its stiffness matrix. Its dimension is 6×6. The element displacement field is given by three us and vs of the three corner nodes.

The total strain energy U in the plate is obtained summing up U_n over all the eight elements in the present case. U can be written as follows.

$$U = \sum_{1}^{n_e} U_n = \frac{1}{2} \{u\}^T [K] \{u\} \quad\quad\quad\quad\quad\quad (5.29)$$

Note that [K] is a matrix of size 18×18 and $\{u\}$ is a column matrix of size 18, and n_e is the total number of elements. In the present case $n_e = 8$. The work done by external forces

$$W = \{u\}^T \{P\}, \tag{5.30}$$

where $\{u\}$ is global displacement vector and $\{P\}$ is global nodal load vector. The potential energy π of the discretized system

$$\pi = U - W = \frac{1}{2}\{u\}^T [K] \{u\} - \{u\}^T \{P\} \tag{5.31}$$

The deformed system is in equilibrium under the action of the external forces. Hence, the potential energy of the system is minimum. By minimizing the potential energy with respect to the displacements u_i, $i = 1, 2, 3 \ldots \ldots 18$, a set of simultaneous equations is obtained.

$$[K]\{u\} = \{P\} \tag{5.32}$$

These are the global nodal equilibrium equations. This set of equations can be solved for after introduction of the displacement boundary conditions, $u_1 = v_1 = u_3 = v_3 = 0$, and, thereby, the displacement field of the whole plate is obtained.

If displacement variation within the element is considered to be quadratic or cubic, the number of nodes per element will have to be increased to six or nine, respectively. Consequently, the strain field within the element will be linear and quadratic, respectively. The element stiffness matrix for the element will have to be obtained through Gauss quadrature or numerical integration.

After obtaining the global displacement vector, the element stresses or Gauss point stresses can be calculated through Eq. (5.27). Based on these displacements and stresses, the SIFs can be calculated through the displacement method, stress method, J integral technique, stiffness derivative procedure, crack closure integral (CCI) technique, and so on. These are discussed subsequently.

5.3.1 Displacement and Stress-Based Methods for Extraction of SIFs

Displacement method For Mode I loading, the x and y displacements are given by

$$u = K_I \sqrt{\frac{2r}{\pi}} \tilde{u}(\theta), \tag{5.33a}$$

$$v = K_I \sqrt{\frac{2r}{\pi}} \tilde{v}(\theta) \tag{5.33b}$$

The exact forms of $\tilde{u}(\theta)$ and $\tilde{v}(\theta)$ are given in Chapter 3 [Eqs. (3.39) and (3.43)]. Generally, values of v are more significant than u around the crack-tip in this mode. By selecting a corner node closer to the crack-tip, hence, small r on the crack face, or $\theta = \pi$, the SIF K_I can be obtained using the above relation for v. It is better to avoid the first corner node, since the FE solution is not that accurate around the crack-tip because of the high stress and strain gradients. It is possible to improve the accuracy of the results by using the appropriate singularity element like the quarter point singularity element (Barsoum 1976; Henshell and Shaw 1975) around the crack-tip.

By selecting a number of nodes on the crack face, a set of K_I can be obtained. Thereby, a variation of K_I with r can be plotted (Fig. 5.11). By extrapolating this variation back to $r = 0$, K_I^* can be determined (Wilson 1973). This appears to be the most accurate value for the Mode I SIF. This graphical procedure helps to avoid the inaccuracy in the SIF due to error in the FE solution close to the crack-tip and the inadequacy of the first term solution of the eigen function expansion away from the crack-tip.

The same procedure can be followed for Mode II problem. In this case, it is more appropriate to use u displacement as the basis for the extraction of the SIF. The extrapolation technique can be employed to enhance the accuracy of the results further.

Stress method   Stresses around the crack-tip are given by

$$\sigma_{ij} = \frac{K_I}{\sqrt{2\pi r}} \tilde{\sigma}_{ij}(\theta) \tag{5.34}$$

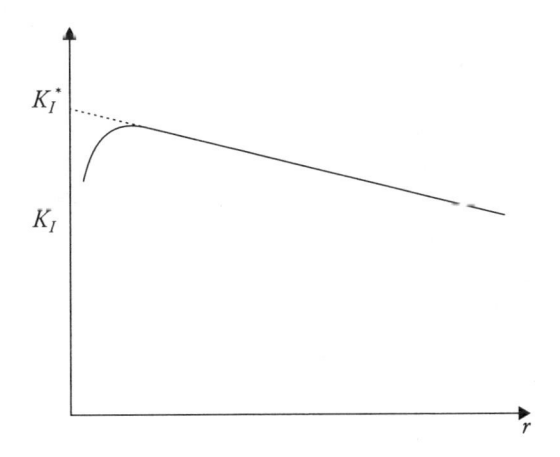

Figure 5.11 Extrapolation method.

The exact form of the functions $\tilde{\sigma}_{ij}(\theta)$ is given in Chapter 2. In Mode I problems, σ_y stresses are more dominant than σ_x.

At the end of FE computations, stresses can be determined at the element Gauss point locations or the nodal locations. More often, the nodal stresses are obtained through extrapolation of the Gauss point data. By selecting a Gauss point on a radial line θ close to, but less than 90°, and comparing σ_y stress, the SIF can be extracted through the above relation. The accuracy can be improved further by resorting to the extrapolation technique. Since the accuracy of stresses is less than displacement in the displacement FE formulation, the results obtained by the stress method are less accurate than the results obtained by the displacement comparison.

In a Mode II problem, τ_{xy} stresses are more significant around the crack line. By selecting θ close to 0° and a small r, the SIF can be extracted. Again, the extrapolation method can be employed to improve accuracy of the results.

5.3.2 Energy-based Methods for Determination of SIFs

Consider a plate with an edge crack a under Mode I loading (Fig. 5.12). Any crack growth will therefore occur in-plane or in a self-similar manner. Under the action of loading, it is possible to calculate the strain energy U_1 stored in the plate. It is possible to calculate the strain energy U_2 corresponding to the same external load and an extended crack of size $a + \Delta a$. The strain energy release rate G is given by

$$G = \lim_{\Delta a \to 0} \frac{U_2 - U_1}{\Delta a} \approx \frac{\Delta U}{\Delta a} \text{ for small } \Delta a. \tag{5.35}$$

where thickness of the plate is taken as unity. Thus, it is possible to calculate the strain energy release rate by two finite element runs. By selecting Δa about 0.1% a (Maiti 1990) and doing computations in extended precision, it is possible to obtain good accuracy. The FE approximation errors in the computation of U_1 and U_2 are of the same order of magnitude because of small Δa. This helps to obtain G, in turn, the SIF, with very good accuracy through this method, which involves $(U_2 - U_1)$. In a sense, this accuracy is obtained at the cost of two separate FE runs.

The requirements of two separate FE runs can be avoided by resorting to the variant of the energy method, for example, J integral method, stiffness derivative procedure, and CCI technique.

J integral method J integral given below can be computed by selecting a contour joining the two points lying on the opposite crack flanks.

$$J = \int_S \left(W dy - T_i \frac{\partial u_i}{\partial x} dS \right) \tag{5.36}$$

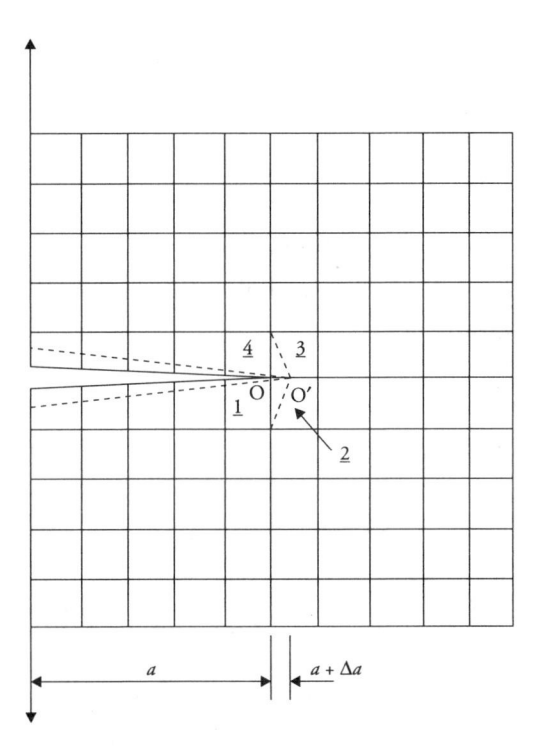

Figure 5.12 Shifting of crack-tip to accommodate small crack extension.

Since J gives the potential energy release rate associated with an in-plane crack extension, it is possible to evaluate the SIF through its computation. It is sometimes convenient to consider the path passing through the element Gauss points. The integration can be carried out through numerical integration by splitting the whole contour S into a number of small segments. The splitting can be done by noting the locations of Gauss points, marked by " in Fig. 5.13, and assuming the values of stresses σ_{ij}, strains ε_{ij}, strain energy density W, and tractions T_j at a Gauss point 1 to be valid over the sector mn. Similarly, the values at Gauss point 2 can be assumed to be valid over the span nr. For the shown contour passing through elements 9 to 16, there are 16 segments. Therefore

$$J = \sum_{k=1}^{16}\left(W\,dy - T_i\frac{\partial u_i}{\partial x}dS\right)_k - \sum_{k=1}^{16}\left[\{W_k\Delta y_k\} - \left\{T_{1k}\left(\frac{\partial u_i}{\partial x}\right)_k + T_{2k}\left(\frac{\partial v_i}{\partial x}\right)_k\right\}\Delta S_k\right], \quad (5.37)$$

where, assuming that all elements around the crack-tip are 8-noded isoparametric elements,

$$W = \frac{1}{2}(\sigma_x\varepsilon_x + \sigma_x\varepsilon_x + \tau_{xy}\gamma_{xy}), \; T_i = T_1\frac{\partial u}{\partial x} + T_2\frac{\partial v}{\partial y},$$

$$T_1 = l\sigma_x + m\tau_{xy}, \; T_2 = l\tau_{xy} + m\sigma_y$$

$$\frac{\partial u}{\partial x} = u_1 \frac{\partial N_1}{\partial x} + u_2 \frac{\partial N_2}{\partial x} + \ldots\ldots\ldots + u_8 \frac{\partial N_8}{\partial x},$$

$$\frac{\partial v}{\partial x} = v_1 \frac{\partial N_1}{\partial x} + v_2 \frac{\partial N_2}{\partial x} + \ldots\ldots\ldots + v_8 \frac{\partial N_8}{\partial x} \qquad (5.38)$$

$\Delta y_k = y_n - y_m$ for a typical Gauss point k, $dS_k = r\, d\theta$

$l, m =$ direction cosines of the outer normal to the contour S at the Gauss point k

$r =$ radius of the contour S.

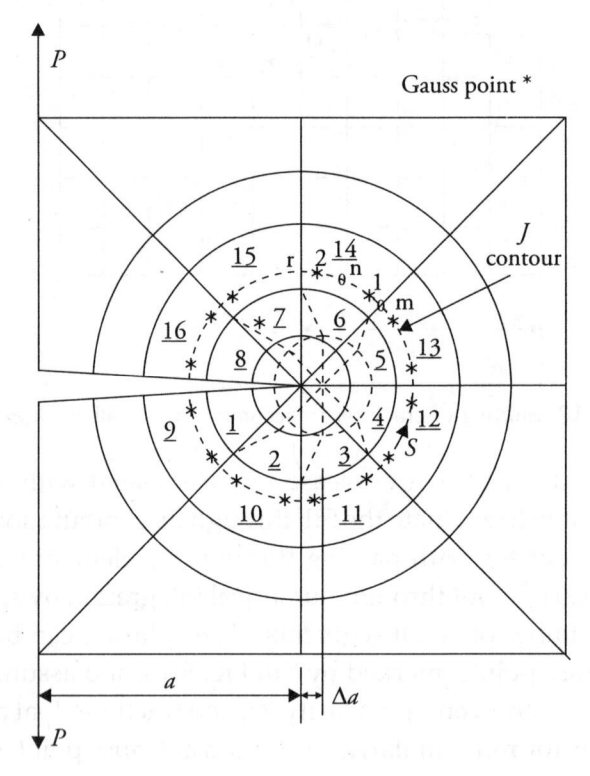

Figure 5.13 Crack-tip shift to accommodate small Δa and contour for J integral.

By using the appropriate relationship between J and K_I, the SIF can be determined. For pure Mode I and Mode II problems, only one-half of the body can be analysed and J can be obtained by integration over elements, for example, 9 to 12 or 13 to 16, and by multiplying by a factor 2.

For a mixed mode problem, J gives the total potential energy release rate. There are schemes whereby the Mode I and Mode II parts of the total energy release rate can be separated (Kitagawa, Okamura and Ishikawa 1976).

Domain integral based evaluation of J Rather than evaluating J through line integral, J can be computed through domain integration (Shih, Moran and

Nakamura 1986; Moran and Shih 1987; Nikishkov and Atluri 1987; Raju and Shivakumar 1990). This conversion is possible using the Green's theorem/formula in two dimensions and the divergence theorem of Gauss in three dimensions (Shih, Moran and Nakamura 1986; Moran and Shih 1987). The domain integral based evaluation of J (Raju and Shivakumar 1990) through two-dimensional finite element analysis is discussed subsequently.

When J is calculated along the contour S_0 or S_1 (Fig. 5.14) by Eq. (5.36), it gives energy flow through the contour in the x direction. Similarly, the energy flow through the contour in the y direction too can be calculated through the following relation.

$$J_y = \int_S \left[-W dx - T_i \frac{\partial u_i}{\partial y} dS \right] \tag{5.39}$$

The two energy expression can be given in the form of a single relation.

$$J_{x_k} = \int_S \left[W n_k - T_i \frac{\partial u_i}{\partial x_k} \right] dS = \int_S \left[W n_k - \sigma_{ij} n_j \frac{\partial u_i}{\partial x_k} \right] dS = \int_S [Q] \, dS, \text{ say.} \tag{5.40}$$

$$J_{x_k} = 1 \int_{S_1} [Q] \, dS - 0 \int_{S_0} [Q] dS$$

$$= 1 \int_{ABC} [Q] \, dS + 1 \int_{CO} [Q] \, dS + 1 \int_{OA} [Q] \, dS$$

$$- 0 \int_{FED} [Q] \, dS - 0 \int_{OF} [Q] \, dS - 0 \int_{DO} [Q] \, dS \tag{5.41}$$

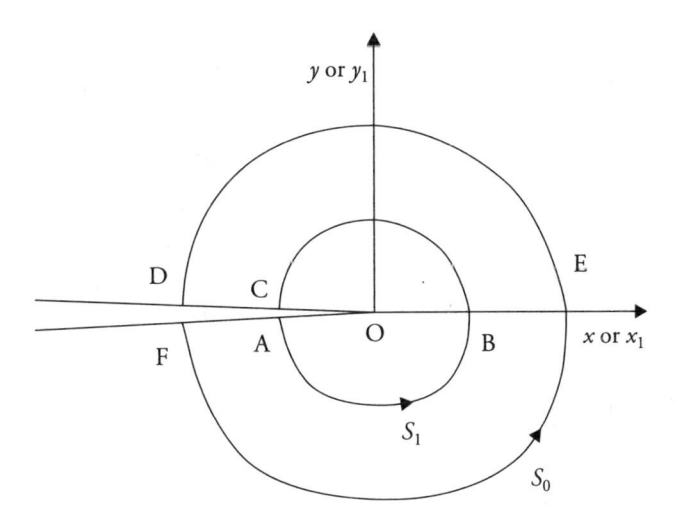

Figure 5.14 Contour for domain integration.

Introducing a function H, which has value 1 along the contour ABC or S_1 and 0 along the contour FED or S_0, the above relation can be written in the following form.

$$J_{x_k} = \int_{ABC} [Q] H \, dS + \int_{CO} [Q] \, dS + \int_{OA} [Q] \, dS - \int_{FED} [Q] H \, dS$$

$$= -\int_{CBA} [Q] H \, dS - \int_{AF} [Q] H \, dS + \int_{CO} [Q] \, dS + \int_{OA} [Q] \, dS$$

$$- \int_{FED} [Q] H \, dS - \int_{DC} [Q] H \, dS + \int_{AF} [Q] H \, dS + \int_{DC} [Q] H \, dS$$

$$= -\int_{FEDCBAF} [Q] H \, dS + \int_{CO} [Q] \, dS + \int_{OA} [Q] \, dS$$

$$+ \int_{AF} [Q] H \, dS + \int_{DC} [Q] H \, dS \tag{5.42}$$

$$J_{x_k} = -\int_{FEDCBAF} [Q] H \, dS + (J_{x_k})_{\text{line}}, \quad Q = \left[W n_k - \sigma_{ij} n_j \frac{\partial u_i}{\partial x_k} \right] \tag{5.43}$$

Considering only the first part,

$$-\int_{FEDCBAF} [Q] H \, dS = -\int_{FEDCBAF} \left[W H n_k - \sigma_{ij} n_j \frac{\partial u_i}{\partial x_k} H \right] dS$$

$$= -\int_A \left[\frac{\partial}{\partial x_k} (W H) - \frac{\partial}{\partial x_j} \left(H \sigma_{ij} \frac{\partial u_i}{\partial x_k} \right) \right] dA$$

$$= -\int_A \left[W \frac{\partial H}{\partial x_k} - \frac{\partial H}{\partial x_j} \left(\sigma_{ij} \frac{\partial u_i}{\partial x_k} \right) \right] dA - \int_A \left[H \frac{\partial W}{\partial x_k} - H \frac{\partial}{\partial x_j} \left(\sigma_{ij} \frac{\partial u_i}{\partial x_k} \right) \right] dA$$

$$= -\int_A \left[W \frac{\partial H}{\partial x_k} - \frac{\partial H}{\partial x_j} \left(\sigma_{ij} \frac{\partial u_i}{\partial x_k} \right) \right] dA - \int_A \left[H \frac{\partial W}{\partial x_k} - H \left\{ \sigma_{ij} \frac{\partial}{\partial x_j} \left(\frac{\partial u_i}{\partial x_k} \right) \right\} \right] dA$$

$$= -\int_A \left[W \frac{\partial H}{\partial x_k} - \frac{\partial H}{\partial x_j} \left(\sigma_{ij} \frac{\partial u_i}{\partial x_k} \right) \right] dA - \int_A \left[H \frac{\partial W}{\partial x_k} - H \left\{ \sigma_{ij} \frac{\partial \varepsilon_{ij}}{\partial x_k} \right\} \right] dA \tag{5.44}$$

since, in the absence of body forces the equilibrium equation is given by $\frac{\partial \sigma_{ij}}{\partial x_j} = 0$, and $\varepsilon_{ij} = \frac{1}{2} \left(\frac{\partial u_i}{\partial x_j} + \frac{\partial u_j}{\partial x_i} \right)$. Finally, the first part of J_{x_k} is as follows.

$$\left(J_{x_k}\right)_{\text{Part I}} = -\int_A \left[W\frac{\partial H}{\partial x_k} - \frac{\partial H}{\partial x_j}\left(\sigma_{ij}\frac{\partial u_i}{\partial x_k}\right)\right] dA - \int_A \left[H\frac{\partial W}{\partial x_k} - H\left\{\sigma_{ij}\left(\frac{\partial \varepsilon_{ij}}{\partial x_k}\right)\right\}\right] dA \tag{5.45}$$

In matrix form

$$\begin{Bmatrix} J_{x_1} \\ J_{x_2} \end{Bmatrix}_{\text{Part I}} = -\int_A \left[\begin{Bmatrix} W\frac{\partial H}{\partial x_1} \\ W\frac{\partial H}{\partial x_2} \end{Bmatrix} - \begin{bmatrix} \frac{\partial u_1}{\partial x_1} & \frac{\partial u_2}{\partial x_1} \\ \frac{\partial u_1}{\partial x_2} & \frac{\partial u_2}{\partial x_2} \end{bmatrix} \begin{bmatrix} \sigma_{11} & \sigma_{12} \\ \sigma_{21} & \sigma_{22} \end{bmatrix} \begin{Bmatrix} \frac{\partial H}{\partial x_1} \\ \frac{\partial H}{\partial x_2} \end{Bmatrix} \right] dA$$

$$-\int_A \left[\begin{Bmatrix} H\frac{\partial W}{\partial x_1} \\ H\frac{\partial W}{\partial x_2} \end{Bmatrix} - H \begin{bmatrix} \frac{\partial \varepsilon_{11}}{\partial x_1} & \frac{\partial \gamma_{12}}{\partial x_1} & \frac{\partial \varepsilon_{22}}{\partial x_1} \\ \frac{\partial \varepsilon_{11}}{\partial x_2} & \frac{\partial \gamma_{12}}{\partial x_2} & \frac{\partial \varepsilon_{22}}{\partial x_2} \end{bmatrix} \begin{Bmatrix} \sigma_{11} \\ \sigma_{12} \\ \sigma_{22} \end{Bmatrix} \right] dA \tag{5.46}$$

where $W = \frac{1}{2}[\sigma_{11}\varepsilon_{11} + \sigma_{22}\varepsilon_{22} + \sigma_{12}\gamma_{12}]$ for linear elastic materials. For non-linear elastic material $W = \int \sigma_{ij} d\varepsilon_{ij}$. Both the area/domain integrals and the line integrals involved in J_{x_1} and J_{x_2} can be evaluated by numerical integrations. In the evaluation of the domain integrals, the order in integration can be kept the same as used in the evaluation of the element stiffness matrices.

The first part of expression for integral J_{x_k} (Eq. (5.42)) consists of the domain integral and the second part includes the line integrals. Under pure Mode I and Mode II loading and load-free crack edges, lines integrals involved in J_{x_1} and J_{x_2} are 0. In the case of a general loading, this is not so. For a general mixed mode loading, Mode I and II energy release rates are given by

$$J_{x_1} = J_I + J_{II}, \quad J_{x_2} = -2\sqrt{J_I J_{II}} \tag{5.47}$$

$$J_I = \frac{1}{4}\left[\sqrt{J_{x_1} - J_{x_2}} + \sqrt{J_{x_1} + J_{x_2}}\right]^2, \quad J_{II} = \frac{1}{4}\left[\sqrt{J_{x_1} - J_{x_2}} - \sqrt{J_{x_1} + J_{x_2}}\right]^2 \tag{5.48}$$

For evaluation of the pure Mode I and Mode II parts of J integral, it is better to decompose displacements and stresses into symmetric and anti-symmetric parts and compute only the domain integrals. The decomposition helps to make the line integrals 0.

Considering two points P and P' that are close to the crack-tip and are symmetrically placed about the crack line (Fig. 5.15), the following relations can be written.

$$\begin{Bmatrix} u_{1P} \\ u_{2P} \end{Bmatrix} = \begin{Bmatrix} u_{1S} \\ u_{2S} \end{Bmatrix} + \begin{Bmatrix} u_{1AS} \\ u_{2AS} \end{Bmatrix}, \quad \begin{Bmatrix} u_{1P'} \\ u_{2P'} \end{Bmatrix} = \begin{Bmatrix} u_{1S} \\ -u_{2S} \end{Bmatrix} + \begin{Bmatrix} -u_{1AS} \\ u_{2AS} \end{Bmatrix} \tag{5.49}$$

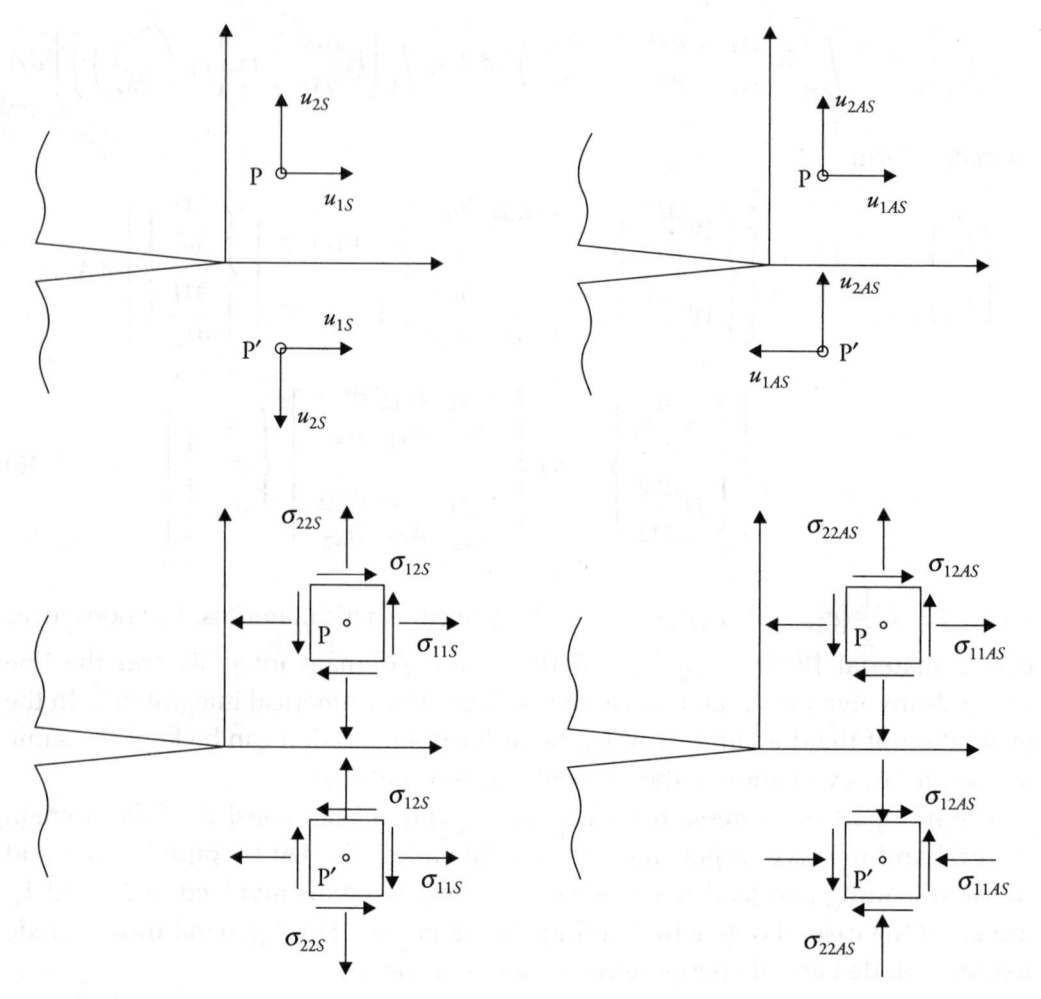

Figure 5.15 Symmetric and anti-symmetric displacements and stresses about crack plane.

These relations give the symmetric and anti-symmetric displacements of P and P' as given below.

$$\begin{Bmatrix} u_1 \\ u_2 \end{Bmatrix}_S = \frac{1}{2} \begin{Bmatrix} u_{1P} + u_{1P'} \\ u_{2P} - u_{2P'} \end{Bmatrix}, \quad \begin{Bmatrix} u_1 \\ u_2 \end{Bmatrix}_{AS} = \frac{1}{2} \begin{Bmatrix} u_{1P} - u_{1P'} \\ u_{2P} + u_{2P'} \end{Bmatrix} \quad (5.50)$$

Similarly, the stresses too can be expressed as follows.

$$\begin{Bmatrix} \sigma_{11} \\ \sigma_{22} \\ \sigma_{12} \end{Bmatrix}_S = \frac{1}{2} \begin{Bmatrix} \sigma_{11P} + \sigma_{11P'} \\ \sigma_{22P} + \sigma_{22P'} \\ \sigma_{12P} - \sigma_{12P'} \end{Bmatrix}, \quad \begin{Bmatrix} \sigma_{11} \\ \sigma_{22} \\ \sigma_{12} \end{Bmatrix}_{AS} = \frac{1}{2} \begin{Bmatrix} \sigma_{11P} - \sigma_{11P'} \\ \sigma_{22P} - \sigma_{22P'} \\ \sigma_{12P} + \sigma_{12P'} \end{Bmatrix} \quad (5.51)$$

The symmetric and anti-symmetric displacements and stresses can be used in Eq. (5.45) or (5.46) to evaluate the four integrals J_{Sx_1}, J_{Sx_2}, J_{ASx_1}, and J_{ASx_2}. It gives rise to $J_{Sx_2} = 0$ and $J_{ASx_2} = 0$ because of the symmetric and anti-symmetric nature of the stress-displacement fields, respectively. Finally,

$$J_I = J_{Sx_1}, \quad J_{II} = J_{ASx_1} \tag{5.52}$$

For the evaluation of the domain integral, a typical zone consisting of elements 1 to 8 or 9 to 16 (Fig. 5.16) can be considered. The function H can be defined to be unity at the inner boundary S_1 of the zone and 0 at the outer boundary S_0 and can be considered to have a linear variation along ζ direction for a constant η. While dealing with integration over each element, the Gauss quadrature can be utilized. The derivatives of H are given by

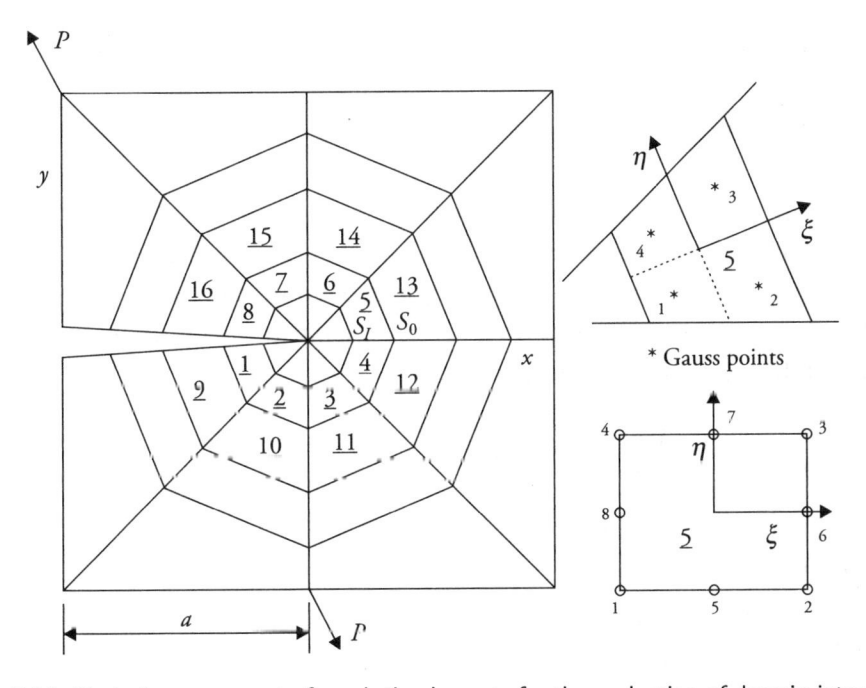

$$\begin{Bmatrix} \dfrac{\partial H}{\partial x_1} \\ \dfrac{\partial H}{\partial x_2} \end{Bmatrix} = \begin{bmatrix} \dfrac{\partial x_1}{\partial \zeta} & \dfrac{\partial x_2}{\partial \zeta} \\ \dfrac{\partial x_1}{\partial \eta} & \dfrac{\partial x_2}{\partial \eta} \end{bmatrix}^{-1} \begin{Bmatrix} \dfrac{\partial H}{\partial \zeta} \\ \dfrac{\partial H}{\partial \eta} \end{Bmatrix}, \quad x_1 = x \text{ and } x_2 = y. \tag{5.53}$$

Figure 5.16 Typical arrangement of crack-tip elements for the evaluation of domain integral and shape of typical element in mapping plane $\zeta - \eta$.

Assuming that all elements in the discretization are 8-noded sub-parametric elements, a typical element, for example, 5, involves eight shape functions for the displacement interpolations and four shape functions associated only with the four corner nodes (1 to 4) for the geometric interpolations. These geometric shape functions are $N_i^g = \frac{1}{4}(1 + \xi\xi_i)(1 + \eta\eta_i)$, where ξ and η are the two natural coordinates, and (ξ_i, η_i) are the coordinates of the corner nodes. Therefore,

$$x = \sum_{i=1}^{4} N_i^g x_i, \; y = \sum_{i=1}^{4} N_i^g y_i, \; \frac{\partial x}{\partial \xi} = \sum_{i=1}^{4} \frac{\partial N_i^g}{\partial \xi} x_i, \; \frac{\partial x}{\partial \eta} = \sum_{i=1}^{4} \frac{\partial N_i^g}{\partial \eta} x_i,$$

$$\frac{\partial y}{\partial \xi} = \sum_{i=1}^{4} \frac{\partial N_i^g}{\partial \xi} y_i, \; \frac{\partial y}{\partial \eta} = \sum_{i=1}^{4} \frac{\partial N_i^g}{\partial \eta} y_i \tag{5.54}$$

where (x_i, y_i) are the coordinates of the associated corner nodes. In order to compute the derivatives, $\frac{\partial W}{\partial x_k}$, $k = 1$ and 2, it is necessary to express the variation of strain energy density W in an appropriate form. For each subparametric element a 4-point Gauss quadrature is good for the evaluation of the element stiffness matrix and all domain integrations. The strain energy density can be evaluated at the four Gauss points. The variation of W within the element can be conveniently considered to be given by the interpolation of the values at the four Gauss points as follows.

$$W = \sum_{i=1}^{4} N_i W_i \tag{5.55}$$

where W_i are the values of strain energy density at the four Gauss points (Fig. 5.16). The same variation, when extrapolated, can also be assumed to be valid in the remaining regions of the element beyond the Gauss point locations. Noting that the element Gauss points are off the natural coordinate axes by $\pm\frac{1}{\sqrt{3}}$, the interpolation functions N_i are given by

$$N_1 = \frac{3}{4}\left(\frac{1}{\sqrt{3}} - \xi\right)\left(\frac{1}{\sqrt{3}} - \eta\right), \; N_2 = \frac{3}{4}\left(\frac{1}{\sqrt{3}} + \xi\right)\left(\frac{1}{\sqrt{3}} - \eta\right),$$

$$N_3 = \frac{3}{4}\left(\frac{1}{\sqrt{3}} + \xi\right)\left(\frac{1}{\sqrt{3}} + \eta\right)$$

$$N_4 = \frac{3}{4}\left(\frac{1}{\sqrt{3}} - \xi\right)\left(\frac{1}{\sqrt{3}} + \eta\right) \tag{5.56}$$

The strain energy derivatives are given as follows.

$$\left\{\begin{array}{c} \frac{\partial W}{\partial x} \\ \frac{\partial W}{\partial x_2} \end{array}\right\} = \left[\begin{array}{cc} \frac{\partial x}{\partial \xi} & \frac{\partial y}{\partial \xi} \\ \frac{\partial x}{\partial \eta} & \frac{\partial y}{\partial \eta} \end{array}\right]^{-1} \left\{\begin{array}{c} \frac{\partial W}{\partial \xi} \\ \frac{\partial W}{\partial \eta} \end{array}\right\}, \; \frac{\partial W}{\partial \xi} = \sum_{i=1}^{4} \frac{\partial N_i}{\partial \xi} W_i, \; \frac{\partial W}{\partial \eta} = \sum_{i=1}^{4} \frac{\partial N_i}{\partial \xi} W_i \tag{5.57}$$

The derivatives of x and y with respect to the natural coordinates ξ and η are given by Eq. (5.54). Finally,

$$\begin{Bmatrix} J_{x_1} \\ J_{x_2} \end{Bmatrix}_{\text{Part I}} = -\sum_{n=1}^{8} \left(\sum_{i=1}^{4} \left[\begin{Bmatrix} Q_{11} \\ Q_{12} \end{Bmatrix} + \begin{Bmatrix} Q_{21} \\ Q_{22} \end{Bmatrix} \right]_i \det [J_{cb}]_i [\omega_{GP}]_i \right)_n \qquad (5.58)$$

where

$$\begin{Bmatrix} Q_{11} \\ Q_{12} \end{Bmatrix} = \left[\begin{Bmatrix} W\dfrac{\partial H}{\partial x_1} \\ W\dfrac{\partial H}{\partial x_2} \end{Bmatrix} - \begin{bmatrix} \dfrac{\partial u_1}{\partial x_1} & \dfrac{\partial u_2}{\partial x_1} \\ \dfrac{\partial u_1}{\partial x_2} & \dfrac{\partial u_2}{\partial x_2} \end{bmatrix} \begin{bmatrix} \sigma_{11} & \sigma_{12} \\ \sigma_{21} & \sigma_{22} \end{bmatrix} \begin{Bmatrix} \dfrac{\partial H}{\partial x_1} \\ \dfrac{\partial H}{\partial x_2} \end{Bmatrix} \right] \qquad (5.59)$$

$$\begin{Bmatrix} Q_{21} \\ Q_{22} \end{Bmatrix} = \left[\begin{Bmatrix} H\dfrac{\partial W}{\partial x_1} \\ H\dfrac{\partial W}{\partial x_2} \end{Bmatrix} - H \begin{bmatrix} \dfrac{\partial \varepsilon_{11}}{\partial x_1} & \dfrac{\partial \gamma_{12}}{\partial x_1} & \dfrac{\partial \varepsilon_{22}}{\partial x_1} \\ \dfrac{\partial \varepsilon_{11}}{\partial x_2} & \dfrac{\partial \gamma_{12}}{\partial x_2} & \dfrac{\partial \varepsilon_{22}}{\partial x_2} \end{bmatrix} \begin{Bmatrix} \sigma_{11} \\ \sigma_{12} \\ \sigma_{22} \end{Bmatrix} \right] \qquad (5.60)$$

$$\det [J_{cb}] = \text{Determinant of Jacobian matrix} = \det \begin{bmatrix} \dfrac{\partial x}{\partial \xi} & \dfrac{\partial y}{\partial \xi} \\ \dfrac{\partial x}{\partial \eta} & \dfrac{\partial y}{\partial \eta} \end{bmatrix} \qquad (5.61)$$

and ω_{GP} is the weightage to be associated with Gauss point i. For a 2×2 integration scheme, the weightage is unity at all the four integration points.

Stiffness derivative procedure This method was proposed by Parks (1974). Considering a crack of size a and loaded as shown in Fig. 5.13, the potential energy of the system is given by

$$\pi = \frac{1}{2} u^T K u - u^T P \qquad (5.62)$$

where u stands for global displacement vector, K is the global stiffness matrix, and P is the global load vector. Since

$$G = -\frac{d\pi}{da}$$

assuming a situation with constant load P and noting that $Ku = P$,

$$G = \frac{du^T}{da}P - \frac{1}{2}\left\{\frac{du^T}{da}Ku + u^T\frac{dK}{da}u + u^TK\frac{du}{da}\right\}$$

$$P = -\frac{du^T}{da}[Ku - P] - \frac{1}{2}u^T\frac{dK}{da}u = -\frac{1}{2}u^T\frac{dK}{da}u$$

$$= -\frac{1}{2}u^T\frac{K_2 - K_1}{\Delta a}u \tag{5.63}$$

where K_1 and K_2 are the global stiffness matrices corresponding to crack lengths a and $a + \Delta a$, respectively. The crack length $a + \Delta a$ can be very easily accommodated by shifting the crack-tip as shown (Fig. 5.13). Δa is recommended as 0.1% a of the initial crack size a (Maiti 1990). It is also recommended that the computation of the product be done in double precision.

It has been shown (Maiti 1990) that it is not necessary to calculate the global stiffness matrices corresponding to the two crack sizes. It is possible to calculate G from the total strain energies of the eight elements 1 to 8 surrounding the crack-tip. That is,

$$G = \frac{\sum_1^8 (U_i)_a - \sum_1^8 (U_i)_{a+\Delta a}}{\Delta a}, \tag{5.64}$$

$$(U_i)_a = \frac{1}{2}(u_i^T)_a (k_i)_a (u_i)_a,$$

$$(U_i)_{a+\Delta a} = \frac{1}{2}(u_i^T)_a (k_i)_{a+\Delta a} (u_i)_a$$

where u_i is displacement matrix and k_i is the stiffness matrix of the element i. Note that k_i has to be evaluated separately for the two crack sizes. Since the displacement is not going to change appreciably due to the extended crack length in the presence of the same external load, the displacements corresponding to the extended crack length $a + \Delta a$ can be approximated by those for the original crack size a.

Crack closure integral technique Consider a simple discretization with 16 elements and crack size a (Fig. 5.17). The external load is P. The crack extends in-plane, or in a self-similar manner, to size $a + \Delta a$ under constant load. The crack opening is v'_{2y} at the original crack-tip position 2. If the crack closure force just before the onset of crack extension is P_{2y}, according to Irwin, the crack closure work is given by

$$W = \frac{1}{2}P_{2y}v'_{2y} \tag{5.65}$$

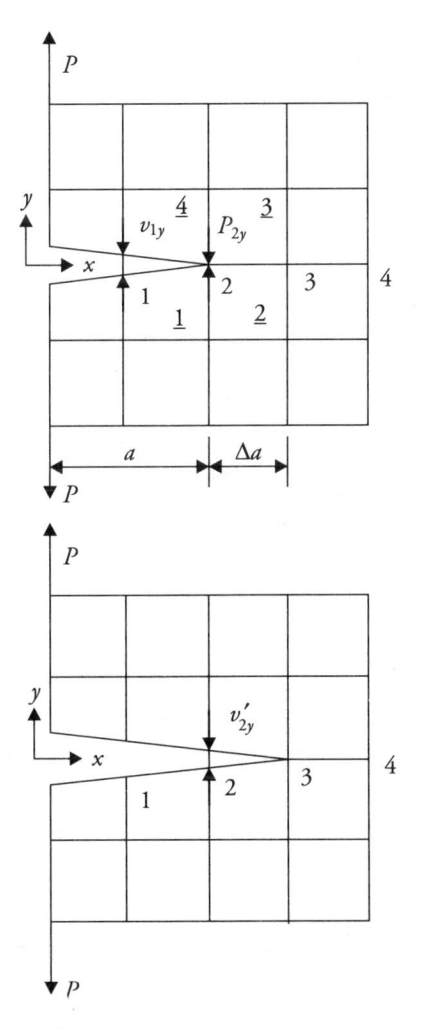

Figure 5.17 Typical crack-tip discretization.

P_{2y} can be obtained from the nodal forces of the elements 3 and 4 or 1 and 2. For infinitesimally small crack extension Δa, v'_{2y} can be approximated by v_{1y} (Fig. 5.17). Therefore

$$W = \frac{1}{2} P_{2y} \, v_{1y} \tag{5.66}$$

Thereby, the crack closure work can be calculated through a single finite element run. The potential energy release rate is given by

$$G = \frac{W}{B \, \Delta a} = \frac{P_{2y} \, v_{1y}}{2 \, B \, \Delta a} \tag{5.67}$$

With a fine discretization around the crack-tip and ensuring that Δa is also very small, G can be very accurately obtained through this method. Again by using the relation between G and K_I, the SIF can be computed.

For pure Mode II loading, closure force at node 2 in the x direction and the corresponding sliding displacement at node 1 must be considered. That is,

$$G_{II} = \frac{P_{2x}\, v_{1x}}{2\, B\, \Delta a} \tag{5.68}$$

To determine the corresponding SIF, the relationship between G_{II} and K_{II} can be utilized.

When elements around the crack-tip are 8-noded quadrilaterals (Fig. 5.18(a)), the potential energy release rate can be computed through single finite element run using the following relationship.

$$G = \frac{W}{B\, \Delta a} = \frac{P_{3y}\, v_{1y} + P_{4y}\, v_{2y}}{2\, B\, \Delta a} \tag{5.69}$$

As before, closure forces P_{3y} and P_{4y} can be obtained from the element nodal forces of elements 3 and 4 or 1 and 2 (Fig. 5.18). An equivalent relation for Mode II can be written replacing the forces and displacements parallel to the crack plane or x direction.

If quarter point singularity elements (Fig. 5.18(b)) are employed around the crack-tip (Henshell and Shaw 1975; Barsoum 1976), the crack closure forces (P_{3y} and P_{4y}) are to be obtained at the nodes 3 and 4, and the corresponding displacements are obtained from the opening displacement at node location 1. Computation of the opening displacement to be associated with P_{4y} is done using the fact that displacement along the crack edges varies as $r^{\frac{1}{2}}$, where r stands for radial distance from the crack-tip. That is, the required displacement $v = \sqrt{\frac{3}{4}}\, v_{1y}$, which is equal to the opening displacement at $r = \frac{3}{4}\, \Delta a$.

To facilitate CCI calculations, in general, element sizes ahead and behind the crack-tip are kept the same. The closure work can also be computed by considering simultaneous opening of the crack over a length more than one-element span Δa. This will involve computing the closure forces and corresponding opening displacements at a number of nodal points.

Practical problems may involve in addition to remote loading, crack edge pressure loading, for example, fluid pressure, thermal loading, and residual stresses. The calculation of CCI and SIFs for such situations is given by Maiti (1992a).

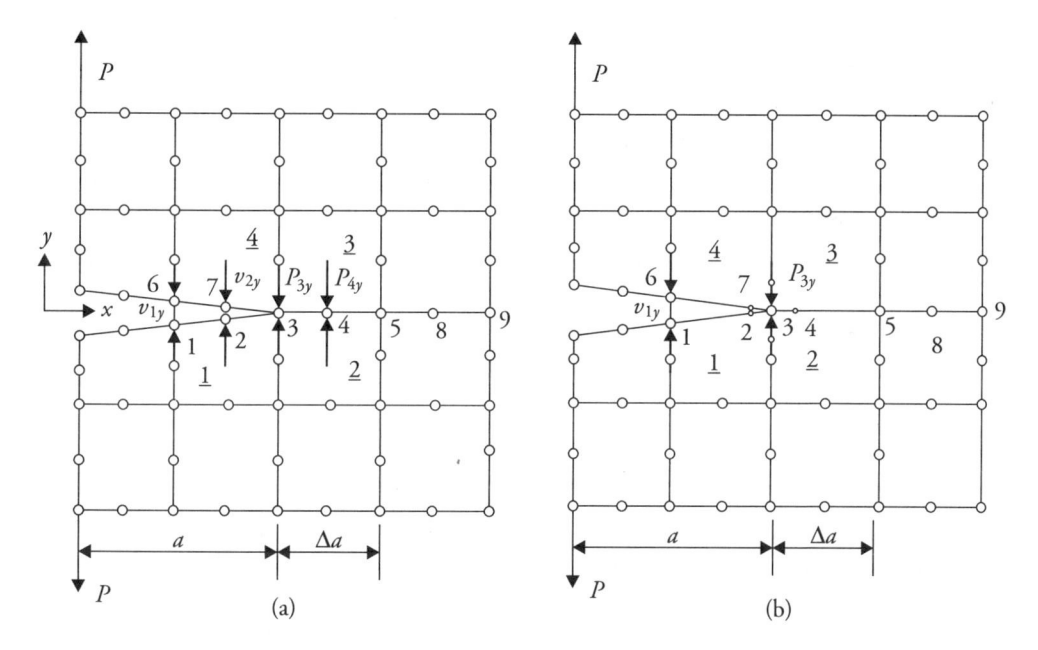

Figure 5.18 (a) Quadratic and (b) quarter point singular elements around crack-tip.

Local smoothing for improving accuracy of CCI-based calculations Generally, the crack closure forces and opening displacements obtained through finite element calculations do not give rise to variations as stipulated by the crack-tip singularity field. It is established (Krishnamurthy et al. 1985; Ramamurthy et al. 1986; Sethuraman and Maiti 1988) that by local smoothing these two quantities utilizing the output data on nodal forces and displacements, the accuracy of determining the strain energy release rates can be improved. Ramamurthy et al. (1986) assumed a variation of the crack closure forces different from the variation assume by Sethuraman and Maiti (1988). The final relations obtained by Sethuraman and Maiti are simpler.

The derivations based on Sethuraman and Maiti (1988) are given subsequently. In keeping with the displacement-based finite element formulations, they assumed a variation of the crack opening displacements as dictated by the crack-tip elements. If the crack-tip elements are quadratic, the displacements vary quadratically; if the crack-tip elements are quarter point singularity elements, the displacements vary directly with square-root of distance from the crack-tip; and so on. The closure stresses are assumed to vary linearly and inversely with square-root of distance from the crack-tip in the two cases respectively.

In the case of the quarter point singularity elements around the crack-tip, the crack opening displacements is assumed to vary as the square-root of distance $|x|$ from the crack-tip 3 (Fig. 5.19a).

$$v\left(\zeta'\right) = v_2\left(1+\zeta'\right) \tag{5.70}$$

noting that $\dfrac{\Delta a}{4}(1+\zeta')^2 = \Delta a - x$, where v_2 stands for y displacement of node 2, and ζ' is a natural coordinate with values 0 at node 2, -1 at node 3, and 1 at node 1. The closure stresses over span 345 (Fig. 5.19(a)) ahead of the crack-tip 3 are assumed to vary as $1/\sqrt{x}$. It can be written in the form

$$\sigma_y\left(\zeta\right) = B_o + \dfrac{B_1}{(1+\zeta)} \tag{5.71}$$

where $\dfrac{\Delta a}{4}(1+\zeta)^2 = x$ and ζ is a natural coordinate with values 0 at node 4, -1 at node 3, and 1 at node 5. The arbitrary constants B_0 and B_1 are obtained by equating the work done by the distributed opening stresses σ_y on y displacements over the span 345 with the work done by nodal forces on the corresponding nodal displacements.

$$F_{3y}\,v_3 + F_{4y}\,v_4 + F_{5y}\,v_5 = \int_0^{\Delta a} v\,\sigma_y\,dx$$

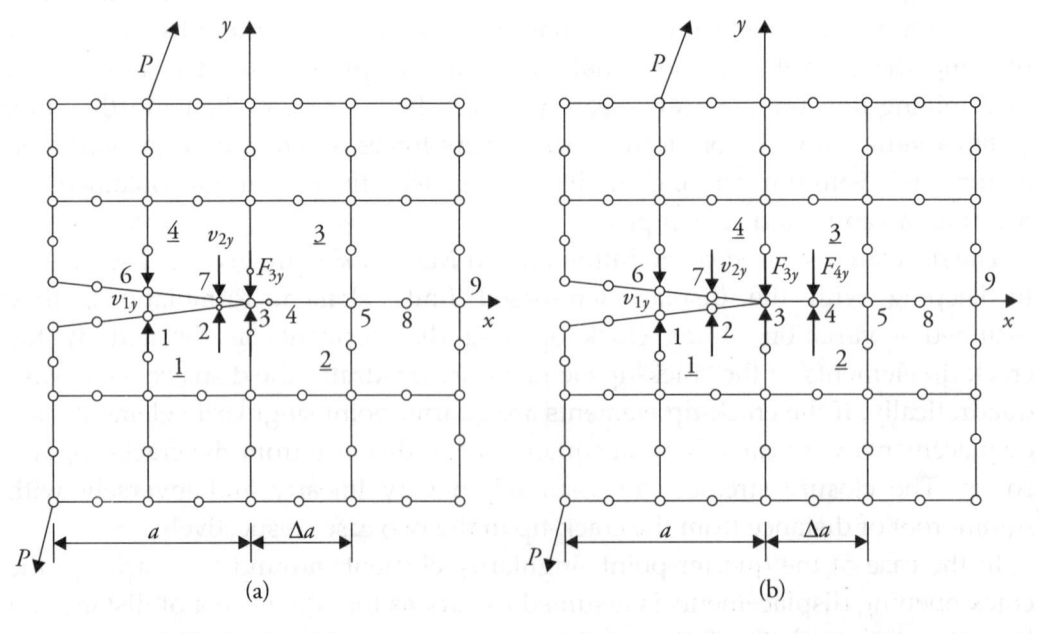

$$= \int_{-1}^{1} [N_3\,v_3 + N_4\,v_4 + N_5\,v_5]\left[\left(B_o + \dfrac{B_1}{1+\zeta}\right)\right]\left[\dfrac{\Delta a}{2}(1+\zeta)\,d\zeta\right] \tag{5.72}$$

Figure 5.19 Crack-tip element arrangements. (a) Quarter point singularity elements. (b) Quadratic elements.

Note that v varies as $\sqrt{x}$ along span 345. F_{3y}, F_{4y}, and F_{5y} are the closure forces in y directions at nodes 3, 4, and 5, respectively and are given by

$$\begin{Bmatrix} F_{3y} \\ F_{4y} \\ F_{5y} \end{Bmatrix} = \int_{-1}^{1} \begin{Bmatrix} N_3 \\ N_4 \\ N_5 \end{Bmatrix} \left(B_o + \frac{B_1}{1+\xi}\right) \left[\frac{\Delta a}{2}(1+\xi)\,d\xi\right] \tag{5.73}$$

where $N_3 = -\frac{\xi}{2}(1-\xi)$, $N_4 = (1-\xi^2)$ and $N_5 = \frac{\xi}{2}(1+\xi)$. Thereby, the two arbitrary constants are obtained.

$$B_o = \frac{3}{2\Delta a}(F_{4y} - 2F_{3y}), \quad B_1 = \frac{6}{\Delta a}F_{3y} \tag{5.74}$$

Finally, strain energy release rate in the opening mode is obtained through

$$G_I = \lim_{\Delta a \to 0} \frac{1}{2\Delta a} \int_0^{\Delta a} \sigma_y(\xi)\,v_y(\xi')\,dx$$

$$= \lim_{\Delta a \to 0} \frac{1}{2\Delta a} \int_0^{\Delta a} \left(B_o + \frac{B_1}{1+\xi}\right) v_{2y}(1+\xi')\,dx$$

$$= \lim_{\Delta a \to 0} \frac{1}{2\Delta a} \int_0^{\Delta a} \left(B_o + \frac{\sqrt{\Delta a}}{2}\frac{B_1}{\sqrt{x}}\right) v_{2y} \frac{2}{\sqrt{\Delta a}}\sqrt{\Delta a - x}\,dx$$

$$= \frac{v_{2y}}{\Delta a}[F_{4y} + (1.5\pi - 4)F_{3y}] \tag{5.75}$$

where v_{2y} is the total crack opening at the nodal location 2 behind the crack-tip and Δa is assumed to be very small. Δa can be taken as 3–4% of given crack size a. Similarly, for the shearing mode or Mode II problem,

$$G_{II} = \frac{v_{2x}}{\Delta a}[F_{4x} + (1.5\pi - 4)F_{3x}] \tag{5.76}$$

where v_{2x} indicates the total crack sliding, or crack 'opening', in x direction at the nodal location 2 and F_{3x} and F_{4y} are the closure forces acting in x direction at nodes 3 and 4, respectively.

The expressions due to Ramamurthy et al. (1986) are as follows.

$$G_I = \frac{1}{2\Delta a}[(C_{11}F_{3y} + C_{12}F_{4y} + C_{13}F_{5y})v_{2y} + (C_{21}F_{3y} + C_{22}F_{4y} + C_{23}F_{5y})v_{1y}] \tag{5.77a}$$

$$G_{II} = \frac{1}{2\Delta a}[(C_{11}F_{3x} + C_{12}F_{4x} + C_{13}F_{5x})v_{2x} + (C_{21}F_{3x} + C_{22}F_{4x} + C_{23}F_{5x})v_{1x}] \tag{5.77b}$$

where $C_{11} = 33\frac{\pi}{2} - 52$, $C_{12} = 17 - 21\frac{\pi}{4}$, $C_{13} = 21\frac{\pi}{2} - 32$, $C_{21} = 14 - 33\frac{\pi}{8}$,

$$C_{22} = 21\frac{\pi}{16} - \frac{7}{2}, \quad C_{23} = 8 - 21\frac{\pi}{8}.$$

For 8-noded quadrilaterals (Fig. 5.19(b)) y displacement varies quadratically along 321 and opening stresses vary along 345 linearly.

$$v = v_2 \left(1 - \xi^2\right) + v_1 \frac{1}{2} \xi \left(1 + \xi\right) \tag{5.78}$$

$$\sigma_y = B_0 + B_1 \xi \tag{5.79}$$

Following the steps as given in the case of the quarter point singularity elements, the following results are obtained.

$$B_0 = \frac{3}{2 \Delta a} F_{4y}, \tag{5.80a}$$

$$B_1 = \frac{3}{2 \Delta a} F_{4y} - \frac{6}{\Delta a} F_{3y} \tag{5.80b}$$

$$G_I = \frac{1}{2 \Delta a} \left(F_{3y} v_{1y} + F_{4y} v_{2y}\right), \quad G_{II} = \frac{1}{2 \Delta a} \left(F_{3x} v_{1x} + F_{4x} v_{2x}\right) \tag{5.81}$$

These results are the same as those obtained by Krishnamurthy et al. (1985).

In the case of 4-noded quadrilaterals, displacement has a linear variation along a crack face, and the closure stress is constant along the crack line ahead of the crack-tip. This gives

$$G_I = \frac{1}{2 \Delta a} F_{3y} v_{1y}, \quad G_{II} = \frac{1}{2 \Delta a} F_{3x} v_{1x} \tag{5.82}$$

where F_{3i} and v_{1i}, $i = x$ or y, stands for closure force at node 3 and total opening at nodal location 1, respectively.

Problem 5.6

A square plate 80 mm× 80 mm of uniform thickness 1 mm contained a central crack of size 8 mm. The plate boundary was maintained at 100°C and the crack edges were at 0°C. Through FE analysis of the problem with arrangements of elements (51–54) and nodes (177–191) around crack-tip A as shown (Fig. 5.20), the following forces and displacements were obtained: $F_{191y} = -0.10048 \times 10^{-2}$ N, $F_{186y} = -0.43123 \times 10^{-2}$ N, $v_{185} = 0.011003$ mm, and $v_{190} = 0.005560$ mm. Calculate the SIF by the displacement method, CCI technique, and CCI technique with local smoothing. Use modulus of elasticity $E = 1$ MPa and Poisson's ratio $\nu = 0.3$. Assume plane strain condition.

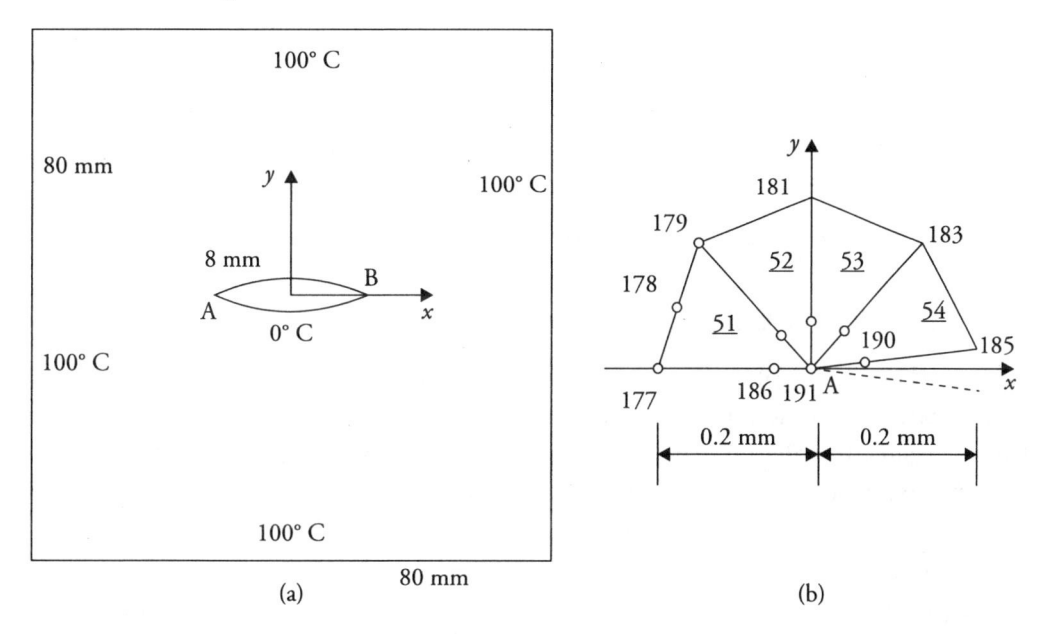

Figure 5.20 (a) Thermo-elastic problem and (b) crack-tip discretization.

Solution
(a) Displacement method
Using v for $\theta = 180°$ and v_{185}

$$0.011003 = \frac{K_I}{\mu}\sqrt{\frac{2r}{\pi}}(1-\nu)$$

Noting that $r = 0.2$ mm, $h = 1$ mm, and $\mu = E/[2(1+\nu)]$, $E = 1$ MPa, and $\nu = 0.3$,

$$K_I = 0.01694 \text{ Nmm}^{-3/2} \text{ (Ans.)}$$

(b) CCI technique

$$G_I = \frac{P_{191y}\, v_{185} + P_{186y}\sqrt{\frac{3}{4}}\, v_{185}}{h\Delta a}$$

$$= \frac{0.0010048 \times 0.011003 + 0.0043123\sqrt{\frac{3}{4}} \times 0.011003}{0.2} \text{ N/mm}$$

$$= (1-\nu^2)K_I^2/E$$

This gives $K_I = 0.01692$ Nmm$^{-3/2}$ (Ans.).

(c) CCI technique with local smoothing
Using relation (5.75)

$$G_I = \frac{2\,v_{190y}}{\Delta a}\left[F_{186y} + (1.5\pi - 4)\,F_{191y}\right]$$

$$= \frac{2 \times 0.00556}{0.2}[0.0043123 + (1.5\pi - 4)]\,0.0010048] = (1-\nu^2)K_I^2/E$$

This gives $K_I = 0.01752\,\text{Nmm}^{-3/2}$ (Ans.).

5.4 FEM-Based Calculation of G Associated with Kinking of Crack

An angled crack in a plate (Fig. 5.21) under tension does not lead to in-plane or self-similar extension when loaded to a critical level. To understand such an extension, it is necessary to calculate the potential/strain energy release rate with such out-of-plane extension. It is possible to calculate this sort of energy release rate by applying the methods, which are available for the in-plane extension (Maiti 1990). Some of the methods are discussed below.

It is possible to calculate the strain energy U_a (Fig. 5.21(a)) of the system with the given crack a and loading. It is then possible to calculate the strain energy U_{a+l} corresponding to an extended crack $a+l$ with an extension l in θ direction. The required energy release rate G_θ is given by

$$G_\theta = \lim_{l \to 0}\frac{U_{a+l} - U_a}{l} = \frac{U_{a+l} - U_a}{l} \text{ for small } l. \tag{5.83}$$

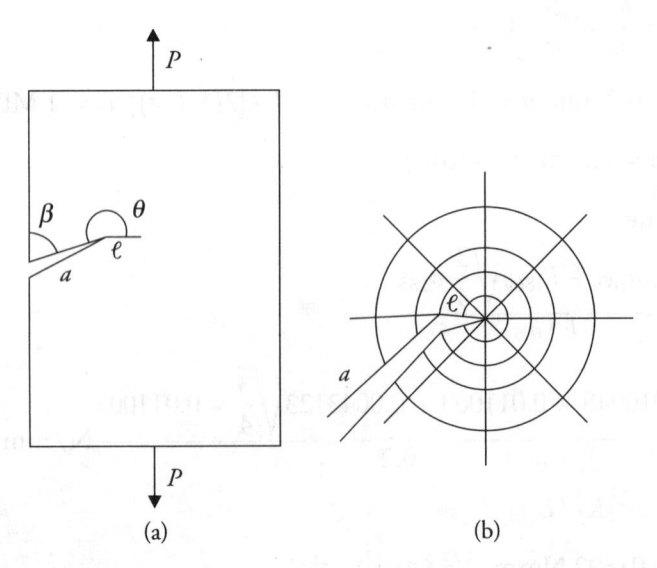

Figure 5.21 (a) Angled crack. (b) Discretization around tip and knee.

assuming thickness to be unity. This procedure will require two FE runs. While computing U_{a+l}, it is possible to consider a branch length l up to 4% of a.

It is also possible to calculate G_θ by considering J integral procedure, stiffness derivative method, CCI technique, and the continuation argument. As per the continuation argument, the energy release rate associated with an out-of-plane extension in the direction θ is equal to the energy release rate associated with the in-plane extension of the branch l when the branch length $l \to 0$. This means that it is necessary to analyse a problem with the given crack plus a branch of small length l in θ direction and compute the energy release rate associated with extension of the branch l in its own plane. While analysing the kinked crack geometry with crack $a + l$, the discretization must be done, recognizing the fact that there is a stress singularity at the knee. The order of singularity depends on the knee angle θ (Williams 1957). This calls for a very refined discretization over the span l, square-root singularity elements at the crack-tip, and variable order singularity elements at the knee (Maiti 1992b) to take care of the variation in the kink angle. Alternatively, it is possible to use multi corner variable order singularity element (Dutta, Maiti and Kakodkar 1990; Maiti 1992c) between the crack-tip and the knee, square-root singularity elements at the crack-tip, and variable order singularity elements at the knee. The calculation of G_θ can be done by adopting the steps that are given earlier for J integral, or stiffness derivative procedure or CCI technique for the in-plane extension (Maiti 1990).

5.5 Other Numerical Methods

Other numerical methods requiring computer-based solution include boundary element method (BEM), extended FEM (XFEM), and meshless/meshfree methods. In the BEM, the boundary alone is discretized and the nodal degrees of freedoms, displacements, and tractions, associated with nodes, are interconnected by the reciprocal theorem of elasticity. These relations can also be obtained by the Galerkin method of weighted residuals. This step gives rise to a set of simultaneous equations. After solving these equations, the SIFs can be obtained through the comparison of displacements and stresses, and energy methods or their variant like the CCI method. Since in this method, displacements and nodal tractions are treated as independent nodal variables, the use of special singularity elements to ensure strain singularity does not automatically guarantee the stress singularity. This calls for use of separate shape functions for displacements and tractions (Mukhopadhyay, Maiti and Kakodkar 2000). The methods to extract the SIFs with good accuracy are given in references (Maiti, Mukhopadhyay and Kakodkar 1997; Mukhopadhyay, Maiti and Kakodkar 1998).

In the FEM, the accuracy of computed SIFs can be increased by representing the crack-tip field over some region around the crack-tip in terms of a few of the terms of the Williams' eigenfunction expansion. The strain energy for this region can be calculated in terms of the arbitrary constants associated with this expansion. The energy for the remaining part of the body can be calculated in terms of the nodal displacements. Thus, the strain energy for the whole body is obtained. After obtaining the potential energy of the body, it can be minimized with respect to the displacements and the arbitrary constants. This will give rise to a set of linear simultaneous equations. Solving these equations, the SIFs can be obtained directly. The details of the method can be found in references (Moës, Dolbow and Belytschko 1999; Stazi et al. 2003; Liu, Xiao and Karihaloo 2004; Xiao and Karihaloo 2007; Rabczuk 2013). This method is termed as XFEM.

To overcome the problems associated with discretization in the FEM and BEM, the meshless methods have evolved (Belytschko et al. 1996; Atluri and Zhu 1998; Rabczuk 2013). The method formulation can be based on different forms of the classical Galerkin method. The two important forms are Petrov-Galerkin and element-free Galerkin methods. In this case, too, the SIFs can be extracted directly by any of the method, for example, the displacement technique, stress method, domain integral method, CCI method (Muthu et al. 2014), and so on, as given earlier in the case of FEM.

5.6 Experimental Methods

5.6.1 Strain Gauge Technique

The technique based on the measurement of strain can be utilized to determine the SIF experimentally. Because of the high stress/strain gradient close to the elastic crack-tip, reduction in the influence of the singularity term of the Williams' stress function expansion away from the crack-tip, and finite dimension of the strain gauge, it is difficult to obtain good accuracy unless some care is exercised in placing the gauges. Dally and Sanford (1987) started with a crack-tip stress field representation using truncated Williams' eigenfunction expansions for the two functions associated with the Airy stress function and proposed a scheme whereby the Mode I SIF could be determined with a good accuracy using a single strain gauge (Fig. 5.22(a)). They ingeniously selected the orientations α and radial line positioning θ_S for the gauge to reduce the number of unknown parameters in the two truncated series. The orientation α, assuming a state of plane stress, is given by

$$\cos 2\alpha = -\kappa = \frac{1-\nu}{1+\nu}, \qquad (5.84)$$

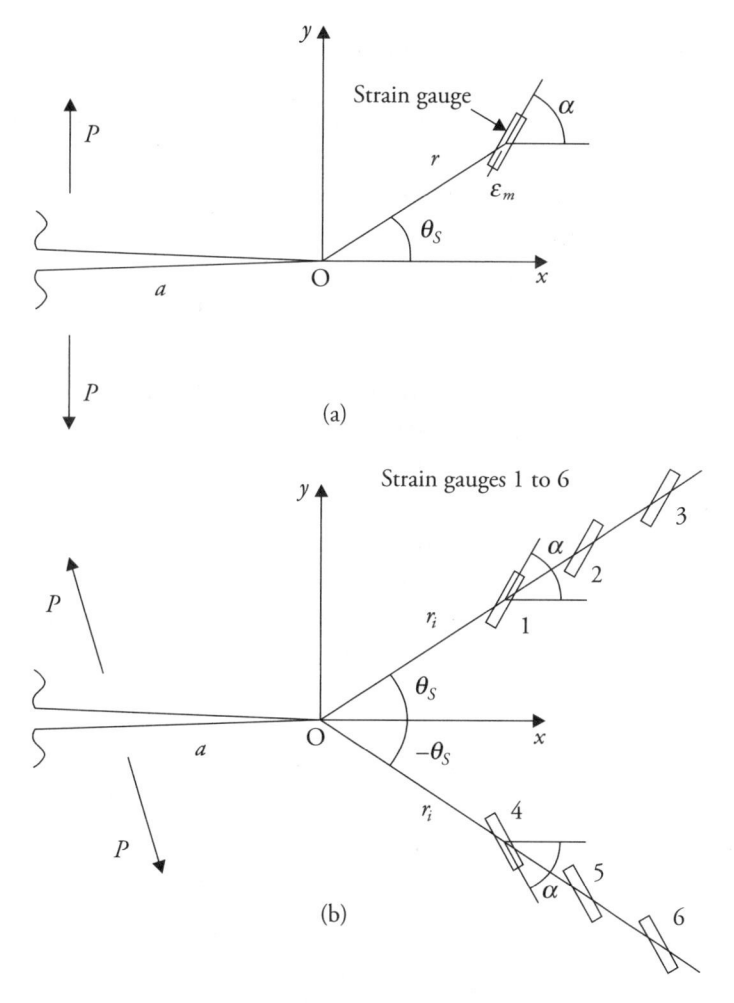

Figure 5.22 Strain gauge arrangement for (a) Mode I and (b) mixed mode.

where ν is Poisson's ratio. The radial line orientation θ_S is given by

$$\tan\frac{\theta_S}{2} = -\cot 2\alpha. \tag{5.85}$$

The measured strain ε_m by the strain gauge (Fig. 5.22(a)) is related to the Mode I SIF by the following relation.

$$2\mu\varepsilon_m = \frac{K_I}{\sqrt{2\pi r}}\left[\kappa\cos\frac{\theta_S}{2} - \frac{1}{2}\sin\theta_S\sin\frac{3\theta_S}{2}\cos 2\alpha + \frac{1}{2}\sin\theta_S\cos\frac{3\theta_S}{2}\sin 2\alpha\right], \tag{5.86}$$

where μ is rigidity modulus. Replacing ν by $\dfrac{\nu}{1-\nu}$ in Eq. (5.84), it is possible to obtain a relationship valid for a state of plane strain. Sarangi, Murthy and

Chakraborty (2010a, 2010b, 2013) recommend that the minimum radial distance r should be greater than half the thickness of the sheet and less than certain maximum, depending on the geometric configuration of the specimen. A distance greater that half the thickness is recommended to exclude the effects of crack-tip local plasticity. They provide a methodology (Sarangi, Murthy and Chakraborty 2010b) for determining the maximum permissible distance r for the location of strain gauge without affecting the accuracy of the SIF obtained.

The problem of mixed mode was first examined by Dally and Berger (1986). For this case, six gauges 1 to 6 (Fig. 5.22(b)) are required. The gauges are placed symmetrically about the crack line. Angles α and θ_S are again given by Eqs. (5.84) and (5.85), respectively. The gauges are not required to be spaced equally over the span 1 to 3 or 4 to 6. However, radial distances of the gauges 1, 2, and 3 should be equal to the distances of gauges 4, 5, and 6, respectively. Sarangi, Murthy and Chakraborty (2012) again recommend that minimum radial distance r should be greater than half the thickness of the sheet and less than certain maximum. They provide a methodology for finding the maximum permissible distance for the furthest strain gauges (3 and 6 in Fig. 5.22(b)) from the crack-tip.

5.6.2 Photoelasticity

Photoelasticity (Dally and Riley 1991) is a well-known method for finding elastic stress distribution. In this method, model conforming to the given geometry is made out of birefringent material and tested in circular polariscope to obtain the fringe pattern. The fringe pattern can be analysed to obtain the required stresses at a point. In the case of an object with a crack, the stress data around the crack-tip can be processed to determine the SIF. A typical fringe pattern around the crack-tip is shown (Fig. 5.23). Each fringe indicates the locus of a point of constant shear stress, whose magnitude is known. That is,

$$\tau_{max} = \frac{\sigma_1 - \sigma_2}{2}$$

$$= \frac{1}{2\sqrt{2\pi r}} \left[(K_I \sin\theta + 2K_{II} \cos\theta)^2 + (K_{II} \sin\theta)^2 \right]^{1/2}, \tag{5.87}$$

where σ_1 and σ_2 are the principal stresses at a point. Angle θ is measured with respect to the crack axis x. Selecting a particular fringe, it is possible to locate the points where the fringe intersects the x and y axes. Thereby, say, r_1 and r_2, respectively are obtained. This gives,

$$\tau_{max} = \frac{K_{II}}{\sqrt{2\pi r_1}} = \frac{1}{2\sqrt{2\pi r_2}} [K_I^2 + K_{II}^2]^{1/2} \tag{5.88}$$

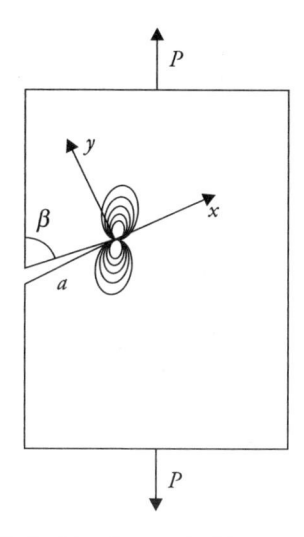

Figure 5.23 Typical isochromatic fringes around crack-tip.

Thus, both the SIFs can be determined in a mixed mode problem.

The fringes get very crowded near the crack-tip because of extremely high stress gradient due to stress singularity. It is very difficult to identify the individual fringes. To overcome this problem, far field fringe data in conjunction with multi parameter approximations of the crack-tip stress field (Sanford and Dally 1979) or the digital photoelasticity techniques (Ramesh 2000) can be employed for an accurate determination of the SIFs.

Exercise

5.1 Solve Problem 5.6 assuming all data to correspond to plane stress condition. [Ans. 0.01542 and 0.01614 Nmm$^{-3/2}$]

5.2 Determine the SIF at the right hand crack-tip at $x = a$ for the loading shown. p represents load per unit thickness. [Ans. $K_I = 0.7086p\sqrt{\pi a}$, $K_{II} = \frac{p}{2}\sqrt{\pi a}$]

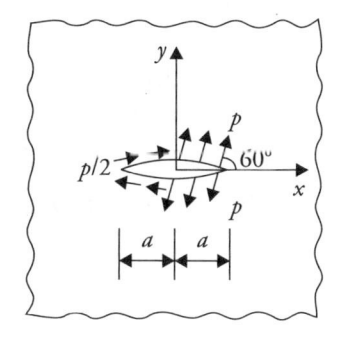

Figure Q.5.2

5.3 Determine the level of safety for the high strength low alloy steel (HSLA) plate (Fig. Q.5.3) with a crack ($2a = 15$ mm) at the end of a rivet hole as shown. Given $K_{IC} = 47.7$ Pa$\sqrt{m}$, 0.2% proof stress $\sigma_Y = 1640$ MPa, and plate thickness $= 25$ mm. [Ans. 3.275]

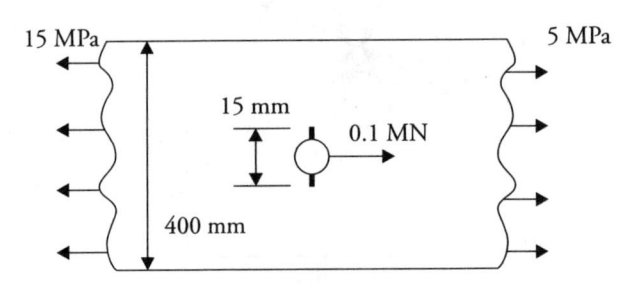

Figure Q.5.3

5.4 Determine the SIFs for the case shown. Crack at each end of the hole is of size 2 mm. If you make any assumption to obtain the solution, indicate it.
[Ans. $K_I = 0.210\sigma$ MPa$\sqrt{m}$, $K_{II} = 0.133\sigma$ MPa$\sqrt{m}$; σ in MPa]

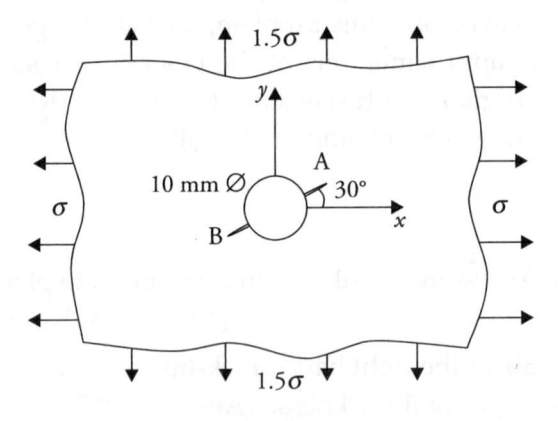

Figure Q.5.4

5.5 Determine the SIFs at the two crack-tips for the four cases shown. In case Fig. Q.5.5c loading p acts over a span a around the centre of crack.
[Ans. (a) K_I, right $= 0.7086p\sqrt{\pi a}$, K_{II}, right $= 0.4092p\sqrt{\pi a}$. (b) K_{II}, left $= 0.4092p\sqrt{\pi a}$, K_{II}, right $= 0.0908p\sqrt{\pi a}$, (c) $K_I = \dfrac{P}{2B\sqrt{\pi a}} + \dfrac{P}{6}\sqrt{\pi a}$, (d) $K_I = \dfrac{2\sigma_0 a}{\sqrt{\pi a}}\left(\dfrac{\pi}{2} - 1\right)$]

Figure Q.5.5(a)

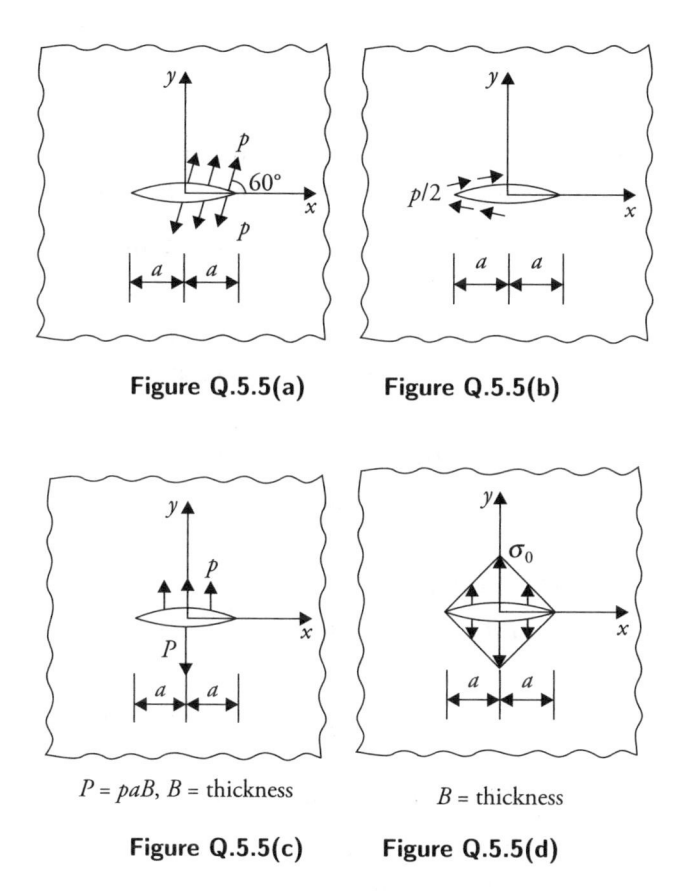

Figure Q.5.5(b)

$P = paB$, B = thickness

Figure Q.5.5(c)

B = thickness

Figure Q.5.5(d)

5.6 Determine the SIFs at the crack-tip for the case shown (Fig. Q.5.6) when the crack size is AB = 20 mm, $\sigma_0 = 200$ MPa, and crack inclination with x axis is 45°. [Ans. $K_I = K_{II} = 25.06$ MPa$\sqrt{m}$]

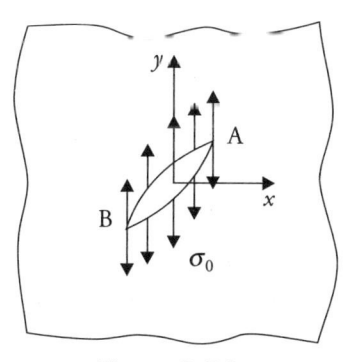

Figure Q.5.6

References

5.1 Aliabadi, M.H., D.P. Rooke and D.J. Cartwright. 1987. 'An Improved Boundary Element Formulation for Calculating Stress Intensity Factors: Application to Aerospace Structures.' *Journal of Strain Analysis* 22: 203–07.

5.2 Atluri, S.N., ed. 1986. *Computational Methods in the Mechanics of Fracture*. Amsterdam: Elsevier Science Publishers B. V.

5.3 Atluri, S.N., A.S. Kobayashi and M. Nakagaki. 1975. 'An Assumed Hybrid Finite Element Model for Linear Fracture Mechanics.' *International Journal of Fracture* 11: 257–71.

5.4 Atluri, S.N. and T. Zhu. 1998. 'A New Meshless Local Petrov-Galerkin (MLPG) Approach in Computational Mechanics.' *Computational Mechanics* 22(2): 117–27.

5.5 Barsoum, R.S. 1976. 'On the Use of Isoparametric Finite Elements in Linear Fracture Mechanics.' *International Journal of Numerical Methods in Engineering* 10: 25–37.

5.6 Belytschko, T., Y. Krongauz, D. Organ, M. Flemming and P. Krysl. 1996. 'Meshless Methods: An Overview of Recent Developments.' *Computer Methods in Applied Mechanics and Engineering* 139: 3–47.

5.7 Bowie, O.L. 1973. 'Solutions of Crack Problems by Mapping Techniques.' In *Methods of Analysis and Solutions of Crack Problems, Mechanics of Fracture*, Vol. 1, ed. Sih, G.C., 1–55. Leyden: Noodhoff International Publishing.

5.8 Bueckner, H.F. 1970. 'A novel principle for the computation of stress intensity factors'; *Zeitschrift Angewandte für Mathematik und Mechanik*, 50, 9, 529-546.

5.9 Byskov, E. 1970. 'The Calculation of Stress Intensity Factors Using the Finite Element Method with Cracked Elements.' *International Journal of Fracture* 6: 159–67.

5.10 Cook, R.D., D.S. Malkus, M.E. Plesha and R.J. Witt. 2002. *Concepts and Applications of Finite Element Analysis*. New York: John Wiley and Sons.

5.11 Cruse, T.A. 1996. 'BIE Fracture Mechanics Analysis: 25 Years of Developments.' *Computational Mechanics* 18: 1–11.

5.12 Dally, J.W. and J.R. Berger. 1986. 'A Strain Gage Method for Determining K_i and K_{ii} in a Mixed Mode Stress Field', 603–12. *Proceedings of the 1986 SEM Spring Conference on Experimental Mechanics*, New Orleans, LA.

5.13 Dally, J.W. and W.F. Riley. 1991. *Experimental Stress Analysis*. Singapore: McGraw Hill.

5.14 Dally, J.W. and R.J. Sanford. 1987. 'Strain Gage Methods for Measuring the Opening Mode Stress Intensity Factor.' *Experimental Mechanics* 27: 381–88.

5.15 Dutta, B.K., S.K. Maiti and A. Kakodkar. 1990. 'On the Use of One Point and Two Points Singularity Elements in the Analysis of Kinked Cracks.' *International Journal of Numerical Methods in Engineering* 29: 1487–99.

5.16 Hartranft, R.J. and G.C. Sih. 1973. 'Alternating Method Applied to Edge Crack and Surface Crack Problems.' In *Methods of Analysis and Solutions of Crack Problems, Mechanics of Fracture*, Vol. 1, ed. Sih, G.C., 179–238. Leyden: Noodhoff International Publishing.

5.17 Henshell, R.D. and K.G. Shaw. 1975. 'Crack-tip Finite Elements are Unnecessary.' *International Journal of Numerical Methods in Engineering* 9: 495–507.

5.18 Kitagawa, H., H. Okamura and H. Ishikawa. 1976. 'Application of *J*-integral to Mixed Mode Crack Problems.' *Transaction of Japan Society of Mechanical Engineers* 760: 46–48.

5.19 Kobayashi, A.S., ed. 1975. *Experimental Techniques in Fracture Mechanics*, SESA Monograph No. 2. Iowa: Iowa State University Press.

5.20 Krishnamurthy, T., T.S. Ramamurthy, V. Vijyakuimar and B. Duttaguru. 1985. 'Modified Crack Integral Closure Integral Method for Higher Order Finite Elements', 89–900. *Proceedings of the International Conference on Finite Elements in Computational Mechanics*, FEICOM, Mumbai, India.

5.21 Liu, X.Y., Q.Z. Xiao and B.L. Karihaloo. 2004. 'XFEM for Direct Evaluation of Mixed Mode SIFs in Homogenous and Bi-materials.' *International Journal for Numerical Methods in Engineering* 59: 1103–18.

5.22 Maiti, S.K. 1990. 'Finite Element Computation of the Strain Energy Release Rate for Kinking of a Crack.' *International Journal of Fracture* 43: 161–74.

5.23 ———. 1992a. 'Finite Element Computation of Crack Closure Integrals and Stress Intensity Factors.' *Engineering Fracture Mechanics*: 41(3): 339–48.

5.24 ———. 1992b. 'A Finite Element for Variable Order Singularities Based on the Displacement Formulation.' *International Journal of Numerical Methods in Engineering* 33: 1955–74.

5.25 ———. 1992c. 'A Multi corner Variable Order Singularity Triangle to Model Neighbouring Singularities.' *International Journal of Numerical Methods in Engineering* 35: 391–408.

5.26 Maiti, S.K., N.K. Mukhopadhyay and A. Kakodkar. 1997. 'Boundary Element Method Based Computation of Stress Intensity Factor by Modified Crack Closure Integral.' *Computational Mechanics* 19: 203–10.

5.27 Maiti, S.K. and R.A. Smith. 1983. 'Theoretical and Experimental Studies on the Extension of Cracks Subjected to Concentrated Loading Near their Faces to Compare the Criteria for Mixed Mode Brittle Fracture.' *Journal of Mechanics and Physics of Solids* 31: 389–403.

5.28 Moës, N., J. Dolbow and T. Belytschko. 1999. 'A Finite Element for Crack Growth Without Remeshing.' *International Journal for Numerical Methods in Engineering* 46: 131–50.

5.29 Moran, B. and C.F. Shih. 1987. 'A General Treatment of Crack-tip Contour Integrals.' *International Journal of Fracture* 25: 295–310.

5.30 Mukhopadhyay, N.K., S.K. Maiti and A. Kakodkar. 1998. 'BEM Based Evaluation of SIFs Using Modified Crack Closure Integral Technique under Remote and/or Crack Edge Loading.' *Engineering Fracture Mechanics* 61: 655–71.

5.31 ———. 2000. 'A Review of SIF Evaluation and Modeling of Singularities in FEM.' *Computational Mechanics* 25: 358–75.

5.32 Murakami, Y. (Editor-in-Chief). 1987. *Stress Intensity Factors Handbook*, Vols. I and II. Oxford: Pergamon Press.

5.33 Muthu, N., B.G. Falzon, S.K. Maiti and S. Khoddam. 2014. 'Modified Crack Closure Integral Technique for Extraction of SIFs in Meshfree Methods.' *Finite Element Analysis and Design* 78: 25–39.

5.34 Nikishkov, G.P. and S.N. Atluri. 1987. 'Calculation of Fracture Mechanics Parameters for an Arbitrary 3-Dimensional Crack, by the Equivalent Domain Integral Method.' *International Journal of Numerical Methods in Engineering* 24: 1801–21.

5.35 Parker, A.P. 1981. *The Mechanics of Fracture and Fatigue – An Introduction*, 60. London: E. & F.N. Spon Ltd.

5.36 Parks, D.M. 1974. 'A Stiffness-derivative Finite Technique for Determination of Crack-tip Stress Intensity Factors.' *International Journal of Fracture* 10(4): 487–502.

5.37 Rabczuk, T. 2013. 'Computational Methods for Fracture in Brittle and Quasi-Brittle Solids: State-of-the-Art Review and Future Perspectives.' *ISRN Applied Mathematics*, Article ID 849231, http://dx.doi.org/10.1155/2013/849231.

5.38 Raju, I.S. and K.N. Shivakumar. 1990. 'An Equivalent Domain Integral Method in the Two-Dimensional Analysis of Mixed Mode Crack Problems.' *Engineering Fracture Mechanics*: 37: 707–25.

5.39 Ramamurthy, T.S., T. Krishnamurthy, K. Badri Narayana, V. Vijyakuimar and B. Duttaguru. 1986. 'Modified Crack Closure Integral Method with Quarter Point Elements.' *Mechanics Research Communications* 13: 179–86.

5.40 Ramesh, K. 2000. *Digital Photoelasticity: Advanced Techniques and Applications*, 1st edn. Heidelberg: Springer.

5.41 Rao, A.K., I.S. Raju and A.V. Krishna Murthy. 1971. 'A Powerful Hybrid Method in Finite Element Analysis.' *International Journal of Numerical Methods in Engineering* 3: 389–403.

5.42 Rice, J.R. 1972. 'Some Remarks on Elastic Crack-tip Stress Fields.' *International Journal of Solids and Structures* 8(6): 751–58.

5.43 Rooke, D.P. and D.J. Cartwright. 1976. *Compendium of Stress Intensity Factors*. Her Majesty's Stationery Office.

5.44 Rosakis, A.J. and A.T. Zehnder. 1985. 'On the Method of Caustics: An Exact Elastic Analysis Based on Geometrical Optics.' *Journal of Elasticity* 15: 347–67.

5.45 Sanford, R.J. and J.W. Dally. 1979. 'A General Method for Determining Mixed Mode Stress Intensity Factors from Isochromatic Fringe Patterns.' *Engineering Fracture Mechanics* 11: 621–33.

5.46 Sarangi, H., K.S.R.K. Murthy and D. Chakraborty. 2010a. 'Radial Locations of Strain Gages for Accurate Measurement of Mode I Stress Intensity Factor.' *Materials and Design* 31: 2840–50.

5.47 ———. 2010b. 'Optimum Strain Gage Location for Evaluating Stress Intensity Factors in Single and Double Ended Cracked Configurations.' *Engineering Fracture Mechanics* 77: 3190–3203.

5.48 ———. 2012. 'Optimum Strain Gage Locations for Accurate Determination of the Mixed Mode Stress Intensity Factors.' *Engineering Fracture Mechanics* 88: 63–78.

5.49 ———. 2013. 'Experimental Verification of Optimal Strain Gage Locations for Accurate Determination of Mode I Stress Intensity Factors.' *Engineering Fracture Mechanics* 110: 189–200.

5.50 Sethuraman, R. and S.K. Maiti. 1988. 'Finite element based computation of strain energy release rate by modified crack closure integral.' *Engineering Fracture Mechanics* 30: 227–31.

5.51 Sih, G.C. 1973a. *Handbook of Stress Intensity Factors*. Pennsylvania: Lehigh University.

5.52 ———, ed. 1973b. *Methods of Analysis and Solution of Crack Problems, Mechanics of Fracture*, Vol. 1. Leyden: Noodhoff International Publishing.

5.53 Shih, C.F., B. Moran and T. Nakamura. 1986. 'Energy Release Rate along a Three-dimensional Crack Front in a Thermally Stressed Body.' *International Journal of Fracture* 30: 79–102.

5.54 Sneddon, I.N. 1973. 'Integral Transform Methods.' In *Methods of Analysis and Solutions of Crack Problems, Mechanics of Fracture*, Vol. 1, ed. Sih, G.C., 315–67. Leyden: Noodhoff International Publishing.

5.55 Sneddon, I.N. and M. Lowengrub. 1969. *Crack Problems in the Classical Theory of Elasticity*. Glassgow: Wiley International.

5.56 Stazi, F.L., E. Budyn, J. Chessa and T. Belytschko. 2003. 'An Extended Finite Element Method with Higher-Order Elements for Curved Cracks.' *Computational Mechanics* 31: 38–48.

5.57 Stern, M. 1979. 'Families of Consistent Conforming Elements with Singular Derivative Fields.' *International Journal of Numerical Methods in Engineering* 14: 409–21.

5.58 Tada, H., P.C. Paris and G.R. Irwin. 2000. *The Stress Analysis of Cracks Handbook*, 3rd edn. New York: ASME Press.

5.59 Theocaris, P.S. 1981. 'Elastic Stress Intensity Factors Evaluated by Caustics'. In *Mechanics of Fracture*, Vol. VII, ed. Sih, G.C. The Netherlands: Sijthoff and Noordhoff.

5.60 Theocaris, P.S. and E.E. Gdoutos. 1976. 'Surface Topography by Caustics.' *Applied Optics* 15 (6): 1629–38.

5.61 Tong, P. and T.H.H. Pian. 1973. 'On the Convergence of Finite Element Method for Problems with Singularity.' *International Journal of Solids and Structures* 9: 313–21.

5.62 Tracey, D.M. 1971. 'Finite Elements for Determination of Stress Intensity Factors.' *Engineering Fracture Mechanics* 3: 255–65.

5.63 Tracey, D.M. and T.S. Cook. 1977. 'Analysis of Power Type of Singularities Using Finite Elements.' *International Journal of Numerical Methods in Engineering* 11: 1225–33.

5.64 Williams, M.L. 1957. 'On Stress Distributions at the Base of a Stationary Crack.' *Journal of Applied Mechanics, Transactions of ASME* 24: 109–14.

5.65 Wilson, W.K. 1973. 'Finite Element Methods for Elastic Bodies Containing Cracks.' In *Methods of Analysis and Solutions of Crack Problems, Mechanics of Fracture*, Vol. 1, ed. Sih, G.C., 484–515. Leyden: Noodhoff International Publishing.

5.66 Wu, X.-R. and A.J. Carlsson. 1991. *Weight Functions and Stress Intensity Factor Solutions*. Oxford: Pergamon Press.

5.67 Xiao, Q.Z. and B.L. Karihaloo. 2007. 'Implementation of Hybrid Crack Element on a General Finite Element Mesh and in Combination with XFEM.' *Computer Methods in Applied Mechanics and Engineering* 196: 1864–73.

5.68 Zienkiewicz, O.C. and R.L. Taylor. 2000. *The Finite Element Method*, Vols 1 and 2. Oxford, Butterworth–Heinemann.

Mixed Mode Brittle Fracture

6.1 Introduction

Griffith considered crack subjected to opening mode of loading and gave the condition for its extension. Thereby, it was possible to determine the critical load capacity of a component made of a brittle material containing a crack. Irwin considered three fundamental modes of loading of a crack and prescribed the condition of their in-plane extension. In reality, it is very difficult to assume that a crack even under pure Mode II loading will extend in its own plane. Furthermore, when a crack is loaded in mixed mode (Fig. 6.1), both the direction of crack extension and the load at which crack extension begins are unknowns. Therefore two questions arise. What governs the direction of crack extension, and what decides the onset of the extension, in all combinations of mode mixity.

For brittle materials, more often catastrophic fracture is observed. The crack extends over a considerable span immediately upon extension, as if the whole course of extension is governed by the state of stress/deformation existing just at

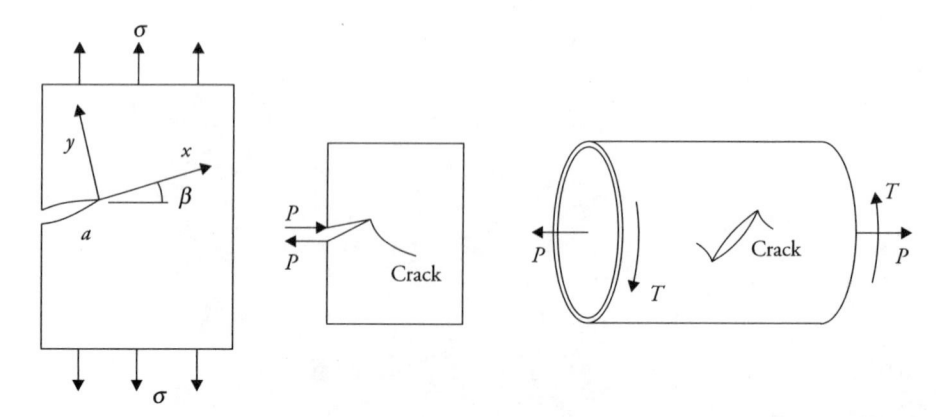

Figure 6.1 Cracks loaded in mixed mode.

the point of onset of extension. If such crack paths can be predicted, it may be possible to take some preventive measures. Even in manufacturing operation like bar shearing, the profile of the sheared edges is dependent on the path of propagation of tool edge cracks. In order to exercise some control on quality of the sheared edges, the prediction of crack path can be very helpful. In rock drilling, the propagation of cracks can help the course of drilling substantially. Any possibility of prediction of crack path in variety of situations can be gainfully exploited.

6.2 Theory based on Potential Energy Release Rate

Erdogan and Sih (1963) proposed that crack extends in the direction θ_c (Fig. 6.2(a)) corresponding to the maximum release rate of potential/strain energy. Similar considerations were also proposed by Palaniswamy and Knauss (1972) and Hussain, Pu and Underwood (1974). The extension occurs when this release rate reaches a critical value. This critical energy release rate is a material property in the same sense as the fracture toughness K_{IC} in Mode I. To apply this criterion, it is necessary to obtain the variation of G with θ (Fig. 6.2(b)). This can not be easily determined analytically and came in the way of application of the theory to practice. The finite element method eased the problem to a certain extent. The difficulty associated with the application of this theory motivated the developments of the other theories. These are presented in this chapter.

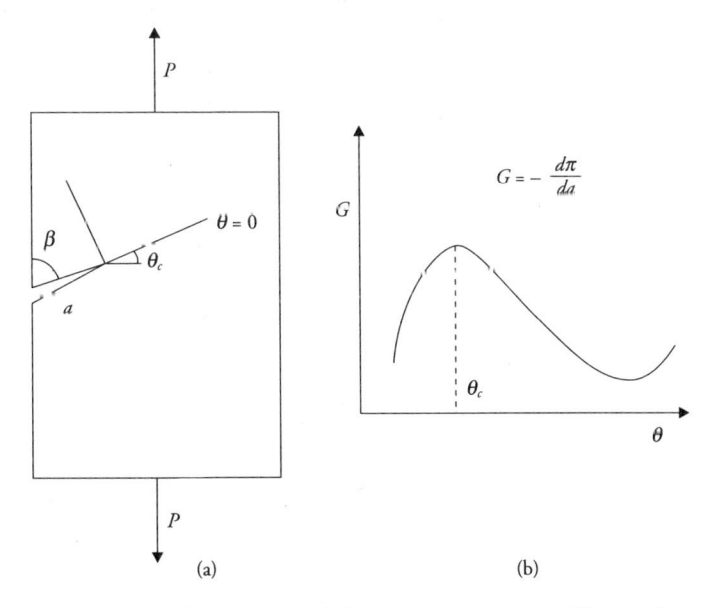

Figure 6.2 (a) Loading. (b) Typical variation of G with θ.

6.3 Maximum Tangential Stress Criterion

According to maximum tangential stress (MTS) criterion, the crack extends in the radial direction from the crack-tip, corresponding to the maximum of tangential stress on a circle of finite radius from the crack-tip. The extension occurs when this maximum reaches a critical value, which can be treated as material constant (Erdogan and Sih 1963; Maiti and Smith 1983a,b, 1984). The crack path can be obtained by joining such points of MTS on circles of different radii. In fact, this criterion is the same as the classical maximum principal stress theory of Rankine (Timoshenko 1986; Srinath 2003). Since the stresses are infinite at the crack-tip, these are evaluated at a finite distance from the crack-tip but given by the first term of the eigenfunction expansion. It is possible to work without specifying this distance.

For a two-dimensional problem, crack-tip stress field involves only Mode I and Mode II SIFs. The stresses in polar coordinates with crack-tip as the origin (Fig. 6.3) are given by

$$\sigma_r = \frac{1}{\sqrt{2\pi r}} \cos\frac{\theta}{2} \left[K_I \left(1 + \sin^2\frac{\theta}{2}\right) + \frac{3}{2}K_{II}\sin\theta - 2K_{II}\tan\frac{\theta}{2} \right]$$
$$+ r^0 f_1(\theta) + r^{\frac{1}{2}} f_2(\theta) + \ldots \quad (6.1a)$$

$$\sigma_\theta = \frac{1}{\sqrt{2\pi r}} \cos\frac{\theta}{2} \left[K_I \cos^2\frac{\theta}{2} - \frac{3}{2}K_{II}\sin\theta \right] + r^0 g_1(\theta) + r^{\frac{1}{2}} g_2(\theta) + \ldots \quad (6.1b)$$

$$\tau_{r\theta} = \frac{1}{2\sqrt{2\pi r}} \cos\frac{\theta}{2} \left[K_I \sin\theta + K_{II}(3\cos\theta - 1) \right] + r^0 h_1(\theta) + r^{\frac{1}{2}} h_2(\theta) + \ldots \quad (6.1c)$$

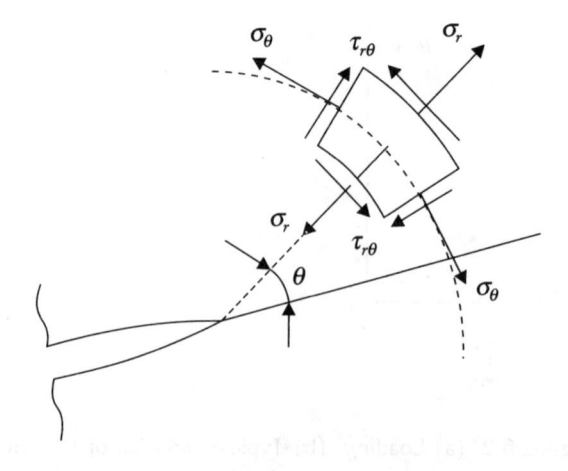

Figure 6.3 Stresses in polar coordinates.

where $f_1(\theta)$, $f_2(\theta)$, etc., $g_1(\theta)$, $g_2(\theta)$, etc., and $h_1(\theta)$, $h_2(\theta)$, etc. are the functions of θ. Reference may be made to Eqs. (2.18) and (2.22) in Chapter 2 for exponents of r.

The criterion indicates that θ_c corresponds to

$$\frac{\partial \sigma_\theta}{\partial \theta} = 0, \quad \frac{\partial^2 \sigma_\theta}{\partial \theta^2} < 0 \tag{6.2}$$

Physically, this means that the MTS is the highest tensile stress on the circle under consideration. The crack passes through the zone dominated by tensile stress in the tangential direction. And the critical stress σ_c or load P_c correspond to

$$(\sigma_\theta)_{max} = (\sigma_\theta)_c \tag{6.3}$$

Considering that the tangential stress σ_θ is given only by the singularity term

$$\frac{\partial \sigma_\theta}{\partial \theta} = -\frac{3}{4\sqrt{2\pi r}} \cos \frac{\theta}{2} [K_I \sin \theta + K_{II}(3\cos\theta - 1)] \tag{6.4}$$

Therefore, $\frac{\partial \sigma_\theta}{\partial \theta} = 0$ corresponds to $\tau_{r\theta} = 0$. That is, the direction of crack extension θ_c corresponds to a principal direction, and it is a solution of the following equation.

$$[K_I \sin\theta + K_{II}(3\cos\theta - 1)] = 0 \tag{6.5}$$

Hence, the corresponding tangential stress is a principal stress. The critical stress or load capacity is given by

$$(\sigma_\theta)_{max} = \frac{1}{\sqrt{2\pi r}} \cos \frac{\theta_c}{2} \left[K_I \cos^2 \frac{\theta_c}{2} - \frac{3}{2} K_{II} \sin\theta_c \right] = \sigma_{cr} \tag{6.6}$$

σ_{cr} can be easily decided by considering a mode I problem.

Mode I

For an internal crack of size $2a$ in an infinite sheet with remote loading σ, $K_I = \sigma\sqrt{\pi a}$ and $K_{II} = 0$. Therefore, θ_c is given by $\sin\theta = 0$. That is, $\theta_c = 0°$. At the onset of fracture

$$(\sigma_\theta)_{max} = \frac{1}{\sqrt{2\pi r}} \cos \frac{\theta_c}{2} \left[K_I \cos^2 \frac{\theta_c}{2} \right] = \frac{K_{IC}}{\sqrt{2\pi r}} = \sigma_{cr} \tag{6.7}$$

Provided the same distance r is considered for all the applications, the same σ_{cr} can be employed for the determination of σ_c.

Mode II

For an internal crack of size $2a$ in an infinite sheet with remote shear loading τ, $K_{II} = \tau\sqrt{\pi a}$ and $K_I = 0$. Therefore, θ_c is given by

$$3\cos\theta - 1 = 0 \tag{6.8}$$

That is, $\theta_c = \pm\cos^{-1}(\frac{1}{3}) = \pm 70.5°$. Incidentally, $\theta_c = -70.5°$ only gives $\frac{\partial^2 \sigma_\theta}{\partial \theta^2} < 0$. Assuming the same distance ahead of the crack-tip as in the case of Mode I, the failure is triggered, when

$$(\sigma_\theta)_{max} = \frac{K_{IC}}{\sqrt{2\pi r}} = \sigma_{cr} \tag{6.9}$$

If it is assumed that the failure in this case is triggered when Mode II SIF reaches the critical value K_{IIC}, substituting K_{II} and θ in the expression of tangential stress

$$(\sigma_\theta)_{max} = \frac{K_{IC}}{\sqrt{2\pi r}} = (\sigma_\theta)_c = \frac{1}{\sqrt{2\pi r}}\left\{\cos\frac{\theta}{2}\left[-\frac{3}{2}K_{IIC}\sin\theta\right]\right\}_{\theta=-70.5°} \tag{6.10}$$

This gives the following relationship between the two fracture toughnesses.

$$K_{IIC} = 0.866\, K_{IC} \tag{6.11}$$

This means that K_{IIC} is not an independent material property, and all two-dimensional fractures are governed by only one fracture resistance K_{IC}.

Considering a problem (Fig. 6.4(a)) with a crack of size $2a$ in an infinite sheet, from Eq. (6.5), we obtain

$$\sin^2\beta\,\sin\theta + \sin\beta\cos\beta\,(3\cos\theta - 1) = 0 \tag{6.12}$$

noting that $K_I = \sigma\sqrt{\pi a}\sin^2\beta$ and $K_{II} = \sigma\sqrt{\pi a}\sin\beta\cos\beta$. For a given β, it is possible to determine θ_c. Using Eq. (6.6) and this critical fracture angle, it is possible to obtain the critical load σ_c. Thereby, we get the corresponding combination of K_I and K_{II} that triggers fracture. Through this approach, it is possible to get all the combinations of the two SIFs that lead to fracture.

A typical plot of these combinations is shown in Fig. 6.4(b). The plot is known as failure locus. The failure locus separates the safe and unsafe zones. All the combinations of K_I and K_{II} that lie on the locus and outside will lead to catastrophic fracture.

The path of unstable extension of crack can be determined by locating the points of MTS on circles of different radii from the crack-tip and joining these points by a smooth curve as illustrated in Fig. 6.5.

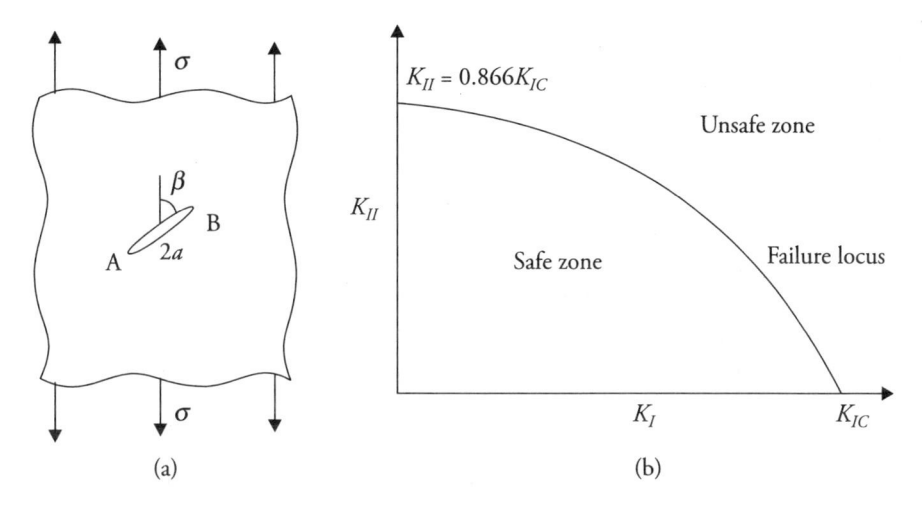

Figure 6.4 Mixed mode crack and failure locus. (a) Angled crack problem. (b) Failure locus.

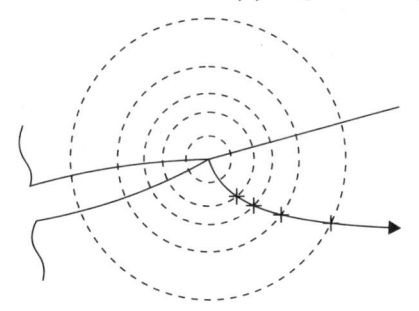

Figure 6.5 Unstable crack path.

6.4 Maximum Tangential Principal Stress Criterion

As per the underlying considerations of the MTS criterion, $\dfrac{\partial \sigma_\theta}{\partial \theta} = 0$ gives a direction, which coincides with the direction $\tau_{r\theta} = 0$. This happens only because the crack-tip stress field is given by the first or singularity term of the eigenfunction expansion. This is alright so long as r is small. It has been observed by Williams and Ewing (1972) that better predictions for both the initial direction of crack extension θ_c and fracture load σ_c are possible by considering stresses at a finite distance from the crack-tip so as to include the effects of the second and the higher order terms of the Williams' (1972) eigenfunction expansion of the crack-tip stress field. With the inclusion of the nonsingular terms, $\dfrac{\partial \sigma_\theta}{\partial \theta} = 0$ does not give rise to a direction, which corresponds to $\tau_{r\theta} = 0$. With reference to relations (6.1b) and 6.1c)

$$\frac{\partial \sigma_{\theta(ST)}}{\partial \theta} = -\frac{3}{2} \tau_{r\theta(ST)} \tag{6.13}.$$

but

$$\frac{dg_1(\theta)}{d\theta} \neq h_1(\theta), \quad \frac{dg_2(\theta)}{d\theta} \neq h_2(\theta), \quad \ldots\ldots\ldots \quad (6.14)$$

where $\sigma_{\theta(ST)}$ and $\tau_{r\theta(ST)}$ represent the singular terms of σ_θ and $\tau_{r\theta}$, respectively [Eqs. (6.1b and 6.1c)]. There are two distinct locations P and Q, and directions θ_{c1} and θ_{c2} (Fig. 6.6) corresponding to $\frac{\partial \sigma_\theta}{\partial \theta} = 0$ and $\tau_{r\theta} = 0$, respectively. The tangential stress at location P is the MTS, but it is not a principal stress ($\tau_{r\theta} \neq 0$). The tangential stress at location Q is a principal stress, but it is not the MTS $\left(\frac{\partial \sigma_\theta}{\partial \theta} \neq 0\right)$. In general, the two tangential stresses differ in magnitude slightly. There are, therefore, two distinct radial directions θ_{c1} and θ_{c2} (Fig. 6.6). This difference was the basis to define the MTPS criterion (Maiti 1980; Maiti and Prasad 1980; Maiti and Smith 1983a, b, 1984).

According to the MTPS criterion, crack extends in a radial direction corresponding to the maximum tangential principal stress (MTPS), and the extension begins when this maximum reaches a critical value. The crack path is given by the smooth curve passing through such points on circles of different radii from the crack-tip.

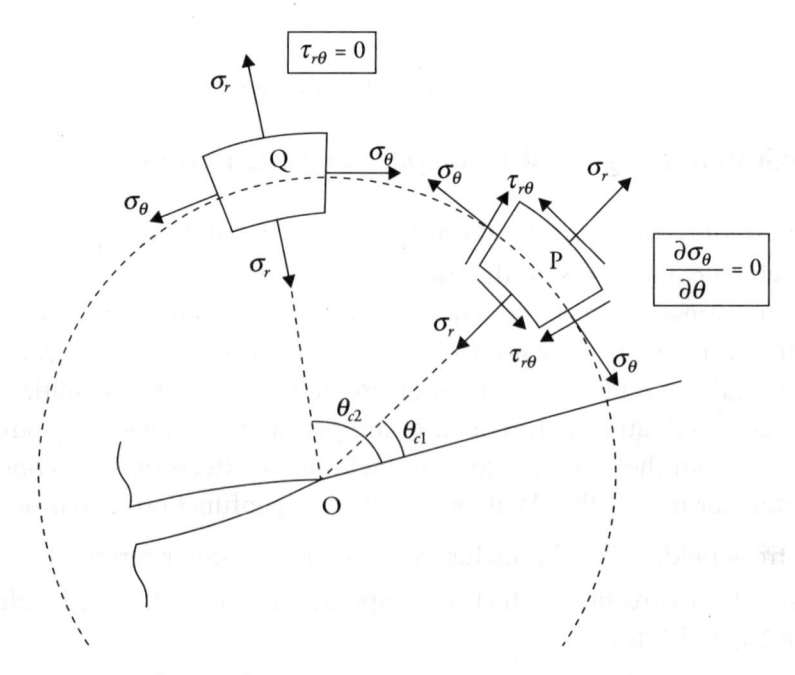

Figure 6.6 Directions θ_{c1} and θ_{c2} corresponding to maximum tangential stress and maximum tangential principal stress.

In the angled crack problems, the difference between the MTS and MTPS criteria increases for crack orientations β greater than 75°. It is observed that predictions based on the MTPS criterion are more close to the experimental observations (Maiti and Smith 1983a, b, 1984). In fact, MTPS criterion is the classical Rankine criterion. The direction of crack growth as per MTPS criterion is not associated with any shear stress. Hence, the crack is driven locally by Mode I type of loading and it grows in a self similar manner.

In general, for any problem with a complex geometry, it is convenient to go in for finite element analysis with circular arrangement of elements around the crack-tip (Fig. 5.13). From the finite element results, it is possible to plot the variations of σ_θ and $\tau_{r\theta}$ on a circle of finite radius passing through element Gauss points. Thereby, both θ_{c1} and θ_{c2} can be obtained (Fig. 6.7). The permissible load capacity can be determined by equating MTPS $(\sigma_{\theta P})_{max}$ with σ_θ that exists at radius r in Mode I for $\theta = 0°$ when the applied load level corresponds to $K_I = K_{IC}$. r is the same radius of circle used for plotting in both the cases. For good predictions, r can be taken up to 5% of crack size (Maiti and Smith 1983). $(\sigma_\theta)_{cr}$ in Mode I can be approximated as $\dfrac{K_{IC}}{\sqrt{2\pi r}}$.

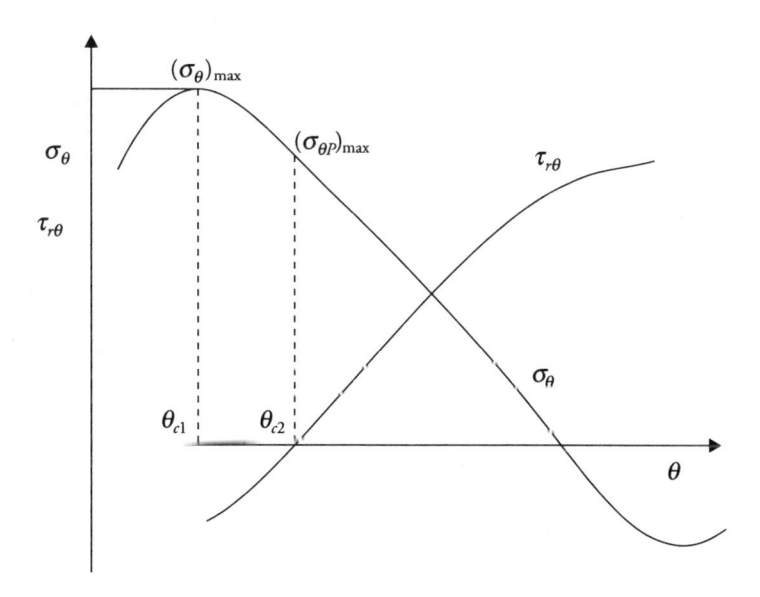

Figure 6.7 Graphical determination of θ_{c1} and θ_{c2}.

6.5 Strain Energy Density Criterion

This criterion has been proposed by Sih (1973). According to this criterion, crack extends in the radial direction from the crack-tip corresponding to the minimum of strain energy density (SED) on a circle of finite radius (Fig. 6.8) from the

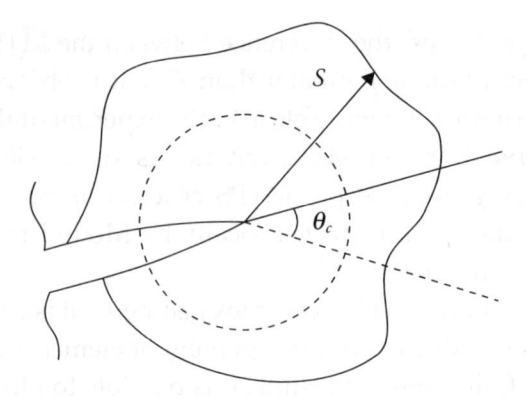

Figure 6.8 Circumferential variation of S.

crack-tip, and it extends when this minimum reaches a critical value, which is a material parameter. The crack path is obtained by joining such points on circles of different radii by a smooth curve (Kipp and Sih 1975). In case, there are more than one minimum on a particular radius, Swedlow (1976) suggested that the one with highest tensile tangential stress should be considered.

The SED is given by

$$SED = \frac{1}{2E}\left(\sigma_x^2 + \sigma_y^2 + \sigma_z^2\right) - \frac{\nu}{E}\left(\sigma_x\sigma_y + \sigma_y\sigma_z + \sigma_z\sigma_x\right) + \frac{1}{2\mu}\left(\tau_{xy}^2 + \tau_{yz}^2 + \tau_{zx}^2\right)$$

$$= \frac{1}{r}\left(a_{11}K_I^2 + 2a_{12}K_IK_{II} + a_{22}K_{II}^2 + a_{33}K_{III}^2\right) = \frac{S}{r} \qquad (6.15)$$

where

$$a_{11} = \frac{1}{16\mu\pi}[(3 - 4\nu - \cos\theta)(1 + \cos\theta)], \quad a_{12} = \frac{1}{16\mu\pi}2\sin\theta[\cos\theta - (1 - 2\nu)]$$

$$a_{22} = \frac{1}{16\mu\pi}[4(1 - \nu)(1 - \cos\theta) + (1 + \cos\theta)(3\cos\theta - 1)], \quad a_{33} = \frac{1}{4\mu\pi}. \qquad (6.16)$$

μ is modulus of rigidity and ν is Poisson's ratio.

The direction of crack extension corresponds to

$$\frac{\partial S}{\partial \theta} = 0, \quad \frac{\partial^2 S}{\partial \theta^2} > 0 \qquad (6.17)$$

and the extension begins when

$$S_{\min} = S_{cr} \qquad (6.18)$$

In order to provide a physical justification, Sih (1973) opined that under fixed loading, the total potential energy of a system is equal in magnitude with the total strain energy, but opposite in sign. Hence, the 'point of minimum SED' corresponds to the 'point of maximum potential energy density'. Since the state of 'maximum potential energy' is very unstable and dangerous, the crack is likely to pass through such a point. In this argument, the total strain energy and total potential energy and the respective energy densities have been treated at par! Later Sih, suggested that, the total SED can be split into volumetric and deviatoric components (e.g., Sih and Macdonald 1974). The crack can be considered to propagate in the radial direction corresponding to the maximum of volumetric energy density on a circle of finite radius from the crack-tip.

Theocaris and Andrianopoulos (1982) modified the SED criterion and stated that a crack propagates in a radial direction from the crack-tip corresponding to the maximum of volumetric energy density (SED$_v$) on a closed contour around the crack-tip along which the distortion energy density (SED$_d$) is constant. It may be stressed here that both SED$_v$ and SED$_d$ are functions of both radius and angle θ. He further opined that the elastic–plastic boundary around the crack-tip is a very good choice for the closed contour, because the distortion energy density on such a curve is constant. It is often referred to as the T criterion.

While MTS and MTPS are two-dimensional criteria, SED is a three-dimensional criterion. SED has been reported to have failed to predict the crack paths, for example, in the case of edge cracks (Maiti 1980; Maiti and Prasad 1980) under shear type of loading, in the case of angled crack (Fig. 6.2(a)) (Maiti and Smith 1983a,b, 1984) when β is less than 15°, and in the case of internal crack in a plate with its edges separated by concentrated forces acting at an angle to the crack (Maiti and Smith 1983c). Applications of the criterion to some cases are discussed subsequently.

Mode I

For this case $S = a_{11}K_I^2$, since $K_{II} = 0$. The direction of initial crack extension is given by

$$\frac{\partial a_{11}}{\partial \theta} = 0 \tag{6.19}$$

That is, $\theta = 0°$ or $\theta = \cos^{-1}(1 - 2\nu)$. The second value is not acceptable because it does not make S a minimum.

At the onset of crack extension $K_I = K_{IC}$, therefore, the corresponding $S_{\min} = \frac{(1-2\nu)}{4\pi\mu}K_{IC}^2 = S_{cr}$. S_{cr} can be treated as a material property.

Mode II
In this case $S = a_{22} K_{II}^2$, since $K_I = 0$. Therefore, the direction of initial crack extension is given by

$$\frac{\partial a_{22}}{\partial \theta} = 0 \qquad (6.20)$$

This gives $\cos \theta = \frac{1-2\nu}{3}$, $\theta = -\cos^{-1} \frac{(1-2\nu)}{3}$. The angle depends on Poisson's ratio ν, unlike in the case of MTS and MTPS criteria. For $\nu = 0$, $\theta_c = -70.5°$, which is the same as the value of θ_c given by the MTS criterion. For $\nu = 0.25, \theta_c = -80.4°$. By substituting θ_c in Eq. (6.15), it is possible to calculate S_{cr} at the onset of crack extension.

$$S_{cr} = \frac{1}{16\pi\mu}[4(1-\nu)(1-\cos\theta) + (1+\cos\theta)(3\cos\theta - 1)]K_{IIC}^2, \text{ for } \theta = \theta_c.$$

$$= \frac{1}{6\pi\mu} K_{IIC}^2 \text{ for } \nu=0 \text{ and } \theta_c=-70.5°. \qquad (6.21)$$

This value of S_{cr} can be equated with the earlier value in terms of K_{IC}, which gives $K_{IIC} = 1.224 K_{IC}$. This is very different from the value $0.866 K_{IC}$ obtained by the MTS criterion. For $\nu = 0.25, \theta_c = -80.4°$, $K_{IIC} = 1.021 K_{IC}$.

Mixed mode
For an angled crack with an inclination β with the loading direction

$$S = a_{11} K_I^2 + 2 a_{12} K_I K_{II} + a_{22} K_{II}^2 \qquad (6.22)$$

The direction of crack extension is given by

$$\frac{\partial}{\partial \theta}\{K_I^2 [(3 - 4\nu - \cos\theta)(1 + \cos\theta)] + 4 K_I K_{II} \sin\theta [\cos\theta - (1 - 2\nu)] +$$

$$K_{II}^2 [4(1-\nu)(1-\cos\theta) + (1+\cos\theta)(3\cos\theta - 1)]\} = 0 \qquad (6.23)$$

Note that $K_I = \sigma\sqrt{\pi a} \sin^2\beta$ and $K_{II} = \sigma\sqrt{\pi a} \sin\beta\cos\beta$. For any particular β, the direction θ_c for which S is minimum can be obtained. For example, for $\beta = 45°$, the direction of crack extension $\theta_c = -53.75°$ using the Poisson's ratio $\nu = 0.35$. Similarly, for $\beta = 75°$ and $60°$ the direction of crack extensions are $-25.8°$ and $-42°$, respectively, using $\nu = 0.35$.

The load capacity σ_c can be determined from the following relation with $\theta = \theta_c$.

$$\sigma_c^2 \pi a \frac{1}{16\mu\pi} \left[\sin^4\beta \left\{ (3-4\nu) - (\cos\theta)(1+\cos\theta) \right\} \right.$$

$$+ 4\sin^3\beta \cos\beta \sin\theta \left\{ \cos\theta - (1-2\nu) \right\}$$

$$\left. + \sin^2\beta \cos^2\beta \left\{ 4(1-\nu)(1-\cos\theta) + (1+\cos\theta)(3\cos\theta - 1) \right\} \right]$$

$$= \frac{(1-2\nu)}{4\pi\mu} K_{IC}^2 \tag{6.24}$$

Exercise

6.1 Calculate the direction of initial crack extension using MTS and SED criteria for the case (Fig. Q.6.1) shown considering (i) loading due to p only, (ii) loading due to $p/2$ only, and (iii) loading due to p and $p/2$ only. p represents load per unit thickness. Use Poisson's ratio $\nu = 1/3$.
[Ans. (i) $-43°$, $-41.5°$, (ii) $-70.5°$, $-84°$, (iii) $-47.2°$, $-45.8°$]

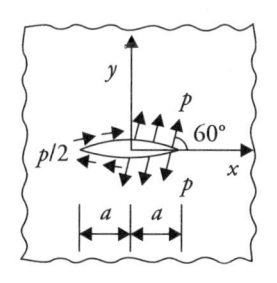

Figure Q.6.1

6.2 Obtain the load capacity $p\sqrt{\pi a}/K_{IC}$ in each case of Q.6.1.
[Ans. (i) 1.041, 1.17, (ii) 2.117, 2.21, (iii) 0.95, 1.09]

6.3 Determine direction of initial crack extension and load capacity v (Fig. Q.6.3) based on the MTS criterion. Crack size at each end of the hole is 2 mm. The given material is 4140 steel with $K_{IC} = 75$ MPa$\sqrt{m}$ and the plate thickness is 25 mm. [Ans. 506 MPa]

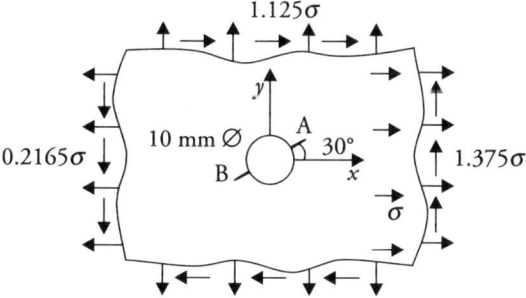

Figure Q.6.3

6.4 What is the direction of initial crack extension and load capacity using the MTS criterion when the crack size is AB = 20 mm, σ_0 = 200 MPa and crack inclination with x axis is 45° (Fig. Q.6.4)? Given that the material is D6ac steel with K_{IC} = 86 MPa$\sqrt{m}$ and thickness 40 mm.

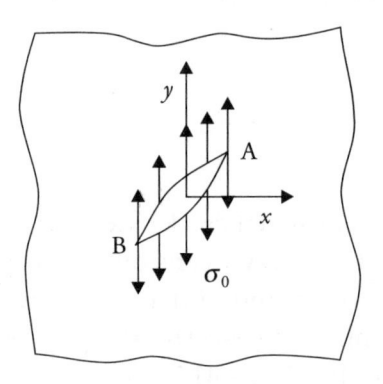

[Ans. $-53.13°, 383.58$ MPa]

Figure Q.6.4

6.5 Estimate the direction of crack extension by MTS and SED criteria for a through-the-thickness crack in a thin cylinder (Fig. Q.6.5) when it is subjected to a torque $T = 50$ kNm and an axial load $P = 1$ MN, which is uniformly distributed over the cross-section. Cylinder has an internal radius $r = 300$ mm and wall thickness $t = 10$ mm. Crack size $2a = 12$ mm. Use $\nu = 1/3$. [Ans. $-18°, -17.6°$]

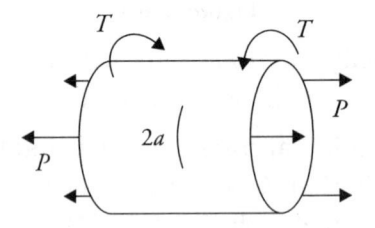

Figure Q.6.5

6.6 Calculate the direction of initial crack extension (Fig. Q.6.6) of CD using the MTS and SED criteria. Determine the load capacity σ_c, considering the load capacity in Mode I as 1, based on the MTS criteria. If the Poisson's ratio is 0.35, compare the load capacity with the SED criteria.

[*Hint*: Treat CD as an inclined crack and solve. Ans. $-51.6°, -51.7°$; Load (MTS) = 1.08, Load (SED) = 1.23, when Mode I load is 1.]

Mixed mode brittle fracture 165

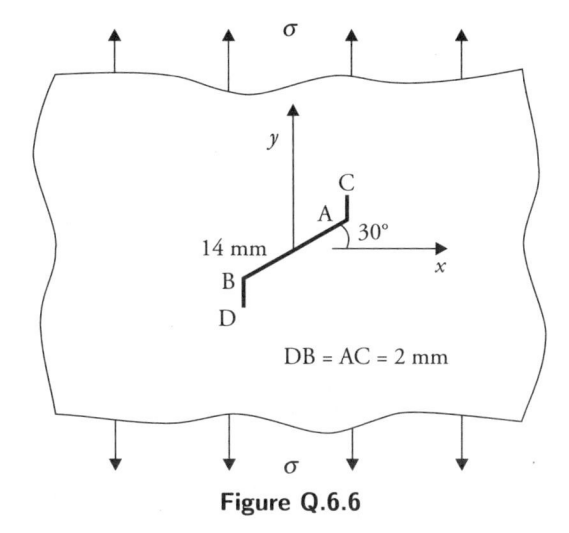

Figure Q.6.6

6.7 Determine the direction of initial crack extension (Fig. Q.6.7) of CD using the MTS and SED criteria. Determine the load capacity when $K_{IC} = 48$ MPa$\sqrt{m}$ based on the MTS criteria. If the Poisson's ratio is 0.35, compare the load capacity with the prediction by the SED criterion.

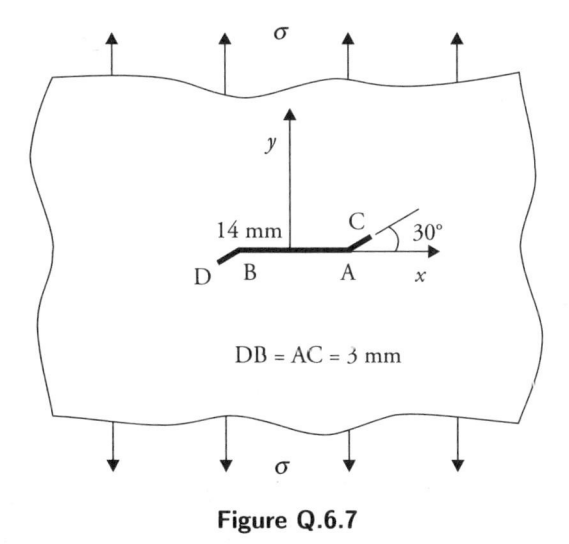

Figure Q.6.7

[*Hint*: Treat CD as an inclined crack and solve. [Ans. $\theta_c = -16.98, -16.7°$ by MTS and SED. Load capacities 271.90 MPa by MTS and 277.85 MPa by SED.]

6.8 For the problem shown (Fig. Q.6.8) $P = 9$ kN and axial load $R = 2$ kN. Calculate the direction of the initial crack extension using thickness 15 mm and Poisson's ratio as 0.30. Try to solve employing both the MTS and SED criteria. [Ans. 3.07°, 3.06°]

Figure Q.6.8

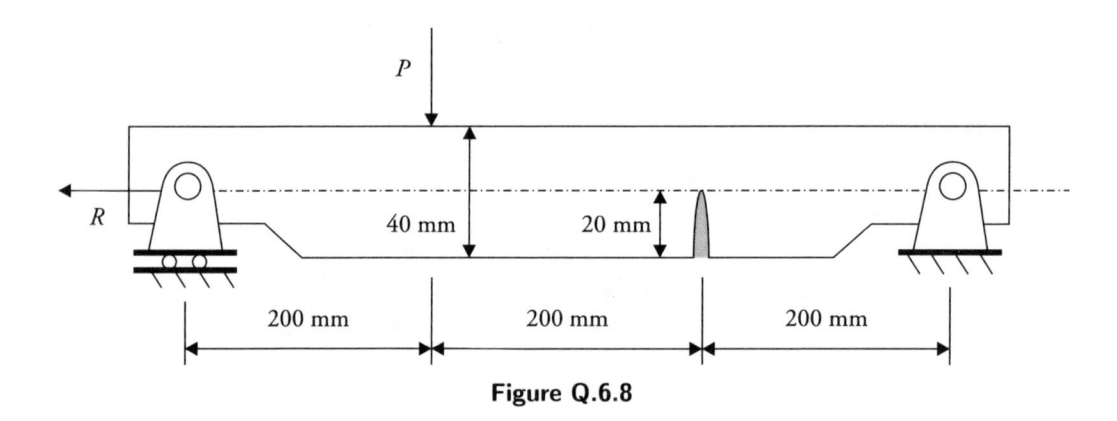

6.9 Determine the load P for the problem shown in Fig. Q.6.8 keeping R constant so as to obtain a direction of initial crack extension of $10°$. Use the MTS criterion. [Ans. -381.2 N]

6.10 Refer to Problem 2.6 (Chapter 2). Calculate the horse power that can be safely transmitted by considering it as a mixed mode problem. [Ans. 72.6 MP]

References

6.1 Erdogan, F. and G.C. Sih. 1963. 'On Extension in Plates Under Plane Loading and Transverse Shear.' *Journal of Basic Engineering, Transactions of ASME* 85: 519–27.

6.2 Hussain, M.A., S.L. Pu and J. Underwood. 1974. 'Strain Energy Release Rate for a Crack under Mode I and Mode II.' In *Fracture Analysis*, 2–28. ASTM Special Technical Publication 560, Philadelphia: American Society for Testing Materials.

6.3 Kipp, M.E. and G.C. Sih. 1975. 'The Strain Energy Density Failure Criteria Applied to Notched Elastic Solids.' *International Journal of Solids and Structures* 11: 153–73.

6.4 Maiti, S.K. 1980. 'Prediction of the Path of Unstable Extension of Internal and Edge Cracks.' *Journal of Strain Analysis* 15: 183–94.

6.5 Maiti, S.K. and K.S.R.K. Prasad. 1980. 'A Study on the Theories of Unstable Crack Extensions for the Prediction of Crack Trajectories.' *International Journal of Solids and Structures* 16: 563–74.

6.6 Maiti, S.K. and R.A. Smith. 1983a. 'Prediction of Initial Direction of crack Extension in a DCB Specimen by Various Criteria.' *International Journal of Fracture* 23: R41–44.

6.7 ———. 1983b. 'Comparison of the Criteria for Mixed Mode Brittle Fracture Based on the Pre-instability Stress–Strain Field, Part I.' *International Journal of Fracture* 23: 281–95.

6.8 ———. 1983c. 'Theoretical and Experimental Studies on the Extension of Cracks Subjected to Concentrated Loading Near their Faces to Compare the Criteria for Mixed Mode Brittle Fracture.' *Journal of Mechanics and Physics of Solids* 31: 389–403.

6.9 ———. 1984. 'Comparison of the Criteria for Mixed Mode Brittle Fracture Based on the Pre-instability Stress–Strain Field, Part II.' *International Journal of Fracture* 24: 5–22.

6.10 Palaniswamy, K. and W.G. Knauss. 1972. 'Propagation of a Crack under General In-plane Tension.' *International Journal of Fracture Mechanics* 8: 114–17.

6.11 Sih, G.C., ed. 1973. 'Methods of Analysis and Solutions of Crack Problems.' In *Mechanics of Fracture*, Vol. 1. Leyden: Moodhoff International Publishing.

6.12 Sih, G.C. and B. Macdonald. 1974. 'Fracture Mechanics Applied to Engineering Problems – Strain Energy Density Fracture Criterion.' *Engineering Fracture Mechanics* 6: 361–86.

6.13 Srinath, L.S. 2003. *Advanced Mechanics of Solids*, p. 113, 2nd edn. New Delhi: Tata McGraw-Hill Publishing.

6.14 Swedlow, J.L. 1976. 'Criteria for Growth of the Angled Crack.' In *Cracks and Fracture*, 506–21. Philadelphia: American Society for Testing and Materials [ASTM STP 601].

6.15 Theocaris, P.S. and N.P. Andrianopoulos. 1982. 'The Mises Elastic–Plastic Boundary as the Core Region in Fracture Criteria.' *Engineering Fracture Mechanics* 16: 425–32.

6.16 Timoshenko, S.P. 1986. *Strength of Materials*, Part 2, 445–446, 3rd edn. USA: Wadsworth Publishing Co, and 1st Indian edn. New Delhi: CBS Publishers.

6.17 Williams, M.L. 1957. 'On Stress Distributions at the Base of a Stationary Crack.' *Journal of Applied Mechanics, Transactions of ASME* 24: 109–14.

6.18 Williams, J.G. and P.D. Ewing. 1972. 'Fracture under Complex Stress – the Angled Crack Problem.' *International Journal of Fracture Mechanics* 8: 441–46.

7

Fatigue Crack Growth

7.1 Introduction

More than 90% components (Dieter 1988) of machines and structures are subjected to cyclically varying loads (Figs. 7.1 and 7.2) in service and fail due to fatigue. The cyclic variations may be due to the change in external loads (arising out of traffic on bridges, waves hitting ships, wind gusts on aircraft, tidal waves hitting offshore structures, pressure fluctuations in pipelines, lift off and landing of aircraft, railway wagons moving on rails, gear tooth getting engaged and disengaged, etc.; due to rotations of shafts of gear boxes, axles, wheels, etc.; and arising out of vibrations of loaded components of automobile running on roads, flow-induced vibration of pipe, etc.).

The fatigue problems can fall into two regimes of fatigue life: low and high cycles. In the case of low cycle fatigue, the problem involves strain level more than the yield strain of the material, and the cumulative damage is considered to be strain controlled. In this regime, the number of cycles to failure is less than about 10^4 cycles (Dieter 1988). These problems are very important for nuclear and

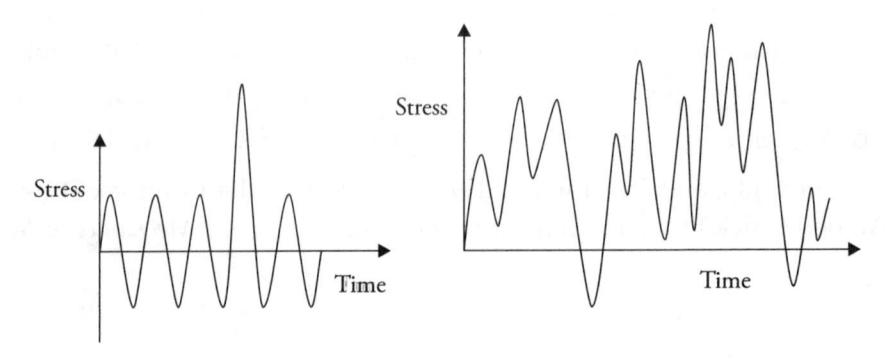

Figure 7.1 Constant amplitude fatigue loading with overload cycle and random fatigue loading.

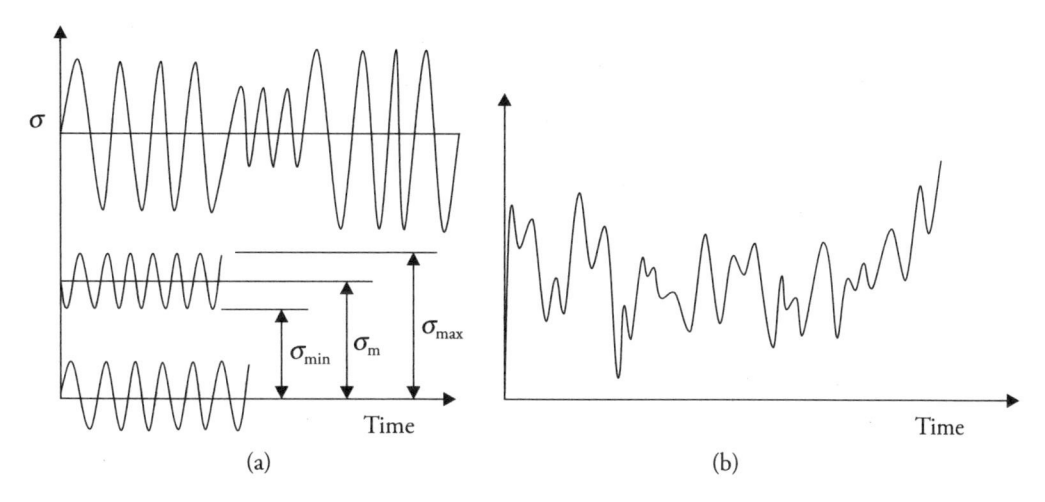

Figure 7.2 Different types of fatigue loading. (a) Time-varying loadings with well defined cycles. (b) Randomly time-varying load without well defined cycles.

thermal power plants. In the case of high cycle fatigue, which is of concern in this chapter, generally failure occurs at stress levels much below the yield stress of the material. In the high cycle regime, the number of cycles to failure is greater than 10^4 cycles.

Mostly machine and structural components are initially crack-free. As time passes, under the action of cyclic loading, damage ensues and it gradually grows to defect like a crack (Fig. 7.3). The crack then grows slowly to a critical size, leading to catastrophic failure. Thus, most of the fatigue crack growth occurs in the sub-critical stage. Since the load levels are low, the crack-tip plastic deformations are small. The growth of a crack under high cycle fatigue loading can be studied within the framework of linear elastic fracture mechanics. A lot of efforts have been devoted to quantify the crack growth in terms of loading and geometry. This has finally given rise to the possibility of calculation of fatigue life of components and machines. Most of the models to study the fatigue crack growth rate (FCGR) are empirical.

Based on an experimental study, Paris and Erdogan (1963) showed that the cyclic crack growth under fatigue loading can be characterized in terms of cyclic stress intensity factor (SIF) range. This paved the way for an estimation of cyclic life of components. This law has been subsequently enlarged in scope to accommodate variable and random amplitude cyclic loading, effects of occasional overloads, etc. Even the scope has been enlarged to take care of crack growth under fatigue together with stress corrosion, or creep, or their combinations.

Varieties of cyclic loading that can come up on a component are schematically shown in Figs. 7.1 and 7.2. A constant amplitude cyclic loading is specified by σ_{max},

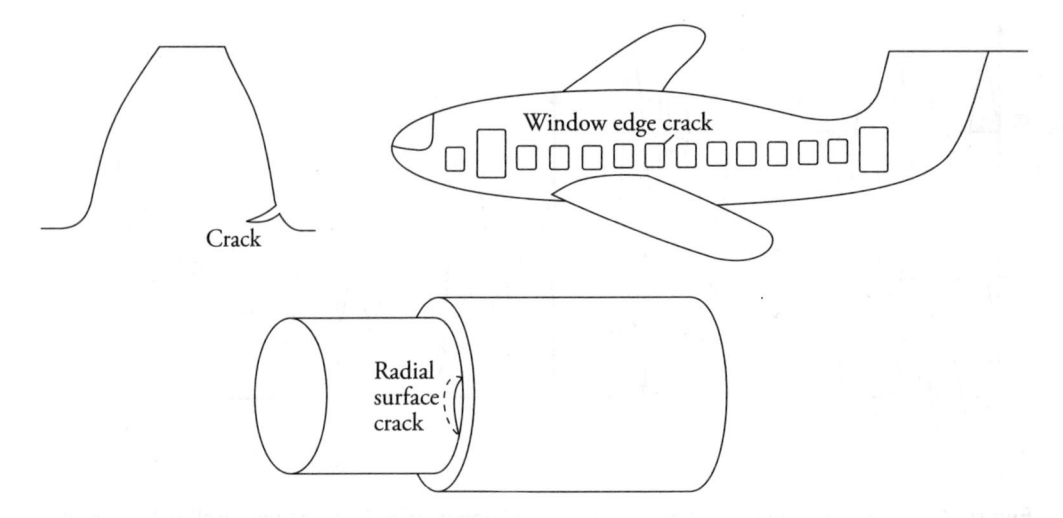

Figure 7.3 Cracks resulting from fatigue loading at the root of a gear tooth, window corner of an aircraft fuselage, and step of a shaft.

$\sigma_{\min}$, and mean stress σ_m. The loading range, $\Delta\sigma = \sigma_{\max} - \sigma_{\min}$, is an important quantity. Sometimes, stress ratio $R = \sigma_{\min}/\sigma_{\max}$ is employed to specify the type of loading. For example, the bottom most loading shown in Fig. 7.2(a), $R = -1$, which is completely alternating load. For the central loading in the same figure, $R > 0$; it has a mean stress σ_m, which is greater than zero. The cyclic loading parameters are the two extreme SIFs and the range of SIF: $K_{I\max} = \sigma_{\max}\sqrt{\pi a}\, Y$, $K_{I\min} = \sigma_{\min}\sqrt{\pi a}\, Y$ and $\Delta K = K_{I\max} - K_{I\min}$. Y is the SIF correction factor.

7.2 Fatigue Crack Growth Rate under Constant Amplitude Loading

Paris and Erdogan (1963) did a wide range of experiments considering specimens with crack subjected to a variety of constant amplitude cyclic loadings and plotted the variation of crack growth rate per cycle with range of SIFs $\Delta K = \Delta\sigma\sqrt{\pi a}\, Y$ on a log–log scale. Typical plots, which are known as sigmoidal plots, are shown in Fig. 7.4. The experimental data generally shows a lot of scatter. The plot can be divided into three distinct stages. Stage I is the crack initiation region and exhibits a fatigue threshold SIF range ΔK_{th}, below which crack does not grow under cyclic stress fluctuations. If the SIF range is slightly higher than this level, crack grows fast.

Stage II is the region of crack growth. In some cases, this is the most dominant region and the crack growth can be represented by the following relation.

$$\frac{da}{dN} = C\,(\Delta K)^m \qquad (7.1)$$

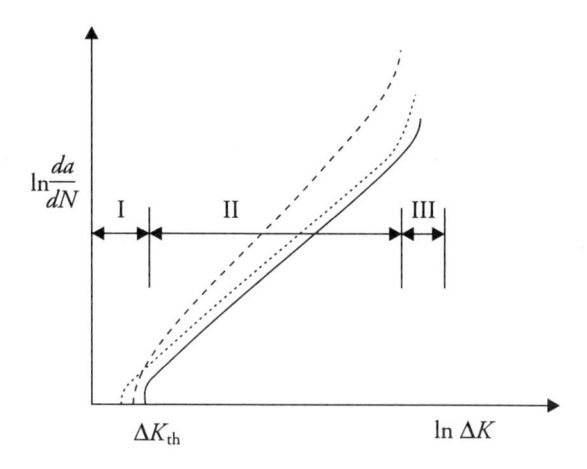

Figure 7.4 Plot of crack growth rate with stress intensity factor range.

Table 7.1 Threshold stress intensity factors.

Material	σ_{ult} (MPa)	Stress ratio R	ΔK_{th} (MPa$\sqrt{m}$)
Mild steel	430	−1.0	6.4
		0.5	4.3
		0.75	3.8
Austenitic steel	685	−1.0	6.0
Low alloy steel	835	−1.0	6.3
	680	0	6.6
18/8 Austenitic steel	685	−1	6.0
	665	0	6.0
Maraging steel	2010	0.67	2.7
Aluminium	77	−1.0	1.0
L65 Aluminium alloy (4.5% Cu)	450	−1.0	2.1
	495	0	2.1
AM503 Magnesium alloy (1.6%Mn) Titanium	165	0	0.99
	540	0.6	2.2
Nickel	430	0	7.9
	455	−1.0	5.9

Source: Pook (1975)

This is an empirical relation and is known as Paris law. C and m are material constants. C varies from 10^{-7} to 10^{-12}, when working in MPa and m units, and m

is dimensionless and varies from 2 to 7. The threshold SIF ΔK_{th} and the two constants depend on the material and stress ratio R. Some experimentally measured values of ΔK_{th} and C and m are presented in Tables 7.1 and 7.2 (Pook 1975), respectively. Some more data on C and m is given in Table 7.3.

In Stage III, crack grows rapidly, leading to catastrophic fracture. $\frac{da}{dN}$ tends to infinity when K_{max} is close to fracture toughness K_{IC} or K_C of the material. To represent the fatigue crack growth behaviour in Stages II and III, Forman proposed the following relations.

$$\frac{da}{dN} = \frac{C\,\Delta K^m}{(1-R)\,K_C - K_{max}} \tag{7.2}$$

where K_C is plane stress fracture toughness. This model includes the effect of stress ratio R. The constants C and m involved here have values and units different from those of the Paris law constants (Anderson 2005).

These relations [Eqs. (7.1) and (7.2)] cannot be applied to Stage I, which is associated with a threshold range ΔK_{th}. Experimental measurements at extremely low levels of the SIF range are difficult, and the results can be dubious. The growth rate in this region has been proposed by Donahue et al. (1972) in the following form.

$$\frac{da}{dN} = C\,(\Delta K - \Delta K_{th})^m \tag{7.3}$$

Like Forman, several investigators have expressed the dependence of the FCGR in Stage II on stress ratio R and proposed relations incorporating this factor. For example, Broek and Schijve (1963) proposed

$$\frac{da}{dN} = C\,K_{max}^3\,(1-R) \tag{7.4}$$

where C is a material constant different from Paris law constant. Walker (1970) proposed a relation of the type

$$\frac{da}{dN} = C\,\Delta K^n\,K_{max}^m \tag{7.5}$$

where C, m, and n are material constants.

None of the relations have been developed from the consideration of basic mechanics. They do not have general applicability. Nevertheless, the Paris law has been most widely applied.

Table 7.2 Typical Paris law constants.

Material	0.1% or 0.2% Proof stress (MPa)	Ultimate strength (MPa)	R	C (MPa, m units)	m
Mild steel	230	325	0.06–0.74	2.427×10^{-12}	3.3
Cold rolled mild steel	655	695	0.07–0.43	2.507×10^{-13}	4.2
			0.54–0.76	3.681×10^{-14}	5.5
			0–0.75	4.624×10^{-12}	3.3
18/8 Austenitic steel	195–225	665	0.33–0.43	3.326×10^{-12}	3.1
Aluminium	95–125	125–155	0.14–0.87	4.56×10^{-11}	2.9
5% Mg–Al alloy	180	310	0.20–0.69	2.811×10^{-12}	2.7
L71 Al alloy (4.5% Cu)	415	480	0.14–0.46	3.920×10^{-11}	3.7
L73 Al alloy (4.5% Cu)	370	435	0.50–0.88	3.821×10^{-11}	4.4
DTD 687A Al alloy (5.5% Zn)	495	540	0.20–0.45	1.261×10^{-11}	3.7
ZW1 Mg alloy (0.5% Zr)	165	250	0	1.230×10^{-9}	3.35
AM503 Mg alloy (1.5% Mn)	107	200	0.5	3.446×10^{-9}	3.35
Cu	26–513	215–310	0.07–0.82	3.384×10^{-12}	3.9
Titanium	440	555	0.08–0.94	6.886×10^{-12}	4.4
5% Al–Titanium alloy	735	835	0.17–0.86	9.558×10^{-12}	3.8
15% Mo–Titanium alloy	995	1160	0.28–0.71	2.138×10^{-11}	3.5
			0.81–0.94	11.666×10^{-12}	4.4

Source: Pook (1975)

Table 7.3 Paris law constants for some more materials.

Material	C (in MPa, m units)	m	Reference
Ferrite–Pearlite steel	6.8×10^{-12}	3.0	Barsom (1971)
Martensitic steel	1.33×10^{-10}	2.25	Barsom (1974)
Austenitic stainless steel	5.5×10^{-12}	3.25	Barsom and Rolfe (1999)
Grey CI (for $R = 0.1$)	2.98×10^{-11}	3.71	Biell IV and Lawrence Jr. (1989)
Al–Si 319 (for $R = 0.1$)	2.05×10^{-11}	3.12	Same as above
7075 T651 Al alloy	6.8×10^{-11}	3.89	Zhao, Zhang and Jiang (2008)

7.3 Factors Affecting Fatigue Crack Propagation

For the correct prediction of fatigue life, the growth rates must be available corresponding to the conditions existing in service. Such data is not always available. Fatigue crack propagation is affected by a number of parameters. Amongst the significant are thickness, anisotropy, heat treatment, cold deformation, temperature, batch-to-batch manufacturing variations, frequency, and environment. The influence of the environment has received a considerable attention. It is realized that the rate of crack growth in wet air can be an order of magnitude higher than that in vacuum. There is no agreement on the influence of environment on the rate of crack growth. Different explanations are likely to be applicable to different materials.

It is very difficult to account precisely all these factors. The FCGR is not a consistent material property as the tensile strength or yield strength of a material. Further, it is influenced by so many factors that it is likely to be a less consistent property than the fracture toughness. This is reflected in the wide scatter bands in the experimental plots of $\frac{da}{dN}$ versus ΔK.

7.4 Crack Closure

When a component with crack is loaded by cyclic loading varying from zero to some maximum amplitude, with the increase in load from the minimum O (Fig. 7.5(a)) crack flanks open. Plastic zone around the crack also grows in size with the increase in the tensile load. The crack opening and the size of the plastic zone continue to increase till the maximum load C is reached. At the maximum

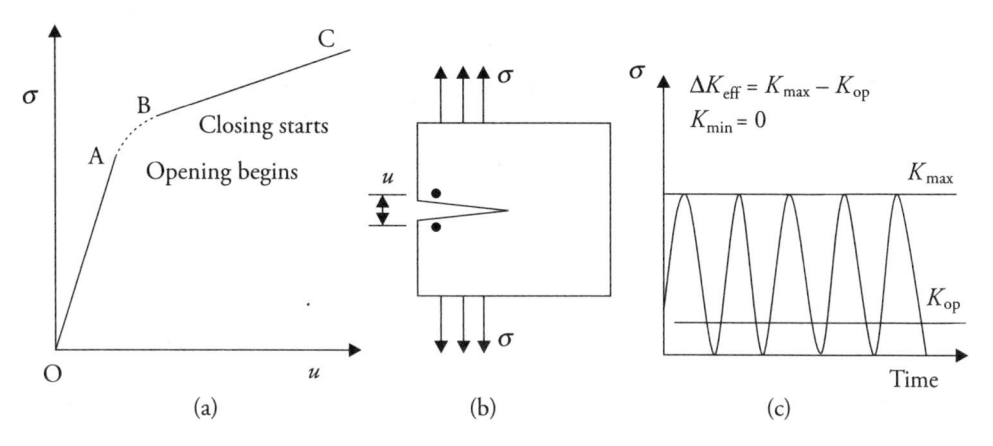

Figure 7.5 Crack closure during cyclic loading. (a) Load-displacement variation. (b) Locations for displacement monitoring. (c) Opening SIF K_{op}.

load, a large plastic zone develops at the crack-tip. Upon reversal of loading, as the load reduces from the maximum, crack flanks close early at B because the material in the neighbourhood of the crack-tip cannot regain the original size and the zone is subjected to compressive load on its boundary (Fig. 7.6(a)). With further reduction in load, the closing becomes complete at A, and the specimen displays crack-free stiffness. In the next loading cycle, crack flanks begin to open at A, and the process continues till point B. Further loading occurs with crack 'fully open'. This process gets repeated in every cycle.

An early closure of crack is induced by crack-tip plastic deformation. Sometimes the crack flanks close due to roughness arising out of zig-zag crack extension and mismatch of crests and valleys (Fig. 7.6(b)) of newly formed crack edges, or debris or metal oxides lodged in between the two flanks (Fig. 7.6(c)).

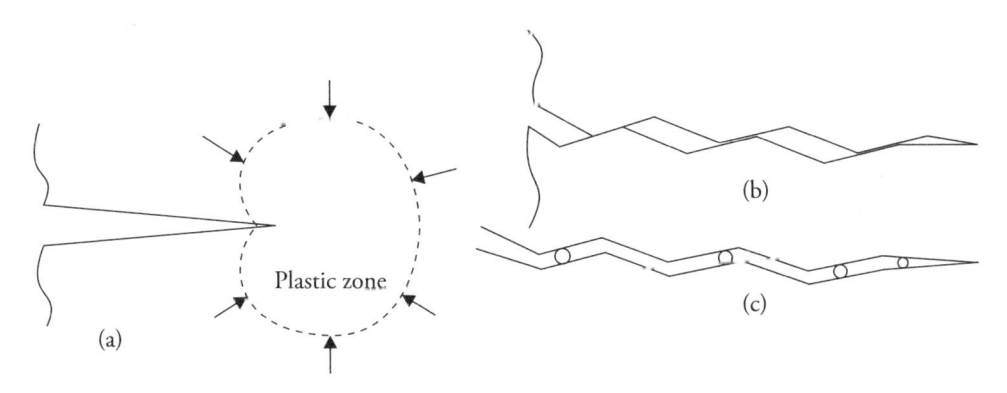

Figure 7.6 (a) Crack-tip plastic zone. (b) Roughness-induced closure. (c) Debris or metal oxide induced closure.

Elber (1970) concluded that the crack extension occurs only due to the part of the cycle from A to C. That is, the effective SIF range responsible for crack extension is $\Delta K_{eff} = K_{max} - K_{op}$. He quantified the crack growth rate in terms the effective stress intensity range ΔK_{eff} in the following form.

$$\frac{da}{dN} = C\left(\Delta K_{eff}\right)^m = C\left(U \Delta K\right)^m \tag{7.6}$$

where $U = \frac{K_{max} - K_{op}}{K_{max} - K_{min}}$. Hence the cyclic crack growth rate reduces due to crack closure. Based on his experimental data for 2024-T3 aluminium sheet material, $U = 0.5 + 0.4R$, where $R = \frac{\sigma_{min}}{\sigma_{max}}$, and R varies from -1.0 to 0.7. Schijve (1981) persented U for the same material differently: $U = 0.55 + 0.33R + 0.12R^2$ for R in the range -1.0 to 0.54.

The opening load factor U, or the opening SIF K_{op}, can be determined experimentally. Fleck, Smith and Smith (1983) give $U = 0.74$ for plane stress and 0.84 for plane strain, when ΔK is in the range 10 to 20 MPa$\sqrt{m}$, for 4360 50B structural steel with Paris law constants $C = 1.48 \times 10^{-8}$ (MPa, m units) and $m = 2.86$.

7.5 Life Estimation Using Paris Law

Knowing FCGR as per the Paris law, it is straightforward to calculate the life expectancy, provided the cyclic loading details and the starting and final crack sizes are known. The starting crack size is mostly dictated by the smallest crack size the non-destructive testing technique employed can detect. The final crack size can be user-specified or it can be taken to correspond the maximum crack size that the component can withstand at σ_{max} of the load cycles. For constant amplitude fatigue loading, the steps in life calculation are as follows.

1. Select the appropriate starting crack size a_0.
2. Determine the final crack size a_f, if it is not specified, from the following relation.
   $\sigma_{max} \sqrt{\pi a_{cr}}\, Y = K_{IC}$ or K_C, as appropriate, where Y is SIF correction factor and $a_f = a_{cr}$.
3. Select the FCGR appropriate for the material: $\frac{da}{dN} = C\left(\Delta K\right)^m$.
4. If Y remains constant over the span a_0 to a_f of crack size, total number of cycles to failure or life N_f is given by integration.

$$N_f = \int_{a_0}^{a_f} \frac{da}{C(\Delta \sigma\, Y\, \sqrt{\pi})^m a^{m/2}} = \frac{a_f^{1-m/2} - a_0^{1-m/2}}{C(\Delta \sigma\, Y\, \sqrt{\pi})^m \left(1 - \frac{m}{2}\right)} \tag{7.7}$$

Problem 7.1

An internal crack in a finite plate (Fig. 7.7) is subjected to cyclic load varying from 0 to 130 MPa. Given $h \gg b$, initial crack size $2a_0 = 10$ mm, $b = 20$ mm. Find the life up to final crack size $2a_f = 12$ mm. For the plate material, the Paris law constants $m = 3.3$ and $C = 2.56 \times 10^{-12}$ (MPa, m units).

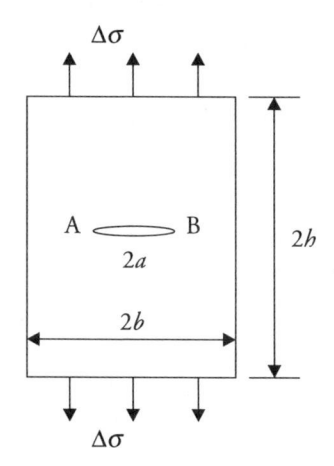

Figure 7.7 Internal crack in plate of constant thickness.

Solution

The SIF correction factor Y in $K = \sigma\sqrt{\pi a}\, Y$ can be taken as $\sqrt{\sec\dfrac{\pi a}{2b}}$. For $a = a_0$, $Y = 1.04038$. Assuming this to remain constant over the whole domain 5–6 mm, using Eq. (7.7) the cyclic life

$$N_f = \frac{a_f^{1-m/2} - a_0^{1-m/2}}{C(\Delta\sigma\, Y\, \sqrt{\pi})^m (1 - \frac{m}{2})}$$

$$= \frac{(0.006)^{1-1.65} - (0.005)^{1-1.65}}{2.56 \times 10^{-12}(130 \times 1.04038\sqrt{\pi})^{3.3}(1 - 1.65)} = 29495 \text{ cycles (Ans.)}.$$

Problem 7.2

Solve the above problem when $2a_0 = 10$ mm and $2a_f = 14$ mm. Consider Y to vary with a.

Solution

Since Y varies, it is better to calculate life through numerical integration.

$$N_f = \int_{a_0}^{a_1} \frac{da}{C(\Delta\sigma\, Y\, \sqrt{\pi})^m a^{m/2}} + \int_{a_1}^{a_2} \frac{da}{C(\Delta\sigma\, Y\, \sqrt{\pi})^m a^{m/2}} + \cdots$$

$$+ \int_{a_n}^{a_f} \frac{da}{C(\Delta\sigma\, Y\, \sqrt{\pi})^m a^{m/2}} \tag{7.8}$$

For an approximate estimate, the full range of crack growth (a_0 to a_f) can be broken into two spans, a_0 to a_1 and a_1 to a_f. ΔK can be calculated at the ends of each span. The corresponding FCGR can also be calculated for the three crack sizes. The average crack growth rate over a span serves as the basis to calculate the number of cycles for each span.

The correction factor Y corresponding to the three crack sizes 5 mm, 6 mm, and 7 mm are

$$Y = \sqrt{\sec\frac{\pi a}{2b}},\ Y_5 = 1.04038,\ Y_6 = 1.0594 \text{ and } Y_7 = 1.083$$

This gives $\Delta K_5 = \Delta\sigma \sqrt{\pi a}\, Y = 130\, \sqrt{\pi \times 0.005}\, Y_5 = 16.95$ MPa$\sqrt{\text{m}}$, $\Delta K_6 = 18.908$ MPa$\sqrt{\text{m}}$ and $\Delta K_7 = 20.878$ MPa$\sqrt{\text{m}}$.

The corresponding crack growth rate $\left(\dfrac{da}{dN}\right)_5 = C\,(\Delta K)^m = 2.914 \times 10^{-8}$ m/cycle, $\left(\dfrac{da}{dN}\right)_6 = 4.18 \times 10^{-8}$ m/cycle, and $\left(\dfrac{da}{dN}\right)_7 = 5.797 \times 10^{-8}$ m/cycle. The crack growth rate over each 1 mm span can be taken as the average of the two terminal growth rates. Therefore the total cyclic life

$$N_f = \frac{10^{-3}}{3.547 \times 10^{-8}} + \frac{10^{-3}}{4.9885 \times 10^{-8}} = 48238 \text{ cycles (Ans.).}$$

7.6 Retardation of Crack Growth Due To Overloads

In a sequence of constant amplitude cyclic loading, if an overload is applied, the cyclic FCGR gets retarded and the cyclic life extends (Wheeler 1972; Willenborg, Engle Jr. and Wood 1971; Fleck 1985; Shin and Fleck 1987; etc.). Conversely, if an underload cycle is applied, the cyclic crack growth rate gets accelerated (e.g., Fleck 1985). The case involving overloads is graphically illustrated in Fig. 7.8. Following the peak of overload applied at A, the crack growth rate drops. This retardation can be explained in terms of plasticity induced crack-tip blunting, crack closure, and residual stresses. The extent of the retardation depends on the cyclic load range $\Delta\sigma$, overload cyclic range $\Delta\sigma_o$, and the span of interval between overloads and the material.

Immediately following the application of an overload, the crack gets blunted due to large plastic deformation. Thereby, the stress concentration at the crack-tip reduces, and the rate of crack growth reduces. The other possibility is that after the overload, the closure occurs early and the crack opening SIF level increases. Thereby, the effective SIF range reduces and so does the crack growth rate. Incidentally, some researchers believe that immediately after the overload, the crack gets blunted, and it increases the crack growth rate, because crack does not close early and the effective SIF range increases. However, after a very small crack growth, the closure begins to occur early, the opening stress increases, and the effective SIF range reduces, which leads to a lower crack growth rate. The reduction in crack growth rate following the application of an overload can also be explained in terms of residual compressive stresses, which develop ahead of the crack-tip, reducing the effective tensile stresses at the crack-tip. In the same light, if an under-load cycle is applied in a sequence of constant amplitude loading, the crack growth rate will increase. These issues make the life estimation under variable amplitude loading difficult.

There are some empirical models available to estimate the crack growth rate following an overload. Wheeler (1972) proposed one such model. According to

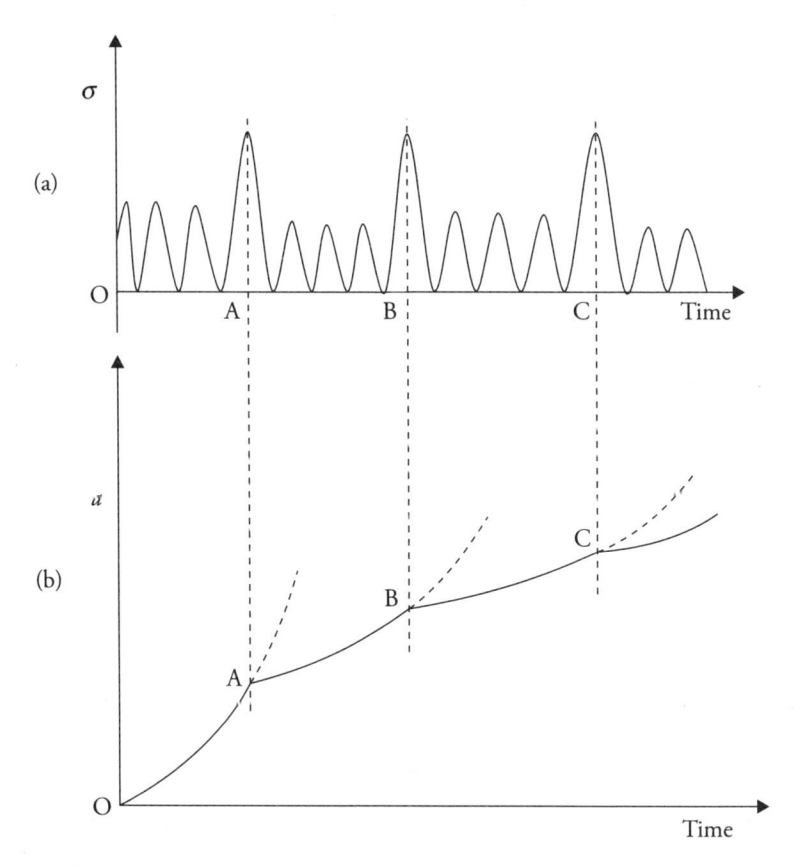

Figure 7.8 (a) Loading. (b) Retardation of crack growth immediately following application of overload.

this model, the FCGR gets retarded to the highest extent immediately following the overload cycle. The retardation effect gradually reduces as the instantaneous plastic zone approaches the overload plastic zone boundary. The retarded crack growth rate depends on two parameters: the instantaneous plastic zone radius r_i and the distance x of the instantaneous crack-tip B (Fig. 7.9(a)) from the overload plastic zone boundary. For $R = 0$,

$$r_i = \frac{\Delta K_i^2}{c\,\sigma_Y^2} \tag{7.9}$$

where $c = 6\pi$ for plane strain and 2π for plane stress. Explicitly, the retarded crack growth rate is given by

$$\left(\frac{da}{dN}\right)_{retarded} = \left(\frac{r_i}{x}\right)^p \left(\frac{da}{dN}\right)_{\Delta\sigma=constant} \tag{7.10}$$

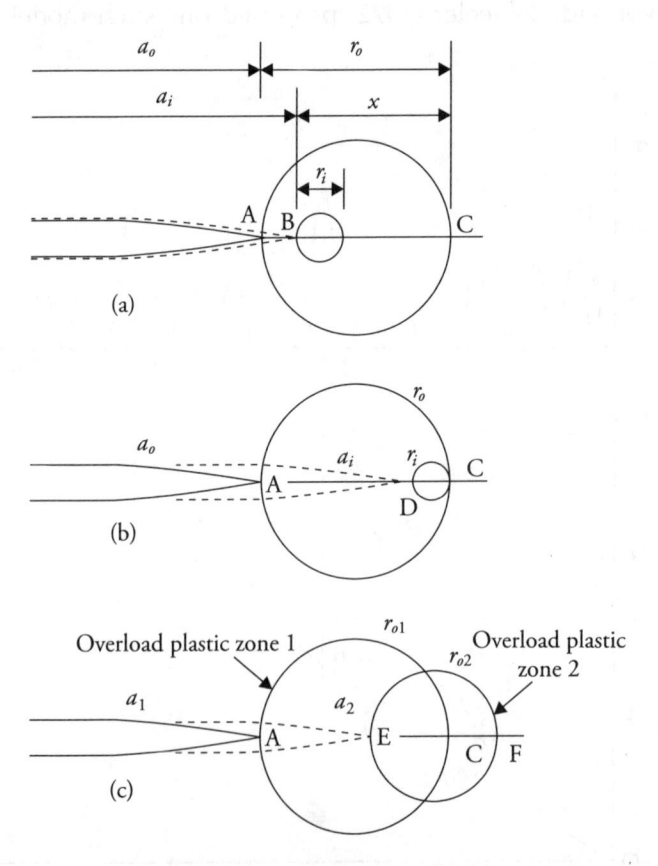

Figure 7.9 (a) Overload and instantaneous plastic zones. (b) Instantaneous plastic zone touching overload plastic zone boundary. (c) New overload plastic zone spreading beyond first overload plastic zone boundary.

where p is an empirical constant and $x = a_o + r_o - a_i$. Wheeler specified that $p = 1.43$ for D_{ac} medium carbon low alloy ultra high strength steel (with yield point 1345 MPa and ultimate limit 1572 MPa) and 3.40 for Ti–6Al–4V titanium alloy (with yield point 930 MPa and ultimate limit 970 MPa). Fleck and Smith (1984) give a value of $p = 0.75$ for BS4360 50B structural steel for which the corresponding Paris law relations are: $\frac{da}{dN} = 6.31 \times 10^{-9}(\Delta K)^{2.98}$ mm/cycle for 3 mm thick specimens and $\frac{da}{dN} = 4.98 \times 10^{-9}(\Delta K)^{3.12}$ mm/cycle for 24 mm thick specimens, when stress ratio R is in the range 0.3 to 0.5. In their relations, ΔK is in MPa$\sqrt{m}$.

When A is the instantaneous crack-tip corresponding to the application of overload, the crack growth rate is the lowest immediately afterwards. As the instantaneous crack-tip B advances from the overload crack-tip A, the FCGR gradually increases. The FCGR picks up the original $\left(\frac{da}{dN}\right)_{\Delta\sigma=\text{constant}}$, when $(a_o + r_o) \leq (a_i + r_i)$. That is to say, when the instantaneous plastic zone radius r_i touches (Fig. 7.9(b)), or spreads beyond, the overload plastic zone boundary C, the crack growth rate becomes equal to the rate corresponding to the constant amplitude cyclic loading $\Delta\sigma$.

In a particular situation, if the first overload occurs at crack length a_1 giving rise to a plastic zone size r_{o1} and before the expiry of the associated retardation effect, a second overload cycle (Fig. 7.9(c)) occurs at crack length a_2, and the corresponding overload plastic zone size r_{o2} spreads beyond the boundary of the first overload plastic zone boundary r_{o1}, fresh calculation of retarded FCGR is recommended considering the new location E of the crack-tip.

Willenborg, Engle Jr. and Wood (1971) gave an empirical model, which considered the existence of residual stresses due to the overload cycle. The residual stresses lead to a reduction of both instantaneous $K_{max,i}$ and $K_{min,i}$ in cycle i. When the crack-tip in the current cycle i is located at a_i (Figs. 7.9(a) and 10), the magnitude of residual stresses correspond to the SIF $K_{res,i} = (K_r - K_{max,i})$, where

$$\frac{K_r^2}{c\,\sigma_Y^2} = x = a_o + r_o - a_i \tag{7.11}$$

and $c = 6\pi$ for plane strain and 2π for plane stress. The effective maximum and minimum SIFs for the current cycle i are $K_{max\,e,i} = K_{max,i} - K_{res,i}$ and $K_{min\,e,i} = K_{min,i} - K_{res,i}$. The effective SIF range $\Delta K_{eff} = K_{max\,e,i} - K_{min\,e,i}$ and the retarded crack growth rate are calculated using the following Forman type relation.

$$\frac{da}{dN} = \frac{C\,(\Delta K_{eff})^m}{(1 - R_{eff})\,K_c - \Delta K_{eff}}; \quad R_{eff} = \frac{K_{min\,e,i}}{K_{max\,e,i}} \tag{7.12}$$

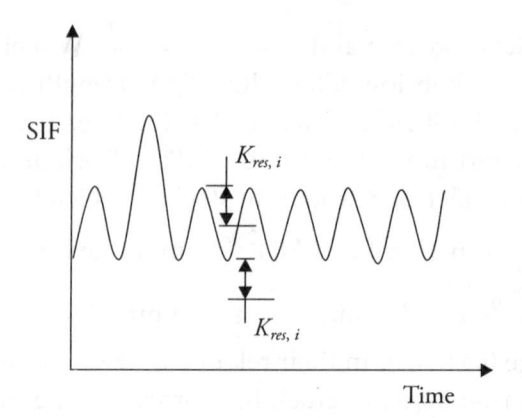

Figure 7.10 Reduction of cyclic stress intensity factors due to residual stresses after overload cycle.

These schemes are useful to calculate fatigue life of components subjected to variety of cyclic loadings with occasional overloads.

7.7 Variable Amplitude Cyclic Loading

Variable amplitude cyclic loading can be of two important varieties. In one case, the load cycles can be well defined but they can occur randomly (Fig. 7.2(a) top). Alternatively, it can be like in the case (Fig. 7.2(b)) where the cycles are not so well defined. The first type of loading will be discussed here; the other case will be discussed in the passing. In the first case, when loads of the same cyclic amplitudes are blocked together and crack growth calculations are done without any interaction effects, the extent of total crack growth can be different depending on the sequence of block-occurrences. This is illustrated in Fig. 7.11. If n_1 cycles of $\Delta\sigma_1$ act first and then n_2 cycles of $\Delta\sigma_2$ (Fig. 7.11(b) top), crack grows by OA_1 followed by A_1B_1 (Fig. 7.11(c)). On the other hand, if n_2 cycles of $\Delta\sigma_2$ act first and then n_1 cycles of $\Delta\sigma_1$ (Fig. 7.11(b) bottom), crack grows by OA_2 followed by A_2B_2 (Fig. 7.11(c)). The total growth is less in the second case. This may not always be the case. It depends on the two constant amplitude crack growth curves, n_1 and n_2.

In the case of second loading sequence, the retardation due to cyclic interactions will come into play. This will further reduce the crack growth in the second case.

If the variation of load amplitude with frequency over the whole life of a component conforms to Gaussian/normal distribution or Rayleigh type distribution (Fig. 7.12(a)), root-mean-square amplitude $\Delta\sigma_{\text{rms}}$ can be determined through the following relation.

$$\Delta\sigma_{\text{rms}} = \sqrt{\frac{\sum_{i=1}^{k} p_i \, \Delta\sigma_i^2}{\sum_{i=1}^{k} p_i}} \tag{7.13}$$

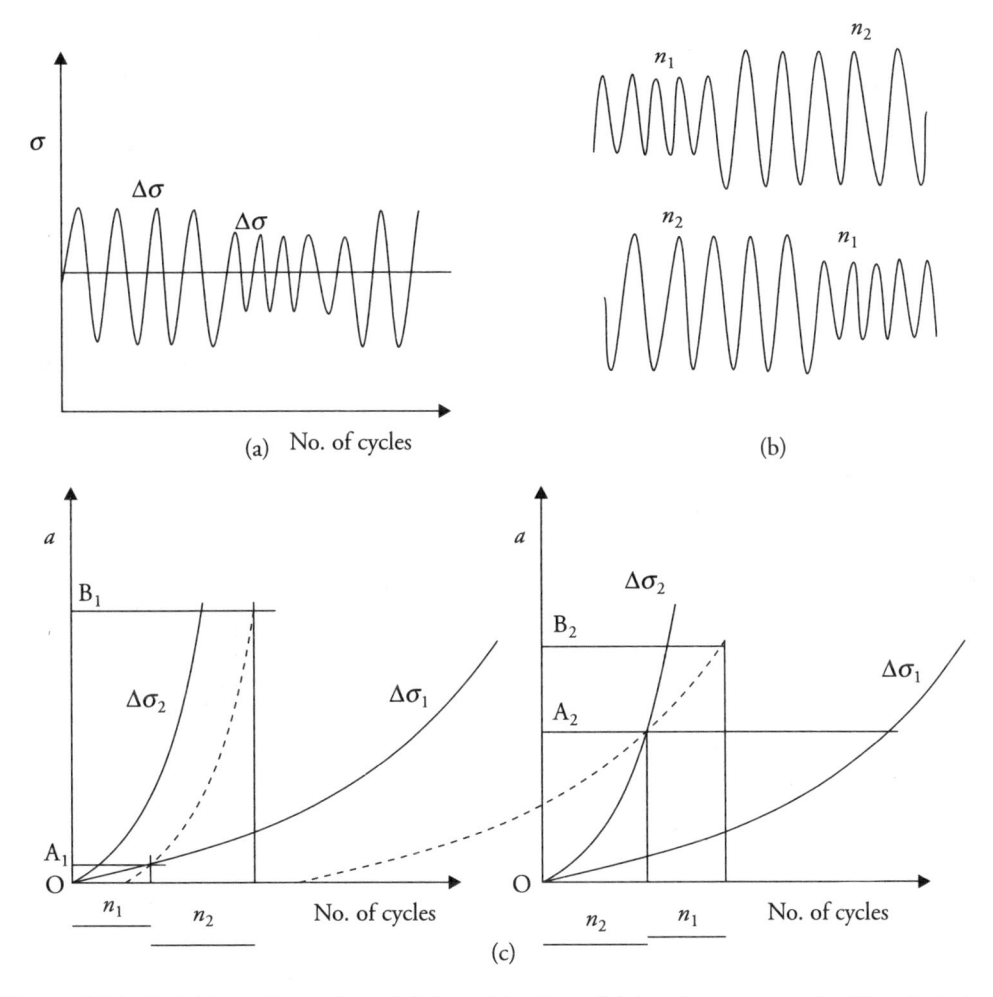

Figure 7.11 Variable cyclic loading. (a) Actual loading. (b) Load sequencing for life calculation. (c) Crack growth due to two loading sequences.

where p_i is the number of cycles, or % occurrences, of amplitude $\Delta\sigma_i$ over the span of life (Barsom and Rolfe 1999). The crack growth life can be calculated using the constant amplitude fatigue loading data and using stress range $\Delta\sigma = \Delta\sigma_{\text{rms}}$.

If the load amplitude occurs repeatedly in blocks, and in each block the cyclic amplitude increases with time (Fig. 7.12(b)), or decreases with time (Fig. 7.12(c)), or increases first and then decreases with time (Fig. 7.12(d)), the life calculation in a block can be done using the root-mean-square approach as in the earlier case (Barsom and Rolfe 1999). In situations of the type shown in Figs. 7.12(c) and 7.12(d), the difference with experimental observations may come up because the crack growth is significantly affected by the cyclic interactions or retardation/acceleration phenomenon. To improve the position, retardation-based calculation can be done.

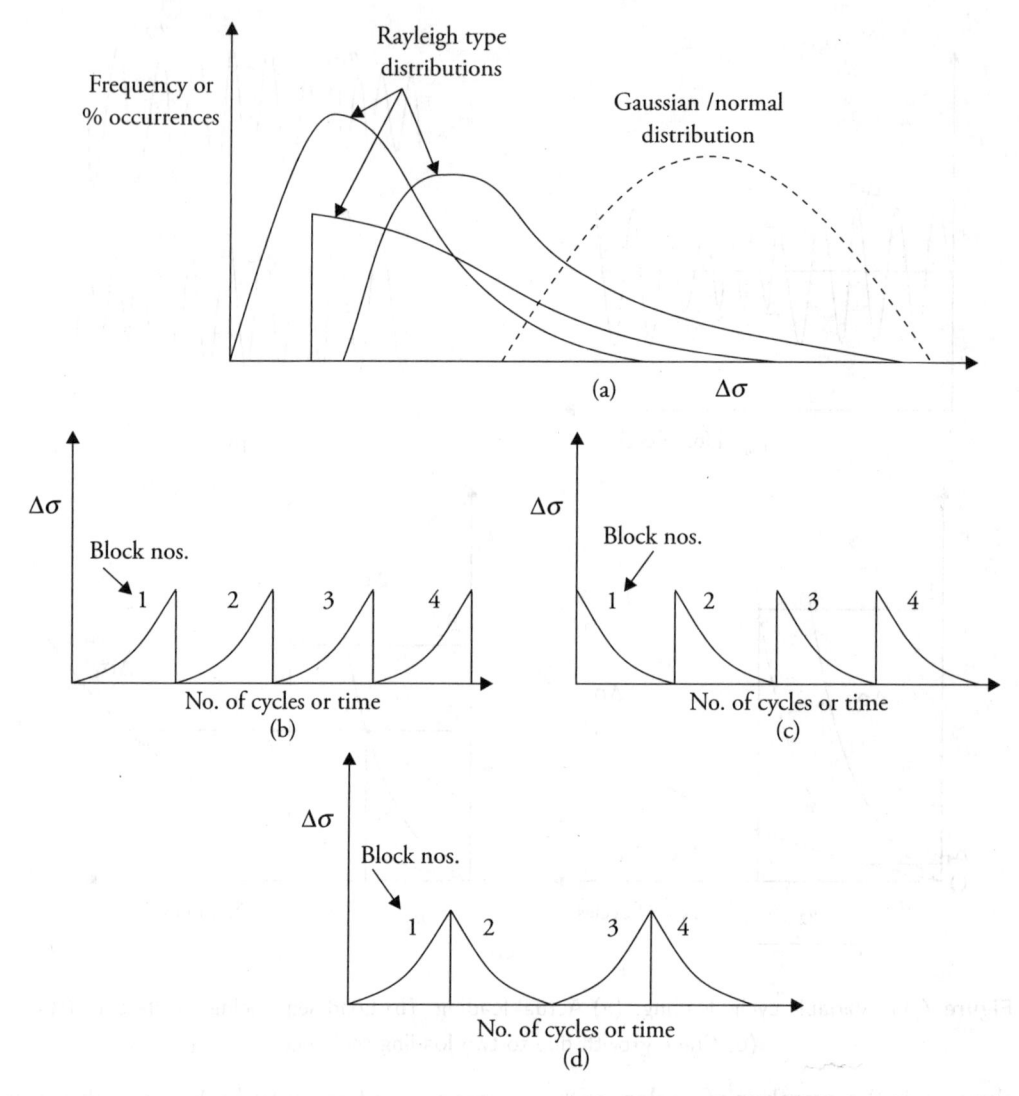

Figure 7.12 Rayleigh-type distribution of range of loading. (a) Frequency variation with stress range. (b) Load range increasing in a block. (c) Load range decreasing in a block. (d) Increasing–decreasing load range in a block.

Problem 7.3 (example on effect of load sequence)
For a steel, FCGR data is: $C = 8.2762 \times 10^{-12}$ (in MPa, m units) and $m = 2.55$. Calculate the final crack size when the starting crack size is 0.010 m and the following cyclic load amplitudes are applied.

1. $\Delta\sigma_1 = 300$ MPa $n_1 = 2000$ cycles followed by $\Delta\sigma_2 = 600$ MPa $n_2 = 1000$ cycles.

2. $\Delta\sigma_1 = 600$ MPa $n_1 = 1000$ cycles followed by $\Delta\sigma_2 = 300$ MPa $n_2 = 2000$ cycles.

The SIF correction factor can be assumed as unity.

Solution

Approximate method

(i) Assuming crack grows at the same rate over 2000 cycles and the rate corresponds to the initial crack size $a_i = 0.010$ m, the final crack size $a_{f1} = 0.010416$ m.

Similarly, assuming that the crack grows in the same fashion over the next 1000 cycles, the final crack size $a_f = 0.011702$ m (Ans.).

(ii) For the second loading, similar calculations give $a_{f1} = 0.0112189$ m and $a_f = 0.0117065$ m.

Thus, the two sequences of loading give rise to different final crack sizes (Ans.).

More accurate method

(i) If the crack growth rate is amended after every cycle of loading, at the end of 2000 cycles of loading of amplitude 300 MPa, the crack attains size $a_{f1} = 0.0104275$ m. This can be easily done by writing a small MATLAB program. A sample FORTRAN program is added in Appendix 7.1 at the end of the chapter and has been used here. If the same procedure is followed for the next 1000 cycles of amplitude 600 MPa starting from size a_{f1}, the final crack size is 0.0118211 m (Ans.).

(ii) For the second sequence of loading, the similar approach gives rise to $a_{f1} = 0.0113198$ m and $a_f = 0.0118211$ m. The final crack size is not affected by the sequence in this approach.

Problem 7.4 (Example on solution by RMS stress range approach)

An edge crack of size $a_0 = 6$ mm in a plate of width $w = 120$ mm was subjected to the following variable load cycles.

Stress range (MPa)	No. of cycles
50	10, 000
100	20, 000
150	30, 000
120	40, 000
40	50, 000

The plate is made of ferrite steel with Paris law constants $C = 6.8 \times 10^{-12}$ (in MPa, m units) and $m = 3.0$. The SIF correction factor Y in $K_I = \sigma\sqrt{\pi a}\, Y$ can be calculated through

$$Y = 1.12 - 0.23r + 10.55r^2 - 21.72r^3 + 30.39r^4, \ 0 < r < 0.6, \ r = \frac{a}{w}.$$

Calculate the final crack size.

Solution

The stress range $\Delta\sigma = \Delta\sigma_{rms} = \sqrt{\dfrac{\sum_{i=1}^{k} p_i \, \Delta\sigma_i^2}{\sum_{i=1}^{k} p_i}} = 101.85$ MPa and total number of cycles is 1, 50, 000.

Calculating the crack growth with continuous amendment of crack size a and ratio $r = \dfrac{a}{w}$ through a small MATLAB program, the final crack size $a_{fb} = 14.744$ mm. If the correction factor Y is taken as 1.12 throughout, the final crack size can be calculated through Eq. (7.7). This gives $a_f = 13.227$ mm.

Problem 7.5

Solve the above problem when $a_0 = 12$ mm and the loading details are as follows.

Stress range (MPa)	No. of cycles
500	10
1000	20
1500	30
1200	40
400	50

Solution

The stress range $\Delta\sigma = \Delta\sigma_{rms} = 1018.5$ MPa, and the final crack size a_f through Eq. (7.7) is 41.40 mm. Through cycle-by-cycle integration using the MATLAB program $a_f = 40.99$ mm assuming $Y = 1.12$ throughout.

Problems involving retardation of crack growth due to overloading

Problem 7.6

There is an initial edge crack of size 8 mm in a plate (width $w = 150$ mm) of Ti–6Al–4V alloy. The material data is as follows: Yield point = 930 MPa, ultimate limit = 970 MPa, the Paris law constants $C = 3.5306 \times 10^{-11}$ (in MPa, m units), and $m = 2.80$, Wheeler retardation constant $p = 3.40$. The plate is in a state of plane stress.

(i) It is subjected to a constant amplitude fatigue loading $\Delta\sigma = 60$ MPa for 1, 00, 000 cycles. Calculate the final crack size.

(ii) After the first 50, 000 cycles of $\Delta\sigma = 60$ MPa, an overload of amplitude 100 MPa is applied. This is followed by the next 50, 000 cycles of the constant amplitude $\Delta\sigma = 60$ MPa. Find out the final crack size. The SIF correction factor can be calculated through

$$Y = 1.12 - 0.23r + 10.55r^2 - 21.72\,r^3 + 30.39\,r^4, \quad r = \frac{a}{w}, \quad 0 < r < 0.6.$$

Solution

(i) Starting from initial crack size $a_i = 8$ mm, the SIF range with the inclusion of the correction factor Y and the cyclic crack growth can be calculated for each cycle. Thereby, crack size at the end of the cycle can be calculated. The process can be repeated for the full span of $1,00,000$ cycles and the final crack size can be obtained. The MATLAB program of Appendix 7.1 is employed to do the calculation. The final crack size a_f is obtained as 11.703 mm. It is noted that the crack size after the first $50,000$ cycles is 9.5822 mm.

(ii) In this case, the growth over the first $50,000$ cycles of amplitude 60 MPa can be calculated in the manner as in the earlier case. The final crack size $a_{f1} = 9.5822$ mm. Since the next cycle is of higher amplitude, the growth of the crack can be determined following the similar procedure. The final crack size after the 50001th cycle is $af_2 = 9.5823$ mm. During the next stages, the cyclic crack growth gets retarded. The retardation factor can be calculated by applying the Wheeler's method. The retardation factor is given by $\left(\frac{r_i}{x}\right)^p$, where r_i is instantaneous plastic zone size, $x = a_o + r_o - a_i$, $a_o = a_{f2}$, and r_o (Fig. 7.9) is the plastic zone size due to the SIF corresponding to the amplitude $\Delta\sigma = 100$ MPa. The instantaneous plastic zone size $r_i = \frac{(\Delta K_i)^2}{2\pi\sigma_Y^2}$. While doing this stage of calculation, crack size can be amended at the end of each cycle. This procedure continues till $(a_o + r_o) = (a_i + r_i)$. This type of growth occurs over only 15,256 cycles, and the crack attains a length $a_{f3} = 9.6285$ mm at the end of 65,257 cycles. After that, crack grows over the remaining cycles without any retardation and it picks up the final size $a_{f4} = 11.0407$ mm after $1,00,001$ cycles. Hence, the total crack growth is 3.0407 mm. It can be noticed that the overload has led to a reduction in the final crack size, for example, from 11.703 mm to 11.0407 mm.

Problem 7.7

In the earlier problem, data is changed as follows. (i) 50 MPa for 50,000 cycles, followed by an overload cycle of 120 MPa and constant amplitude loading of 50 MPa for 50,000 cycles; starting crack size is 10 mm and plate width is 180 mm. (ii) 60 MPa for 50,000 cycles, followed by an overload cycle of 90 MPa and then a constant amplitude loading of 60 MPa for 50,000 cycles; starting crack size is 10 mm and plate width is 180 mm.

Calculate changes in the crack size at the different stages. Assume plane stress.

Solution

(i) The method of calculations remains the same and has been done using the program given in Appendix 7.1. The crack lengths at the end of different stages are now obtained as follows. After first 50, 000 cycles, $a_{f1} = 11.242$ mm and after the overload cycle, $a_{f2} = 11.242$ mm. The growth during the overload cycle is less than 10^{-3} mm. The crack growth is highly retarded subsequently, and it remains the same till the end of 1, 00, 001 cycles.

(ii) In this case, after the first 50, 000 cycles, $a_{f1} = 12.204$ mm and after the overload cycle $a_{f2} = 12.205$ mm. The crack grows with retardation effect over the next 5035 cycles and grows to size $a_{f3} = 12.246$ mm. That is, crack size after 55, 036 cycles is 12.246 mm. After 1, 00, 001 cycles, it grows to the final size $a_{f4} = 14.968$ mm.

In Problem 7.7(ii), if the state of stress conforms to plane strain, after the first 50, 000 cycles, $a_{f1} = 12.204$ mm and after the overload cycle, $a_{f2} = 12.205$ mm. The crack grows with retardation effect over the next 1687 cycles and grows to size $a_{f3} = 12.219$ mm. That is, crack size after 51, 688 cycles is 12.219 mm. After 1, 00, 001 cycles, it grows to the final size $a_{f4} = 15.17$ mm. In this case, the number of retardation cycles reduces, but the final crack size increases slightly.

Problem 7.8

In a plate (width $w = 180$ mm) of Ti–6Al–4V alloy, there is an initial edge crack of size 10 mm. The material data is the same as given earlier. (i) It is subjected to constant amplitude loading of 60 MPa for 50, 000 cycles, followed by an overload cycle of 100 MPa and then a constant amplitude loading of 60 MPa for 50, 000 cycles. (ii) It is subjected to constant amplitude loading of 60 MPa for 50, 000 cycles, followed by an overload cycle of 100 MPa and then constant amplitude loading of 60 MPa for 4000 cycles. This is followed by the second overload cycle of 90 MPa, and then constant amplitude loading of 60 MPa for 46, 000 cycles. Assume plane strain condition and employ material data of Problem 7.6. The SIF correction factor can be calculated through

$$Y = 1.12 - 0.23r + 10.55r^2 - 21.72\, r^3 + 30.39\, r^4, \quad r = \frac{a}{w}, \quad 0 < r < 0.6.$$

Calculate the crack sizes at different stages.

Solution

(i) The solution procedure remains the same and results are obtained using the program. However, the plastic zone size is calculated using the relation $r_p = \dfrac{(\Delta K)^2}{6\pi\sigma_Y^2}$.

Upon solving, it is observed that after the first 50, 000 cycles, $a_{f1} = 12.2048$ mm and after the overload cycle, $a_{f2} = 12.2050$ mm. The crack grows with retardation effect over the next 4368 cycles and grows to a size $a_{f3} = 12.2248$ mm. After 1, 00, 001 cycles, it grows to the final size $a_{f4} = 14.9854$ mm. Comparing the results for this case with the same for solved problem 7.7(ii), it can be seen that crack growth is accelerated in this case.

(ii) The crack growth calculations can be done along the similar lines as in the case of part (i). After the first 50, 000 cycles, crack attains a size $a_{f1} = 12.2048$ mm. At the end of the first overload cycle, it grows to size $a_{f2} = 12.2050$ mm. Over the next 4000 cycles, crack growth gets retarded. At the end of 54, 001 cycles, it grows to size $a_{f3} = 12.2179$ mm. After the second overload cycle, or 54, 002 cycles, it picks up a length $a_{f4} = 12.2180$ mm. Over the next 1685 cycles, the crack growth gets retarded and it grows to size $a_{f5} = 12.232$ mm. That is, after 55, 687 cycles crack attains a size 12.232 mm. Finally, at the end of 1, 00, 002 cycles, its size $a_{f6} = 14.9020$ mm.

Problem 7.9

A plate 25 mm thick and width $w = 250$ mm made of structural steel BS4360 50B has an initial edge crack of size $a_0 = 10$ mm. It is subjected to a constant amplitude fatigue cycling with stress range $\Delta\sigma = 80$ MPa for 7, 50, 000 cycles. Determine the final crack size using the Paris law constants, $C = 6.31 \times 10^{-12}$ (MPa, m units) and $m = 2.98$, and considering the crack closure effect to be given by the factor $U = 0.84$. The plate is in a state of plane strain.

Solution

Assuming the SIF correction factor to remain constant and given by $Y = 1.12$, Eq. (7.7) can be written for the present case as follows.

$$N_f = \frac{a_f^{1-m/2} - a_0^{1-m/2}}{C(U\Delta\sigma\, Y \sqrt{\pi})^m (1 - \frac{m}{2})} \tag{7.14}$$

Substituting the data

$$75 \times 10^4 = \frac{a_f^{1-2.98/2} - (0.010)^{1-2.98/2}}{6.31 \times 10^{-12} (0.84 \times 80 \times 1.12\sqrt{\pi})^{2.98} (1 - \frac{2.98}{2})}.$$

After solving the above relation, the final crack size $a_f = 45.24$ mm.

7.7.1 Rainflow Cycle Counting

When the load records do not show clearly the load cycles as in the case of random cyclic loading, the counting methods as per, for example, American Society for Testing Materials (ASTM E 1049-85 2005), the rainflow counting method (Endo et al. 1974; Downing and Socie 1982), and so on, are very useful to reduce a record of load/stress spectrum into a set of simple stress reversals. In the rainflow method, the component loading record is placed vertically with the time axis pointing downward (Fig. 7.13(a)). The load record looks like the Japanese pagoda. The rainflow technique is applied to calculate each half load cycle. By combining equal, or nearly equal, size half cycles, the complete load cycles are identified. The rules to be followed in identifying each half cycle are as follows.

1. It is necessary to consider water flows downwards starting from each minimum. The half cycles are counted spanning over starting and termination. The termination occurs when either of the following takes place.
   (i) It reaches the end of time history, for example, points I and K.
   (ii) If it comes across a minimum of greater magnitude than its origin, for example, points C and E.
   (iii) It merges with flow that started at an earlier level, for example, flow from H to I.
2. Similar situation is considered with flow starting from each maximum as a source. Steps (i) to (iii) are again followed to count half cycles.
3. Next, half cycles of equal or very nearly equal magnitude but opposite sign are combined to count complete load cycles, for example, A to F and F to I, C to B and B to C, G to H and H to G, and so on. A few half cycles may be left out, for example, I to J and J to K; these two can be paired to form a cycle.

Alternatively, considering the cyclic loading in the sequence in which it is applied on a component and the component material to display kinematic plasticity, it is possible to count the loading cycles as shown in Fig. 7.13(b). This method is unmanageable when the load spectrum records are quite voluminous. The rainflow counting is then the easier way out.

The component life can be calculated through integration of cycle-by-cycle crack growth neglecting cyclic interactions. To account for cyclic interactions, various methods are proposed (Socie 1977; Schijve 1980; Kikukawa, Jono and Kondo 1981; Broek 1984; Fleck and Smith 1984).

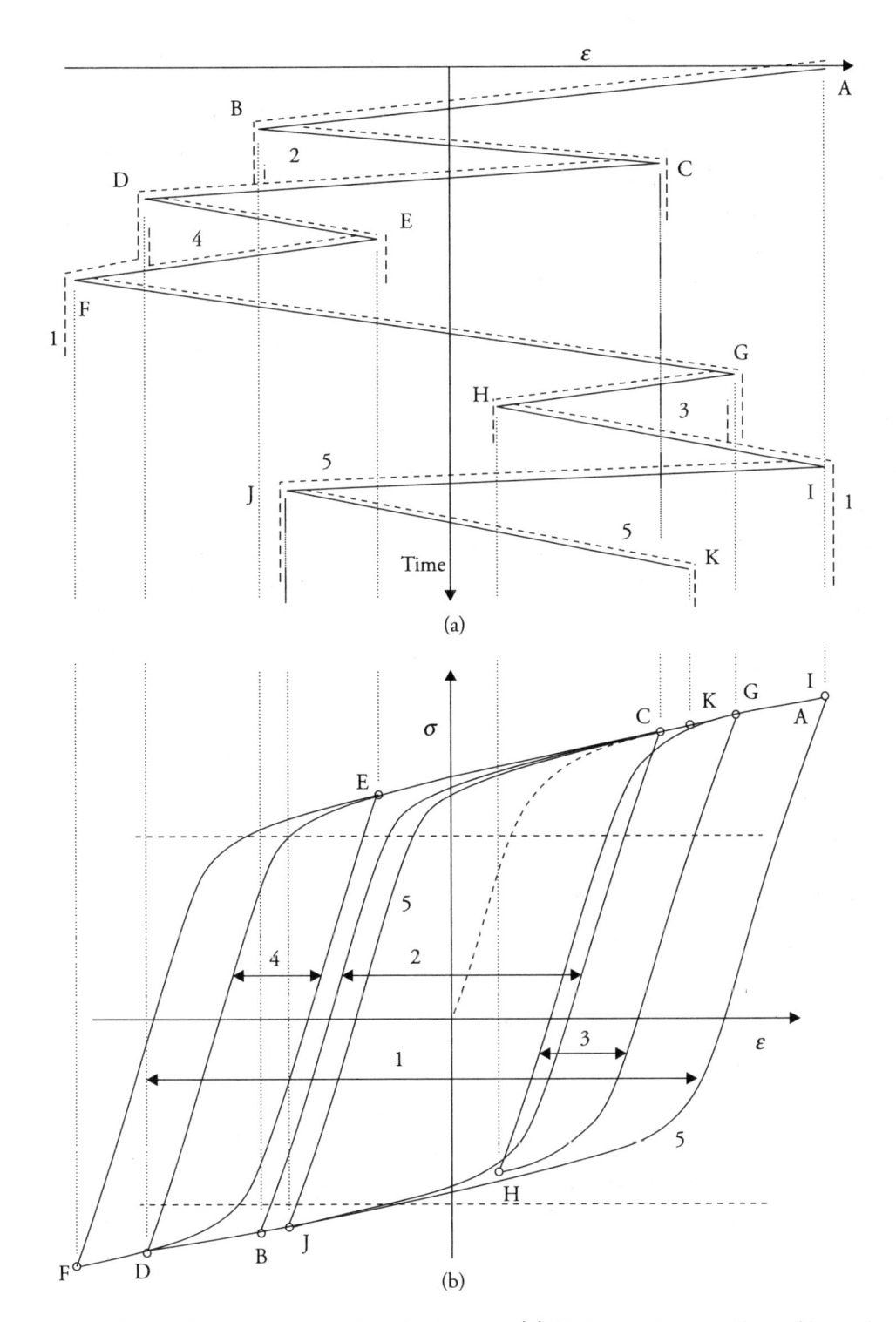

Figure 7.13 (a) Rainflow counting from load record. (b) Cycle counting considering kinematic hardening material properties.

7.8 Closure

Fatigue crack growth encompasses a very vast spectrum. The coverage in this chapter has been centred mainly on the Paris law and linear elastic fracture mechanics (LEFM). There is no coverage of short cracks, low cycle fatigue, fatigue interactions with other subcritical crack growth phenomena like corrosion, creep, and so on. Readers may find the references (Smith 1983; Tanaka and Nakai 1983; Suresh and Ritchie 1984) very useful in the case of short cracks. Reference by Suresh(1998) will be useful in many of these areas. Reference by Larsson (1983) deals with the initial developments concerning stress corrosion and creep.

APPENDIX 7.1

Fortran Program for Crack Growth Calculations

The program, FCGRETARD, listed below can be used for crack growth calculations both with and without the inclusion of any retardation, or cycle-to-cycle interaction, effects. At the end of the listing, a sample dataset for a case study with the interaction effects is included. To carry out the calculations without any interaction effects, replace 1 by 0 in the third line of the dataset.

```
PROGRAM FCGRETARD
C   Date: 25 Jun 2014.
C   Fatigue crack growth calculations.
C   For variable amplitude loading with overloads, retardation
C   calculation according to Wheeler model is included. This program
C   can also be used for calculations without any cyclic interactions.
    DIMENSION DELSIG (200), NFREQ (200), IOLINDX (50)
C   Units for input data: stress and stress range in MPa, crack size in m,
C   Paris law constant in (MPa, m) units,
C   DELSIG = Delsigma = Stress range, NFREQ = No. of cycles, IOLINDX =
C   Index for overload (0 for normal cycles; 1 for overload cycles).
    PI=3.14159265
    OPEN (UNIT = 1, FILE = 'TAPE1', STATUS = 'OLD')
    OPEN (UNIT = 2, FILE = 'TAPE2', STATUS ='NEW')
```

```
      OPEN (UNIT = 5, FILE = 'TAPE5', STATUS = 'NEW')
C   NAMPL = Number of cases of stress range.
C   IPLST = Index for plane stress and plane strain, 0 for
C   plane stress and 1 for plane strain.
      READ (1, *) NAMPL, IPLST
      WRITE (2, *) 'NAMPL, IPLST', NAMPL, IPLST
      READ (1, *) (DELSIG(IA), NFREQ(IA), IOLINDX(IA), IA = 1, NAMPL)
      DO 90 IA = 1, NAMPL
      WRITE (2, *) DELSIG(IA), NFREQ(IA), IOLINDX(IA)
90    CONTINUE
      WRITE (*, *)'GIVE Paris law C & m, initial crack size a0, width w,
1   Yield point, Wheeler p, No.of stress ranges and AFD'
C   AFD Specifies final crack size up to which cycle life is are required.
C   Give w as 1.0E10 for using YC = 1.0 ————
C   Next line lists all input variables: C, m, a0, w, Y.P.,
C   number of stress range in the problem, specified crack
C   size, if any.
      READ (1, *) CP, AM, AI0, W, SIGY, AP, NC, AFD
      WRITE (2, *) 'C, m, ai, w, sigy, p, no.of ampl. cycles'
      WRITE (2, *) CP, AM, AI0, W, SIGY, AP, NC, AFD
C   For AFD = 0.0, DIFF is arbitrarily set at a very high level.
      IF (AFD.EQ.0.0) DIFF = 10000.0
      AOL = 0.0
      ROL = 0.0
      XO = 0.0
      XI = 0.0
      AI = AI0
```

```
        DO 60 I = 1, NC
        NFREQT = NFREQ(I)
C   Crack extension calculation begins.
180     LC=NFREQT
        WRITE (2, *) 'AF, AFD, NC', AF, AFD, NC
        DO 170 JO = 1, LC
C   SIF correction factor YC=1.0, when w = 1.0E10 is given as input.
        IF (W.EQ.1.0E10) YC = 1.0
        IF (W.NE.1.0E10) THEN
        R = AI/W
        YC =4 1.12-0.23*R + 10.55*R**2−21.72*R**3 + 30.39*R**4
        ENDIF
        SIFI = DELSIG(I)*SQRT(PI*AI)*YC
        IF (IPLST.EQ.1) RI = SIFI**2/(6.0*PI*SIGY**2)
        IF (IPLST.EQ.0) RI = SIFI**2/(2.0*PI*SIGY**2)
C       IF (NCHEK.EQ.1.AND.XC.LE.0.0) THEN
        IF (AOL.EQ.0.0) RAT = 1.0
        IF (AOL.GT.0.0) THEN
        XO = AOL + ROL
        XI = AI + RI
        XB = XO−AI
        IF (XO.GT.XI) RAT = (RI/(XO − AI))
        IF (XO.LE.XI) RAT = 1.0
        WRITE (5, *) 'Cycle No., RAT, ai, aol, ri, rol =',
       1 JO, RAT AI, AOL, RI, ROL
        ENDIF
        AINC = CP*(SIFI)**AM*(RAT)**AP
```

```
C       IF (JO.EQ.1) WRITE (*, *) 'I, AINC'
C       IF (JO.LE.15) WRITE (*, *) I, SIFMX, SIFMN, AINC
        AF=AI+AINC
        IF (JO.EQ.1) WRITE (2, *) 'I, JO, SIFI, RAT, AINC, AI, AF'
        IF (JO.EQ.1) WRITE (2, *) I, JO, SIFI, RAT, AINC, AI, AF
C
        IF (AFD.GT.0.0.AND.RAT.LE.0.9999)
      1 WRITE (2, *)I, JO, SIFI, RAT, AINC, AI, AF
        IF (AFD.LE.0.0.AND.RAT.LE.0.9999)
      1 WRITE (2, *)I, JO, SIFI, RAT, AINC, AI, AF
C       IF (AF.GE.AFD) WRITE (2, *) 'JO, AI, AF', JO, AI, AF
        AI = AF
        IF (AFD.GT.0.0) THEN
C       IF (AF.GT.AFD) WRITE (2, *) 'I, JO, AF, AFD', I, JO, AF, AFD
        DIFF = ABS(AFD-AF)
        IF (DIFF.LE.10.0E-6) GO TO 190
        ENDIF
        IF (IOLINDX(I).EQ.1) THEN
        AOL = AF
        ROL = RI
        XO = AOL + ROL
        ENDIF
C       WRITE (2, *) JO, AI, AF, AFD
170     CONTINUE
190     CONTINUE
        WRITE (2, *) I, JO, SIFI, RAT, AINC, AI, AF
        IF (DIFF.LE.10.0E-6) STOP
```

```
C
60      CONTINUE
40      FORMAT (1X, 8(1X, E12.5))
        CLOSE(2, STATUS = 'KEEP')
        CLOSE(5, STATUS = 'DELETE')
        STOP
        END
C —— SAMPLE DATASET ——————————————————————
C   A typical dataset with one overload cycle is given below.
C   There are three stress ranges 60, 120 and 60MPa. The total
C   number of cycles in the three cases are 50000, 1 and 50000 respectively.
C   Last line gives values of C, m, initial crack size, plate width, yield point,
C   Wheeler retardation constant p, number of cases of stress ranges/amplitudes
C   and $a_{fd}$, the crack size for which the printing of number of cycles is needed.
        3 0
        60.0 50000 0
        120.0 1 1
        60.0 50000 0
        3.5306E-11 2.80 0.010 0.180 930.0 3.40 3 0.000
```

Excercise

7.1 Convert Paris law constants C and m given in ksi$\sqrt{\text{in}}$-in units to MPa$\sqrt{\text{m}}$-m units for the following two materials. (i) D_{ac} steel: $C = 0.0022 \times 10^{-6}$ (ksi$\sqrt{\text{in}}$, in units) and $m = 2.55$. (ii) Ti–6Al–4V titanium alloy: 0.00181×10^{-6} (ksi$\sqrt{\text{in}}$, in units) and $m = 2.8$.

[Ans. (i) 4.3937×10^{-11} (MPa$\sqrt{\text{m}}$, m units) and 2.55. (ii) 3.5306×10^{-11} (MPa$\sqrt{\text{m}}$, m units) and 2.8].

7.2 Is there any dependence of $\dfrac{da}{dN}$ on R. How can you accommodate it?

7.3 What is the difference between opening and closing SIFs? Which is important in relation to the Elber crack growth model?

7.4 What is the reason for crack closure?

7.5 Is the cyclic sequence important in fatigue crack growth? If yes, why is it so?

7.6 If the cyclic amplitude distribution does not follow the standard Gaussian distribution, how do you plan to calculate the cyclic crack growth rate?

7.7 Why does fatigue crack growth rate reduce following an overload cycle?

7.8 Give some suggestions for extending the life of a component subjected to fatigue loading.

7.9 In a plate of D6$_{ac}$ steel, there is a starting crack of size 12 mm. Assuming the SIF correction factor as 1.0, calculate the crack sizes in the following two cases. Use material data of Q.7.1. Assume no cycle-to-cycle interaction.
(i) $\Delta\sigma_1 = 250$ MPa $n_1 = 1000$ cycles followed by $\Delta\sigma_2 = 500$ MPa $n_2 = 1000$ cycles.
(ii) $\Delta\sigma_1 = 500$ MPa $n_1 = 1000$ cycles followed by $\Delta\sigma_2 = 250$ MPa $n_2 = 1000$ cycles.
[Ans. Solution through cycle-to-cycle integration: (i) $a_{f1} = 0.012918$ m, $a_f = 0.0205549$ m. (ii) $a_{f1} = 0.018905$ m, $a_f = 0.020555$ m].

7.10 A plate (width $w = 150$ mm) is made of Ti–6Al–4V alloy. The material data are as follows: Yield point = 930 MPa, ultimate limit = 970 MPa, and the Paris law constants $C = 3.5306 \times 10^{-11}$ (in MPa, m units) and $m = 2.80$, Wheeler retardation constant $p = 3.4$. Assume plane strain.
(i) Find out the final crack size after 10, 00, 000 cycles when it is subjected to a constant amplitude fatigue loading $\Delta\sigma = 30$ MPa and the starting edge crack size is 10 mm.
(ii) Find the number of cycles when the starting and final edge crack sizes are 15 mm and 25 mm, respectively, and the cyclic load amplitude $\Delta\sigma = 35$ MPa.
(iii) Find the number of cycles when the starting and final edge crack sizes are 25 mm and 40 mm, respectively, and the cyclic load amplitude $\Delta\sigma = 55$ MPa. What difference is likely to be observed if the state of stress changes to plane stress?
[Hint: Using the MATLAB program given in Appendix 7.1, these problems can be solved. Ans. (i) 19.8048 mm. (ii) 3, 76, 820 cycles. (iii) 56, 334 cycles. No change in number of cycles if the same Paris law constants are considered to be valid.]

7.11 Solve problem Q.7.10(i) when an overload cycle of 60 MPa is applied after the first 400, 000 cycles. Assume a plane state of strain.
[Ans. Final crack size 12.67149 mm after 10,00,001 cycles]

7.12 In a plate (width $w = 220$ mm) of D6$_{ac}$ steel, there is an initial edge crack of size 10 mm. The material data are as follows: Yield point = 1345 MPa, ultimate limit = 1572 MPa, the Paris law constants $C = 4.3937 \times 10^{-11}$ (in MPa, m units) and $m = 2.55$, and Wheeler retardation constant $p = 1.43$. Assume plane strain.

The plate is subjected to a constant amplitude fatigue loading $\Delta\sigma = 120$ MPa for 10,000 cycles. This is followed by an overload cycle of amplitude 360 MPa. Subsequently, a constant amplitude loading of 220 MPa is applied for 12,000 cycles. Find the crack sizes at different stages.

[Ans. Crack sizes are 11.5935 mm after 10,000 cycles, 11.5964 mm after 10,001 cycles, 11.7054 mm after 10,321 cycles, and 30.1329 mm after 22,001 cycles.]

7.13 A three-point bend specimen (Fig. Q.7.13) with an edge crack is made of Ti–6Al–4V alloy and has the following dimensions: starting crack size $a_0 = 14$ mm, specimen depth $w = 30$ mm, support span $L = 120$ mm, and specimen thickness $B = 15$ mm. It is subjected to a cyclic load varying from 1000 N to 3500 N for 1,00,000 cycles. Find the final crack size. Given $C = 3.5306 \times 10^{-11}$ (MPa$\sqrt{m}$, m units) and $m = 2.8$.

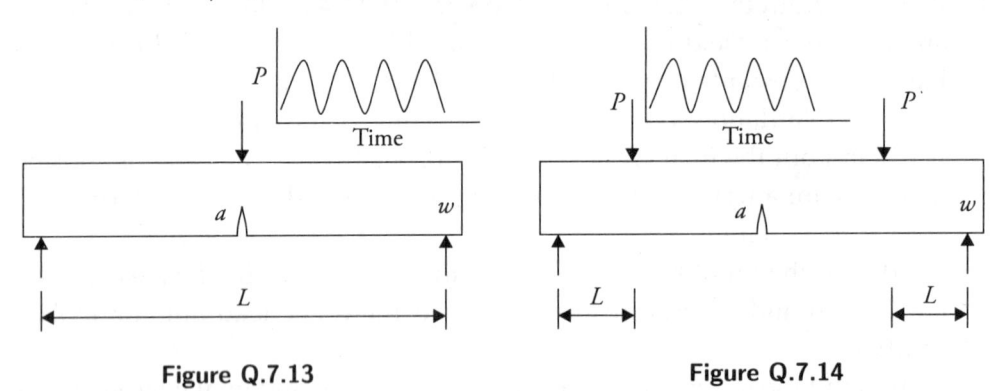

Figure Q.7.13

Figure Q.7.14

[*Hint*: Use the appropriate SIF correction factor and update crack size after every cycle. Ans. Final crack size 16.5333 mm.]

7.14 A four-point bend specimen (Fig. Q.7.14) with an edge crack is made of Ti–6Al–4V alloy and has the following dimensions: starting crack size $a_0 = 10$ mm, specimen depth $w = 30$ mm, load-support span $L = 100$ mm, and specimen thickness $B = 15$ mm. It is subjected to a cyclic load varying from 400 N to 1400 N for 1,00,000 cycles. Find the final crack size. Use properties data of Q.7.10. Assume plane strain.   [Ans. Final crack size 12.2236 mm.]

7.15 Solve problem Q.7.14 when the material is D6$_{ac}$ steel with properties data: yield point = 1345 MPa, ultimate limit = 1572 MPa, the Paris law constants

$C = 4.3937 \times 10^{-11}$ (in MPa, m units) and $m = 2.55$, and Wheeler retardation constant $p = 1.43$. (i) Load range is 1100 N. (ii) Load range is 1450 N. Assume plane strain. [Ans. Final crack size: (i) 11.9249 mm, (ii) 15.6084 mm.]

7.16 A solid shaft made of Ti–6Al–4V alloy has a circumferential crack of depth $a = 4$ mm. Its diameter $D = 60$ mm. It is subjected to a constant amplitude axial fatigue loading of 80 MPa for 25, 000 cycles. It is followed by an overload cycle of amplitude 190 MPa and then constant amplitude loading of 80 MPa for 25, 000 cycles. Find the size of the crack at the different stages. Given $C = 3.5306 \times 10^{-11}$ (MPa$\sqrt{m}$, m units), $m = 2.8$, Wheeler constant $p = 3.40$, and yield point = 930 MPa.
[Hint: Use the appropriate correction factor in terms of a and D. Ans. $a_{f1} = 4.7377$ mm after 25, 000 cycles, $a_{f2} = 4.7381$ mm after 25, 001 cycles, $a_{f3} = 4.7510$ mm after 50, 001 cycles. The full span of growth over the second span of 25000 cycles occurs with retardation.]

7.17 Solve problem Q.7.10(ii) when there is closure effect given by $U=0.75$. [Ans. 1.094 million cycles, assuming $Y=1.12$ over the whole span.]

7.18 A plate of uniform thickness made of Ti–6Al–4V alloy with an 8 mm edge crack is subjected to a constant amplitude cyclic loading of range 60 MPa. Given $C = 5.5306 \times 10^{-11}$ (MPa$\sqrt{m}$, m units) and $m = 2.8$. The plate width is 200 mm. Determine approximately the final crack size after 50, 000 cycles. [Hint: Assume constant SIF correction factor Y and solve. Ans. 9.50 mm.]

References

7.1 Anderson, T.L. 2005. *Fracture Mechanics – Fundamentals and Applications*, 3rd edn. London: Taylor and Francis.

7.2 ASTM (American Society for Testing and Materials) E 1049-85 (Reapproved 2005). *Standard Practices for Cycle Counting in Fatigue Analysis*. Philadelphia: ASTM International.

7.3 Barsom, J.M. 1971. 'Fatigue-crack Propagation in Steels of Various Yield Streng- ths.' *Journal of Engineering for Industry, Transactions of ASME, Series B* 93 (4).

7.4 ———.1974. *Fatigue Behavior of Pressure-vessel Steels*, WRC Bulletin, No. 194. New York: Welding Research Council.

7.5 Barsom, J.M. and S.T. Rolfe. 1999. *Fracture and Fatigue Control in Structures: Application of Fracture Mechanics*, 3rd edn., 194–236. Philadelphia: American Society for Testing and Materials.

7.6 Biell IV, A.J. and F.V. Lawrence Jr. 1989. *The Effect of Casting Porosity on the Fatigue Life of Lost-foam CI and Al–Si 319*, Report No. 150, UILU-ENG-89-3604. Department of Material Science and Engineering, University of Illinois at Urban-Champaign, USA.

7.7 Broek, D. 1984. *Elementary Engineering Fracture Mechanics.* The Hague: Martinus Nijhoff Publishers.

7.8 Broek, D. and J. Schijve. 1963. *The Influence of the Mean Stress on the Propagation of Fatigue Cracks in Aluminium Alloy Sheets.* Report No. TR-M-2111 Amsterdam: National Aerospace Institute.

7.9 Dieter, G.E. 1988. *Mechanical Metallurgy.* London: McGraw-Hill.

7.10 Donahue, R.J., H.M. Clark, P. Atanmo, R. Kumble and A.J. McEvily. 1972. 'Crack Opening Displacement and Rate of Fatigue Crack Growth.' *International Journal of Fracture Mechanics* 8: 209–19.

7.11 Downing, S.D. and D.F. Socie. 1982. 'Simple Rainflow Counting Algorithms.' *International Journal of Fatigue* 4(1): 31–40.

7.12 Elber, W. 1970. 'Fatigue Crack Closure under Cyclic Tension.' *Engineering Fracture Mechanics* 2: 37–45.

7.13 Endo, T., K. Mitsunaga, K. Takahashi, K. Kobayashi and M. Matsuishi. 1974. 'Damage Evaluation of Metals for Random or Varying Load. Three Aspects of Rainflow Method'. *Proceedings of Symposium, Mechanical Behaviour of Materials*, 21–24 August, Kyoto, Japan. Kyoto: Published by Society of Material Science.

7.14 Fleck, N.A. 1985. 'Fatigue Crack Growth Due to Periodic Underloads and Overloads.' *Acta Metallurgica* 33 (7): 1339–54.

7.15 Fleck, N.A. and R.A. Smith. 1984. 'Fatigue Life Prediction of a Structural Steel under Service Loading.' *International Journal of Fatigue* 6(4): 203–10.

7.16 Fleck, N.A., I.F.C. Smith and R.A. Smith. 1983. 'Closure Behavior of Surface Cracks.' *Fatigue of Engineering Materials and Structures* 6(3): 225–239.

7.17 Kikukawa, M., M. Jono and Y. Kondo. 1981. 'An Estimation Method of Fatigue Crack Propagation Rate under Varying Loading Conditions of Low Stress Intensity Level.' In *Advances in Fracture Research*, ed. Francois, D., 1799–1806, Vol. 4. Proceedings of Fifth International Conference on Fracture, 29 March–3 April 1981, Cannes, France. Oxford: Pergamon Press.

7.18 Larsson, L.H., ed. 1983. *Subcritical Crack Growth due to Fatigue, Stress Corrosion and Creep.* London: Elsevier Applied Science Publishing.
https://www.efatigue.com/variable/background/rainflow.html

7.19 Paris, P.C. and F. Erdogan. 1963. 'A Critical Analysis of Crack Propagation Laws.' *Journal of Basic Engineering, Transactions of ASME* 85: 528–34.

7.20 Pook, L.P. 1975. 'Analysis and application of fatigue crack growth data.' *Journal of Strain Analysis* 10(4): 242–50.

7.21 Schijve, J. 1980. 'Prediction Methods for Fatigue Crack Growth in Aircraft Material', 3–34. In *Fracture Mechanics: Twelfth Conference*. Philadelphia: American Society for Testing Materials [ASTM STP 700].

7.22 ———. 1981. 'Some Formulas for the Crack Opening Stress Level.' *Engineering Fracture Mechanics* 14: 461–65.

7.23 Shin, C.S. and N.A. Fleck. 1987.'Overload Retardation in a Structural Steel.' *Fatigue & Fracture of Engineering Materials and Structures* 9(5): 379–93.

7.24 Smith, R.A. 1983. 'Short Fatigue Cracks.' In *Fatigue Mechanisms: Advances in Quantitative Measurement of Physical Damage*, eds. Lankford, J., D.L. Davidson, W.L. Morris and R.P. Wei, 264–79. Philadelphia: American Society for Testing Materials [ASTM STP 811].

7.25 Socie, D.F. 1977. 'Prediction of Fatigue Crack Growth in Notched Members under Variable AmplitudeLoading Histories.' *Engineering Fracture Mechanics* 9: 849–65.

7.26 Suresh, S. 1998. *Fatigue of Materials*, 2nd edn. Cambridge: Cambridge University Press.

7.27 Suresh, S. and R.O. Ritchie. 1984. 'Propagation of Short Cracks.' *International Metal Reviews* 29(1): 445–75.

7.28 Tanaka, K. and Y. Nakai. 1983. 'Propagation and Non-propagation of Short Fatigue Cracks at a SharpNotch.' *Fatigue and Fracture of Engineering Materials and Structures* 6(4): 315–27.

7.29 Walker, E.K. 1970. 'An Effective Strain Concept for Crack Propagation and Fatigue with Specific Application to Biaxial Stress Fatigue', 225–33. *Air Force Conference on Fracture and Fatigue* (1969), AFFDL-TR-70-144.

7.30 Wheeler, O.E. 1972. 'Spectrum Loading and Crack Growth.' *Journal of Basic Engineering, Transactions of ASME* 94: 181–86.

7.31 Willenborg, J., R.M. Engle Jr. and R.A. Wood. 1971. *A Crack Growth Retardation Model Using an Effective Stress Concept*. Report No. AFFL-TM-71-1-FBR Air Force Flight Dynamics Laboratory.

7.32 Zhao, T., J. Zhang and Y. Jiang, 2008. 'A Study of Fatigue Crack Growth of 7075 T651 Al Alloy.' *International Journal of Fatigue* 30: 1169–80.

8

Elastic Plastic Fracture Mechanics

8.1 Introduction

Metals are abundantly used in machine building and civil structures. In the presence of a crack, metals give way to plastic deformation near the crack-tip prior to fracture. The degree of plastic deformation may be higher than the level that can be accommodated in the linear elastic fracture mechanics (LEFM). This has lead to development of elastic plastic fracture mechanics (EPFM) or yielding fracture mechanics (YFM) to cater for ductile metals. In the early stages of the development, focus was on understanding of the crack-tip field and its characterization through parameter like the SIF in LEFM. These developments, measurement of the parameter and application of the concept to practice are discussed in this chapter. Before presenting these developments, some preliminaries of plasticity of metals are given in the following section. Some aspects of plasticity, which may have relevance in the case of fatigue, are included.

8.2 Briefs on Plasticity

During tensile test, a specimen is loaded by an external tensile load P (Fig. 8.1(a)), and the instantaneous deformation is measured from changes in length between the two points spaced by a gauge length l_0 at no load. If the instantaneous spacing is l, the engineering strain $\epsilon = (l - l_0)/l_0$ and the corresponding natural, or true, or logarithmic strain,

$$\epsilon_n = \int_{l_0}^{l} \frac{dl}{l} = \ln \frac{l}{l_0} = \ln\left(1 + \frac{l - l_0}{l}\right) = \ln(1 + \epsilon) \tag{8.1}$$

The engineering stress σ is given by instantaneous load P divided by the original cross-sectional area A_0, that is, $\sigma = P/A_0$. The true stress σ_n is given by

$\sigma_n = \dfrac{P}{A}$, where A is the current cross-section. If small changes in the volume due to elastic deformation are neglected, that is, the material can be assumed plastically incompressible, then $Al = A_0 l_0$.

Hence,

$$\sigma_n = \frac{P}{A} = \frac{Pl}{A_0 l_0} = \sigma(1+\epsilon) \tag{8.2}$$

In the small strain regime, which is roughly up to about yield point of a metal, the two strains are practically the same, and the difference between the nominal and true stresses is negligible. The natural strains offer some advantages (Mendelson 1968; Chakrabarty 1987). The natural strains are additive, but the conventional strains are not. Again, the true stress and true strain diagrams for a material are the same in tension and compression, but they are not the same if the conventional engineering stress and strain measures are employed. Further, the incompressibility conditions in the cases of engineering and natural strain measures are as follows.

$$(1+\epsilon_1)(1+\epsilon_2)(1+\epsilon_3) - 1 = 0, \tag{8.3a}$$

$$\epsilon_{n1} + \epsilon_{n2} + \epsilon_{n3} = 0, \tag{8.3b}$$

where ϵ_i and ϵ_{ni}, $i = 1, 2,$ and 3, are the principal engineering and natural strains at a point. The first equation reduces to $\epsilon_1 + \epsilon_2 + \epsilon_3 = 0$ only when strains are small.

The point of instability, $dP = 0$, can be easily obtained from σ_n versus ϵ_n diagram through the following considerations.

$$P = A\sigma_n \tag{8.4}$$

$$dP = dA\sigma_n + Ad\sigma_n = 0 \tag{8.5}$$

Therefore,

$$\frac{d\sigma_n}{\sigma_n} = -\frac{dA}{A}.$$

Further, $Al = A_0 l_0$ gives $\dfrac{dA}{A} = -\dfrac{dl}{l} = -d\epsilon_n$.

Combining the two, $\dfrac{d\sigma_n}{d\epsilon_n} = \sigma_n$.

The point of instability is given by the location where the slope of the σ_n versus ϵ_n diagram is equal to σ_n. Such location, for example, D, is obtained by drawing a

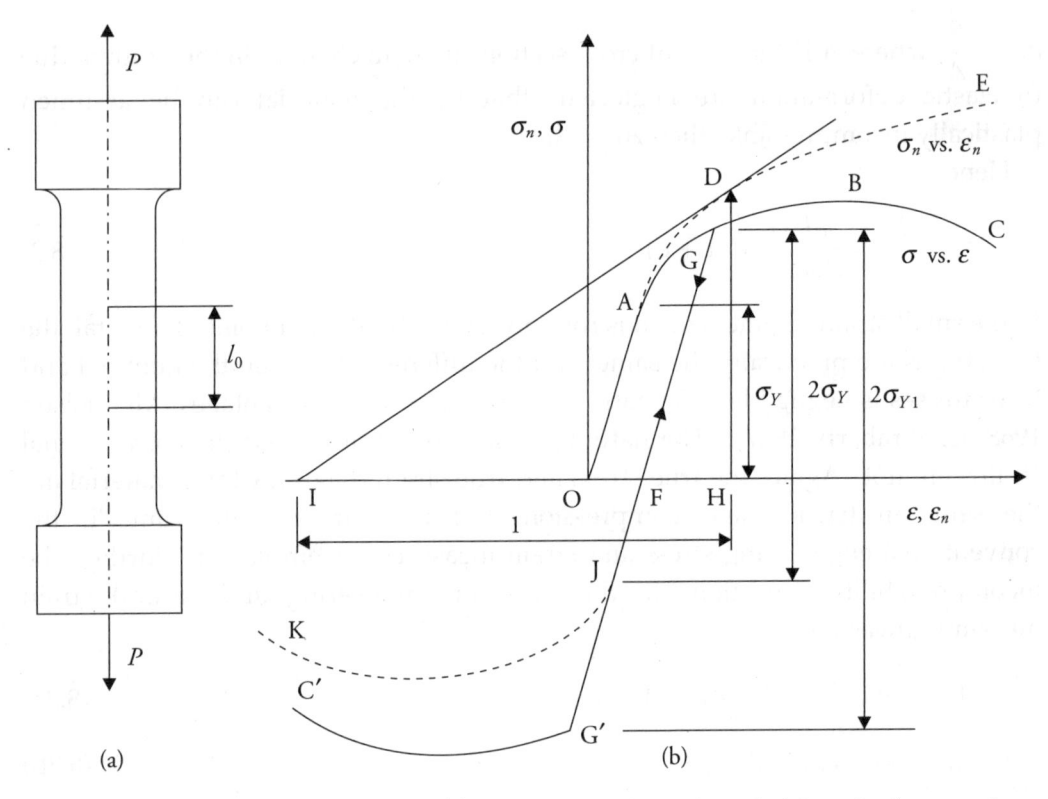

Figure 8.1 (a) Tensile test specimen. (b) Stress–strain plot under uniaxial tension/compression.

tangent to the σ_n versus ϵ_n diagram from I (Fig. 8.1(b)) such that the subtangent is unity.

Beyond the proportional limit, stress–strain behaviour is nonlinear. If the yield point σ_Y of the material is roughly taken to coincide with the point A, and if the material is loaded up to a point like G and then unloaded, the unloading occurs along GF, a straight line parallel to OA. After full unloading, the material shows a residual strain of magnitude OF. If the material is loaded back, it will show a linear behaviour up to G and then stress–strain variation will follow the path GB. The yield point of the material increases from σ_Y to the value corresponding to level G, say, σ_{Y1}. This is known as strain hardening or work hardening. The stress increases with strain up to the point B, at which the instability or necking sets in. The stress corresponding to this point is known as ultimate tensile strength. After point B, further deformation is associated with reduction in load. The specimen breaks at the fracture point C.

In the case of isotropic hardening materials, if it is loaded up to G and then fully unloaded, followed by loading in compression, the yielding will occur when the stress level reaches G' such that vertical distance from G is $2\sigma_{Y1}$. With further increase in compression load, it will show nonlinear variation of stress with strain along G'C'. In case the material demonstrates kinematic hardening, it will start yielding in compression, when the stress reaches a level J such that the vertical distance from the point G is $2\sigma_Y$. This type of reduction in yield strength upon reversal of loading is known as Bauschinger effect.

The conditions for yielding are as follows, assuming the three principal stresses satisfy the condition, $\sigma_1 > \sigma_2 > \sigma_3$.

$\sigma_1 - \sigma_3 = \sigma_Y$, $\sigma_1 = \sigma_Y$, $\sigma_2 = \sigma_Y$ according to Tresca criterion (8.6a)

$(\sigma_1 - \sigma_2)^2 + (\sigma_2 - \sigma_3)^2 + (\sigma_3 - \sigma_1)^2 = 2\sigma_Y^2$ according to von Mises criterion (8.6b)

where σ_Y is the yield strength. In the three-dimensional stress space, according to the von Mises criterion, the yield surface is a cylinder, whose axis is equally inclined with the three stress axes. It is a hexagonal cylinder according to the Tresca criterion. In the two-dimensional stress space, with $\sigma_3 = 0$, the two yield surfaces reduce to an ellipse and a hexagon (Fig. 8.2(a)), respectively. The yield locus in two dimensions expands in the case of isotropic hardening (Figs. 8.2(b) and 8.3(a)); it shifts in the same space in the case of kinematic hardening, depending on the loading path, for example, O to O_1, O_1 to O_2, and O_2 to O_3 (Fig. 8.3(b)).

When a material undergoes elastic deformation, the stresses and strains are related by the Hooke's law. The final stresses and strains are independent of loading path. On the other hand, when it is subjected to elastic–plastic deformation, the elastic part of the total strain is related to the state of stress by Hooke's law, but the plastic part of the strain is governed by the incremental theory of plasticity. The incremental plastic strains and the final state of strains are loading path dependent. That is, for the same state of final stresses, the final total strains are different. This can be illustrated by considering the loading of thin cylinder by an axial load P and a torque T. The material is assumed to undergo

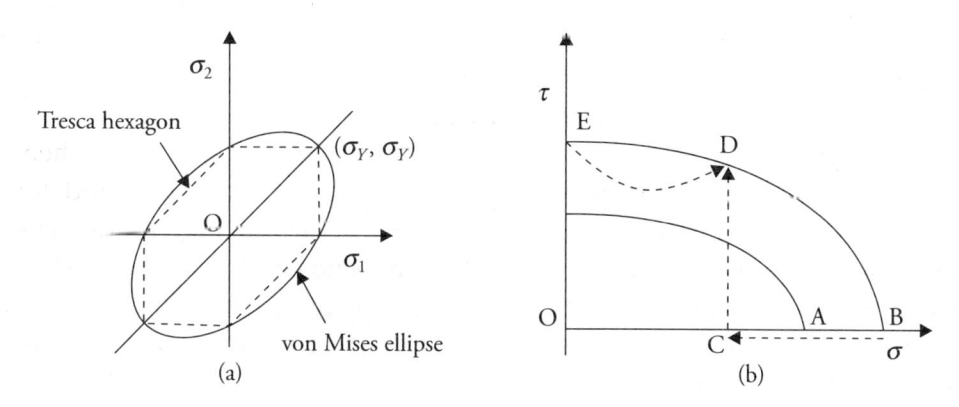

Figure 8.2 (a) Two yield loci in $\sigma_1 - \sigma_2$ plane. (b) von Mises yield locus in $\tau - \sigma$ plane.

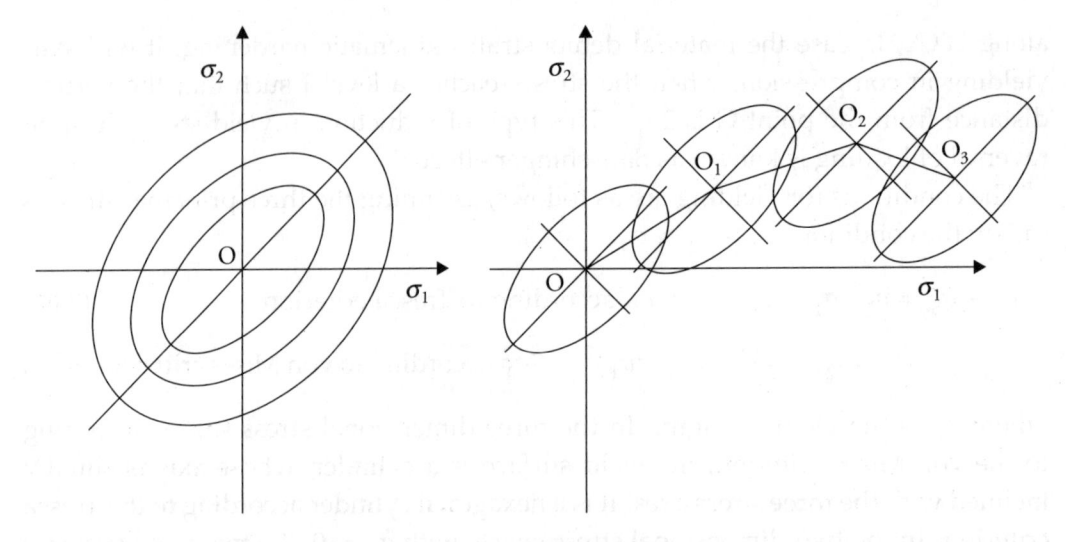

Figure 8.3 (a) Isotropic hardening. (b) Kinematic hardening.

isotropic hardening. The yield locus for the material is shown in Fig. 8.2(b). The cylinder is loaded by the axial load gradually to bring it to initial yielding, that is, state A. Considering the cylinder to be developed in $x - y$ plane and the axial direction is coincident with the x direction, the axial strain can be represented by ϵ_x and shear strain by γ_{xy}. It is then loaded further in the axial direction up to state B, which lies on the expanded yield locus. The strains at this point will consist of some elastic normal strain ϵ_x^e and normal plastic strain ϵ_x^p and no shear strain; that is, $\gamma_{xy}^e = 0$ and $\gamma_{xy}^p = 0$. The cylinder is then unloaded up to the point C. At this point, the strains are: $\epsilon_x^e \neq 0, \epsilon_x^p \neq 0, \gamma_{xy}^e = 0, \gamma_{xy}^p = 0, \epsilon_y^p = \epsilon_z^p = -0.5\epsilon_x^p$. The plastic strain ϵ_x^p will be the same as that at B. The plastic strains in the y and z directions are related ϵ_x^p because of volume consistency or incompressibilty. The cylinder is now subjected to torque gradually to raise the stress levels to point D. Since point D lies on the expanded yield locus, it will undergo deformation from C to D only elastically. At D, it will have ϵ_x^e and ϵ_x^p, ϵ_y^p and ϵ_z^p the same as those at C and shear strains $\gamma_{xy}^e \neq 0$ and $\gamma_{xy}^p = 0$. Consider now a second loading path of the virgin cylinder. Load the cylinder only by torque up to point E. It will have all elastic and plastic components of strains zero except γ_{xy}^e and γ_{xy}^p. Then the cylinder is loaded along the path ED by adjusting the axial load and torque continuously. At this point, the material will have ϵ_x^e, whose magnitude depends on axial stress σ_x at D, γ_{xy}^e, whose magnitude depends on the torque at D, γ_{xy}^p is equal to the plastic shear strain at E, and all the remaining elastic and plastic strain components are zero. Although the state of stress at D in the two cases is the same, the final strains are different. Thus, path dependence of plastic deformation is established.

8.2.1 Incremental Theories of Plasticity

There are two theories, which are employed to relate the incremental strains to stresses. In the first, Levy–Mises theory, the elastic strains are considered negligible. The total strain increments constitute only the plastic part. The strain increments are given by

$$\frac{d\epsilon_x}{S_x} = \frac{d\epsilon_y}{S_y} = \frac{d\epsilon_z}{S_z} = \frac{d\gamma_{xy}}{\tau_{xy}} = \frac{d\gamma_{yz}}{\tau_{yz}} = \frac{d\gamma_{zx}}{\tau_{zx}} = d\lambda \tag{8.7}$$

where S_x, S_y, and S_z are deviatoric stresses and τ_{xy}, τ_{yz}, and τ_{zx} are shear stresses and $d\lambda$ is a material constant dependent on stress state. In the second rule, Prandtl–Reuss theory, the elastic part of the strain increments is not considered negligible. The plastic part of the strain increments is proportional to the corresponding deviatoric stresses. That is,

$$\frac{d\epsilon_x^p}{S_x} = \frac{d\epsilon_y^p}{S_y} = \frac{d\epsilon_z^p}{S_z} = \frac{\gamma_{xy}^p}{\tau_{xy}} = \frac{\gamma_{yz}^p}{\tau_{yz}} = \frac{\gamma_{zx}^p}{\tau_{zx}} = d\lambda \tag{8.8}$$

The constant of proportionality is obtained by tensile test and the following definition of equivalent stress and equivalent plastic strain increment.

$$\bar{\sigma} = \frac{1}{\sqrt{2}}\left[(\sigma_x - \sigma_y)^2 + (\sigma_y - \sigma_z)^2 + (\sigma_z - \sigma_x)^2 + 6\{\tau_{xy}^2 + \tau_{yz}^2 + \tau_{zx}^2\}\right]^{1/2} \tag{8.9a}$$

$$d\bar{\epsilon}^p = \frac{\sqrt{2}}{3}\left[(d\epsilon_x^p - d\epsilon_y^p)^2 + (d\epsilon_y^p - d\epsilon_z^p)^2 + (d\epsilon_z^p - d\epsilon)^2 + 6(d\epsilon_{xy}^{p\,2} + d\epsilon_{yz}^{p\,2} + d\epsilon_{zx}^{p\,2})\right]^{1/2} \tag{8.9b}$$

In the case of uniaxial test, $\bar{\sigma} = \sigma_x = \sigma$, $\sigma_y = \sigma_z = 0$, $d\epsilon_x^p = -2d\epsilon_y^p = -2d\epsilon_z^p$ and $d\epsilon_{xy}^p = d\epsilon_{yz}^p = d\epsilon_{zx}^p = 0$. Therefore,

$$d\lambda = \frac{3}{2}\frac{d\bar{\epsilon}^p}{\bar{\sigma}} = \frac{3}{2}\frac{d\epsilon_x^p}{\sigma_x} = \frac{3}{2}\frac{d\bar{\sigma}}{H'\bar{\sigma}} = \frac{3}{2}\frac{d\sigma}{H'\sigma}, \quad H' = \frac{d\bar{\sigma}}{d\bar{\epsilon}^p} = \frac{d\sigma}{d\epsilon^p} \tag{8.10}$$

The modulus of plasticity is obtained by plotting σ versus ϵ^p noting that $\epsilon^p - \epsilon$ ϵ^e and ϵ is the total uniaxial strain corresponding to stress σ. The slope at any point of the curve A'B'C' (Fig. 8.4) will give the instantaneous plasticity modulus H'.

Finally, the incremental plastic strains can be written as

$$d\epsilon_{ij}^p = \frac{3}{2}\frac{d\bar{\epsilon}^p}{\bar{\sigma}}S_{ij} = \frac{3}{2}\frac{d\bar{\sigma}}{H'\bar{\sigma}}S_{ij} \tag{8.11}$$

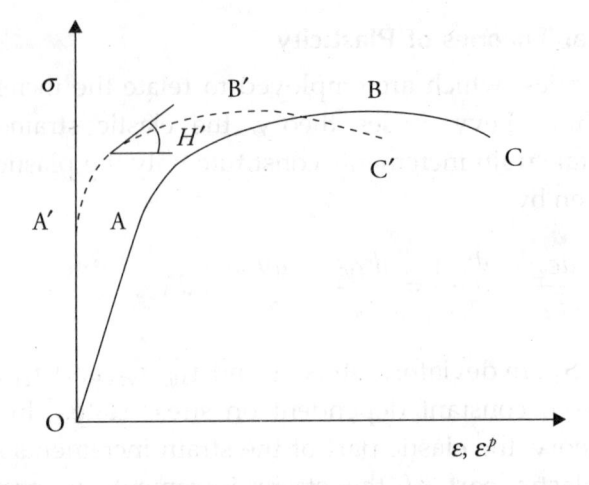

Figure 8.4 Plasticity modulus H'.

In the case of proportional or radial loading, that is, if all the stresses are increasing with time in the same ratio, the incremental theory reduces to deformation theory. $\sigma_{ij} = c\,\sigma_{ij}^0$, where σ_{ij}^0 is a reference state of stress and c is a monotonically increasing function of time. Therefore, $S_{ij} = c\,S_{ij}^0$ and $\bar{\sigma} = c\bar{\sigma}^0$.

Then, at any level of loading

$$d\epsilon_{ij}^p = \frac{3}{2}\frac{d\bar{\epsilon}^p}{\bar{\sigma}^0}S_{ij}^0. \tag{8.12}$$

which can be integrated to give

$$\epsilon_{ij}^p = \frac{3}{2}\frac{\bar{\epsilon}^p}{\bar{\sigma}^0}S_{ij}^0 = \frac{3}{2}\frac{\bar{\epsilon}^p}{\bar{\sigma}}S_{ij} \tag{8.13}$$

So the plastic strain is a function of the current state of stress and is independent of the path of loading. Such deformations are considered to obey the total or deformation theory of plasticity.

8.3 Crack Opening Displacement Criterion

In the presence of plastic deformation, the sharp crack-tip gets blunted and strains increase rapidly than the stresses. Strains manifest in the form of displacements, opening δ at the locations of crack-tip (B and C). Figure 8.5 represents the extent of deformations. Wells (1961) experimented with different metals and found that in many cases, the fracture resistance cannot be defined in terms of the critical SIFs. He also observed that crack opens up considerably at the original physical crack-tip locations, and materials over segments AB and CD get stretched before fracturing.

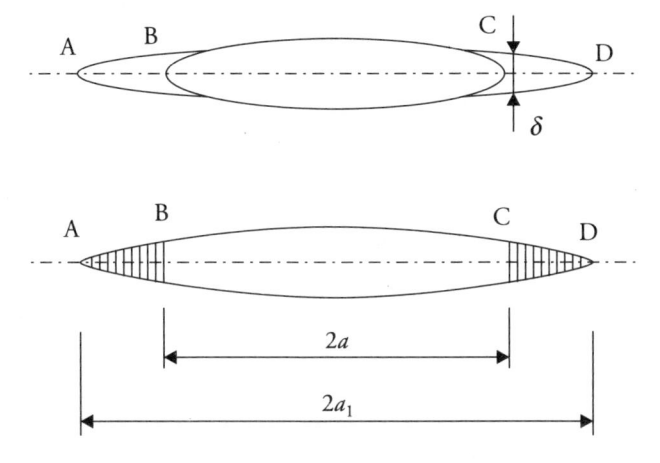

Figure 8.5 Crack opening displacement.

He defined the condition for extension of crack for such materials in terms of this opening. Since then, the crack opening displacement (COD) or crack-tip opening displacement (CTOD) is regarded as a fracture resistance parameter. It has been shown in Chapter 4 that this COD at the crack-tip is related to the Griffith energy release rate through consideration of both Irwin's plastic zone correction factor and Dugdale–Barenblatt model in the case of small-scale plastic deformation.

8.4 Mode III Crack-Tip Field for Elastic-Perfectly-Plastic Materials

Although Wells (1961) gave the condition for Mode I fracture in terms of COD, it was not known if there was any stress or strain singularity at the crack-tip as it was in the case of elastic materials. The crack-tip field for perfectly plastic materials was first obtained under Mode III loading by Hult and McClintock (1957). Readers may also refer McClintock (1971). They considered the material to be perfectly plastic with distinct shear yield strength τ_Y. In the presence of elastic plastic deformation, the actual crack-tip can be replaced by an equivalent elastic crack with tip at O'. The elastic stress-displacement field with origin O' can be written using Eqs. (2.28) and (2.29) as follows.

$$\tau_{\bar{r}z} = \frac{K}{\sqrt{2\pi\bar{r}}} \sin\frac{\bar{\theta}}{2}, \quad \tau_{\bar{\theta}z} = \frac{K}{\sqrt{2\pi\bar{r}}} \cos\frac{\bar{\theta}}{2} \tag{8.14}$$

$$w = \frac{K}{\mu}\sqrt{\frac{2\bar{r}}{\pi}} \sin\frac{\bar{\theta}}{2} \tag{8.15}$$

$\bar{r}$ and $\bar{\theta}$ are referred to O'. The stress field within the elastic-perfectly-plastic zone satisfies the equilibrium equations and the yield conditions.

210 *Fracture mechanics*

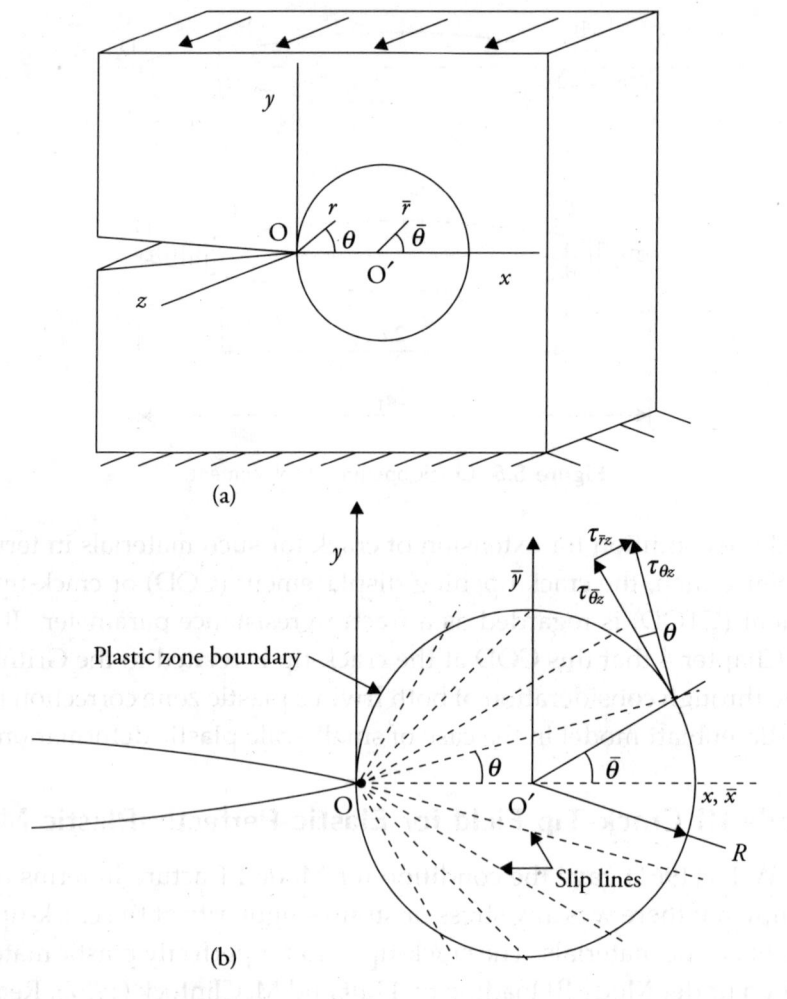

Figure 8.6 (a) Mode III loading and coordinates. (b) Crack-tip elastic plastic field.

$$\frac{\tau_{rz}}{r} + \frac{\partial \tau_{rz}}{\partial r} + \frac{1}{r}\frac{\partial \tau_{\theta z}}{\partial \theta} = 0 \quad \text{or,} \quad \frac{1}{r}\frac{\partial (r\tau_{rz})}{\partial r} + \frac{1}{r}\frac{\partial \tau_{\theta z}}{\partial \theta} = 0 \tag{8.16}$$

$$\tau_{\theta z}^2 + \tau_{rz}^2 = \tau_Y^2 \tag{8.17}$$

Assuming the deformation theory plasticity, the total strains within the plastic zone is given by

$$\gamma_{rz} = \frac{\tau_{rz}}{\mu} + \lambda_1 \tau_{rz} = \lambda \tau_{rz}, \text{ and} \tag{8.18a}$$

$$\gamma_{\theta z} = \frac{\tau_{\theta z}}{\mu} + \lambda_1 \tau_{\theta z} = \lambda \tau_{\theta z} \tag{8.18b}$$

where μ is rigidity modulus, λ_1 is plasticity constant, and $\lambda = \dfrac{1}{\mu} + \lambda_1$. The first part indicates the elastic strain and the second part is the plastic strain component.

The strains are given by displacement function w, which is just a function of r and θ, by

$$\gamma_{rz} = \frac{\partial w}{\partial r} \text{ and } \gamma_{\theta z} = \frac{\partial w}{r \partial \theta} \tag{8.19}$$

since other two displacement components u and v are zero. From the slip line field analysis, it was shown by Hult and McClintock (1957) that $\tau_{rz} = 0$ along the radial lines from O and, in the tangential direction $\tau_{\theta z} = \tau_Y$, which is obtained from the yield condition. This stress state satisfies the equilibrium Eq. (8.16). Therefore, from Eqs. (8.18a and b)

$$\gamma_{rz} = 0, \ \gamma_{\theta z} = \lambda \tau_Y \tag{8.20}$$

From Eqs. (8.18) and (8.19), w is only a function of θ. Assuming $w = f(\theta)$

$$\gamma_{\theta z} = \frac{1}{r} f'(\theta) = \lambda \tau_Y \tag{8.21}$$

From Eq. (8.20), since strain $\gamma_{\theta z}$ is a function of r and θ, λ is a function of r and θ. By imposing the continuity of stress at the elastic plastic boundary,

$$\tau_{\theta z} = \sqrt{\tau_{\theta z}^2 + \tau_{rz}^2} = \tau_Y = \sqrt{\frac{K^2}{2\pi R}} \tag{8.22}$$

Furthermore, as per the elastic field with origin at O'

$$w = \frac{K}{\mu} \sqrt{\frac{2R}{\pi}} \sin \frac{\bar{\theta}}{2} = \frac{K}{\mu} \sqrt{\frac{2R}{\pi}} \sin \theta, \text{ since } \bar{\theta} = 2\theta \tag{8.23}$$

Using Eqs. (8.20), (8.21), and (8.22)

$$\gamma_{\theta z} = \frac{2R \tau_Y}{\mu r} \cos \theta \tag{8.24}$$

Therefore, there is a strain singularity of order -1 at the crack-tip, but there is no stress singularity. The COD δ is given by

$$\delta = w_{\theta=\pi/2} - w_{\theta=-\pi/2} = \frac{4}{\pi \tau_Y} \frac{K^2}{2\mu} \tag{8.25}$$

212 Fracture mechanics

Since strain energy release rate $G = \dfrac{K^2}{2\mu}$, $G = \dfrac{\pi}{4}\tau_Y \delta$. This relation again indicates that if fracture is governed by a critical value of G, it is also governed by a critical value of δ. That is, fracture occurs when $\delta = \delta_c$.

8.5 Relationship between J and COD

Considering Dugdale (1960) type strip yield zone ahead of the crack-tip under Mode I loading, the elastic boundary ABC can be assumed to be under closure stresses σ_Y, where σ_Y is the yield strength of the material. The material is assumed elastic-perfectly-plastic. The material is therefore elastic outside ABC; J integral can be easily evaluated along the contour ABC. In the presence of large-scale yielding (Fig. 8.7) at the crack-tip,

$$J = \int_S \left(W dy - T_i \frac{\partial u_i}{\partial x} dS \right) = \int_{ABC} \left(W dy - T_i \frac{\partial u_i}{\partial x} dS \right)$$

$$= \int_A^B \left(W dy - T_i \frac{\partial u_i}{\partial x} dS \right) + \int_B^C \left(W dy - T_i \frac{\partial u_i}{\partial x} dS \right) \tag{8.26}$$

Since dy is zero along AB and BC,

$$J = \int_A^B \left(-T_i \frac{\partial u_i}{\partial x} dS \right) + \int_B^C \left(-T_i \frac{\partial u_i}{\partial x} dS \right) \tag{8.27}$$

Further,

$$T_i \frac{\partial u_i}{\partial x} = \sigma_{ij} n_j \frac{\partial u_i}{\partial x} = \sigma_{11} n_1 \frac{\partial u_1}{\partial x} + \sigma_{12} n_2 \frac{\partial u_1}{\partial x} + \sigma_{21} n_1 \frac{\partial u_2}{\partial x} + \sigma_{22} n_2 \frac{\partial u_2}{\partial x} \tag{8.28}$$

The direction cosines of the two outer normals on AB and BC are shown. Noting that shear stresses are zero along the two segments, due to symmetry and self-similar crack growth.

$$J = \int_A^B \left(\sigma_{22} \frac{\partial u_2}{\partial x} dx \right) + \int_B^C \left\{ -\sigma_{22} \frac{\partial u_2}{\partial x}(-dx) \right\} = \sigma_Y (v_B - v_A) + \sigma_Y(v_C - v_B)$$

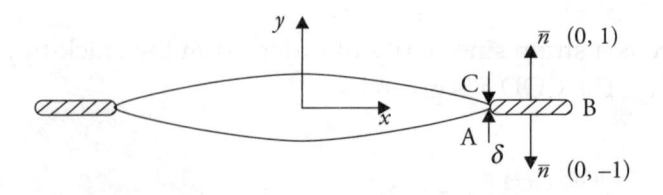

Figure 8.7 Mode I crack with strip yield zones.

$$= \sigma_Y(v_C - v_A) = \sigma_Y \delta \tag{8.29}$$

This relationship is valid for both small-scale and large-scale plastic deformation. For small-scale yielding, it was shown in Chapter 4 that $J = G = \sigma_Y \delta$. It must be emphasized that J loses its significance as a energy release rate quantity, because the part of the energy that goes into deforming material plastically is non-recoverable.

8.6 Fracture Assessment Diagram and R-6 Curve

Based on the linkage between J and δ, it has been possible to arrive at a graphical presentation of failure region, which is very useful in the assessment of safety of components showing plastic deformation before crack extension. These diagrams are known as fracture assessment diagrams in USA and R-6 curve in Europe. This is discussed subsequently.

According to Dugdale strip yield model (1960) or Barenblatt cohesive zone model (1962), COD δ at the crack-tip of a Mode I internal crack in an infinite sheet, is given by Eq. (4.7). Combining this equation with Eq. (8.29), J can be written as

$$J = \sigma_Y \delta = \frac{8\sigma_Y^2 a}{\pi E} \left[\ln \sec \frac{\pi \sigma}{2\sigma_Y} \right] \tag{8.30}$$

At fracture $J = J_C = \dfrac{8\sigma_Y^2 a}{\pi E} \left[\ln \sec \dfrac{\pi \sigma_C}{2\sigma_Y} \right]$, where σ_C is fracture stress. Using $\epsilon_Y = \sigma_Y/E$, Eq. (8.30) can be alternatively written as follows.

$$\frac{J_C}{\sigma_Y \epsilon_Y a} = \frac{8}{\pi} \ln \sec \left(\frac{\pi \sigma_C}{2\sigma_Y} \right) \quad \text{for both small-scale and large-scale yielding.} \tag{8.31}$$

$$= \frac{8}{\pi} \left[\frac{1}{2} \left(\frac{\pi \sigma_C}{2\sigma_Y} \right)^2 \right] \quad \text{for only small-scale yielding.} \tag{8.32}$$

These two relations are plotted in Fig. 8.8. It shows variation of σ_C with a.

Using $K_C = \sqrt{EJ_C}$, and representing the corresponding fracture stress by σ_C, it is possible to write from Eq. (8.30)

$$K_C = \sigma_C \sqrt{\pi a}\, \frac{\sigma_Y}{\sigma_C} \left[\frac{8}{\pi^2} \ln \sec \frac{\pi \sigma_C}{2\sigma_Y} \right]^{\frac{1}{2}} \tag{8.33}$$

For small-scale yielding, the critical SIF $K_{SSY} = \sigma_C \sqrt{\pi a}$. Therefore, using Eq. (8.33)

$$\frac{K_{SSY}}{K_C} = \frac{\sigma_C}{\sigma_Y}\left[\frac{8}{\pi^2}\ln\sec\frac{\pi\sigma_C}{2\sigma_Y}\right]^{-\frac{1}{2}} \tag{8.34}$$

Knowing the fracture stress σ_C during specimen testing and noting $K_{SSY} = \sigma_C\sqrt{\pi a}$, it is possible to determine K_C either from Eq. (8.33) or (8.34). If the whole ligament undergoes plastic deformation, the ratio $\frac{\sigma_C}{\sigma_Y}$ will become 1. Thus, Eq. (8.34) can be used to interpolate between elastic fracture, for which $\frac{K_{SSY}}{K_C} = 1$, to fully plastic collapse, for which $\frac{\sigma_C}{\sigma_Y} = 1$ (Fig. 8.9).

Equation (8.34) can also be used to interpolate between linear elastic fracture, that is, $\frac{\sigma_C}{\sigma_Y}$ much less than 1, and large scale plastic deformation, that is, $\frac{\sigma_C}{\sigma_Y}$ close to 1. For such situations, the yield stress σ_Y is replaced by limit stress σ_l (Harrison, Loosemoore and Milne 1976) and K_C is represented by K_l. The resulting failure assessment curve (FAC) or R-6 curve is given by the following relation with K_{SSY} substituted by K, K_C replaced by K_l, σ_C by σ, and σ_Y replaced by σ_l.

$$\frac{K}{K_l} = \frac{\sigma}{\sigma_l}\left[\frac{8}{\pi^2}\ln\sec\frac{\pi\sigma}{2\sigma_l}\right]^{-\frac{1}{2}} \tag{8.35}$$

The collapse load is obtained by limit analysis. For plate (of unit thickness and width 2W) of elastic-perfectly-plastic material with a centre crack of size 2a

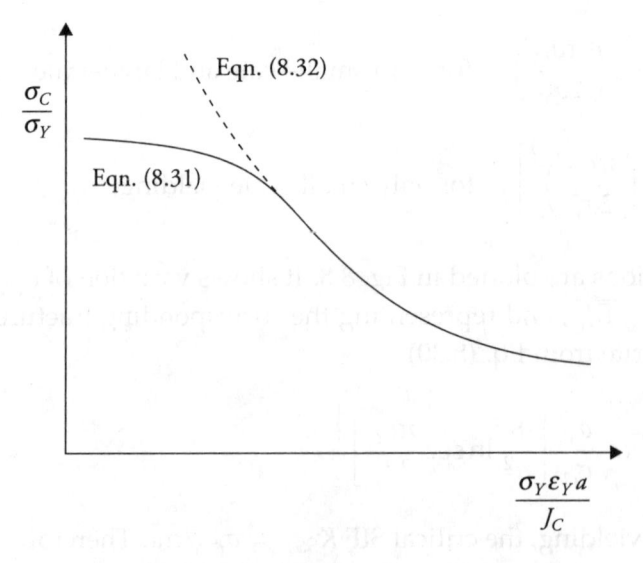

Figure 8.8 Variation of σ_C with a.

$$\sigma_l = \sigma_Y \left(1 - \frac{a}{W}\right) \tag{8.36}$$

The FAC is plotted in $\frac{K}{K_I} - \frac{\sigma}{\sigma_l}$ plane by all the combinations of $\frac{K}{K_I}$ and $\frac{\sigma}{\sigma_l}$ that correspond to failure. A typical FAC or failure assessment diagram (FAD) is shown in Fig. 8.9. Both the ratios $\frac{K}{K_I}$ and $\frac{\sigma}{\sigma_l}$ depend mostly linearly on the external loading for a given crack size a. ABC indicates the failure locus; the combination of the ratios $\frac{K}{K_I}$ and $\frac{\sigma}{\sigma_l}$ that lies on the curve ABC indicates failure. If a given loading of a component with a crack a is represented by point D on the FAD, as the load increases, the point moves along OB. At D, the component is safe and the loading will not lead to any extension of the crack. The available factor of safety is given by OB/OD. The FAD is the basis for R-6 curve, which is widely considered for safety analysis. In general, the FAD varies with specimen geometry (Kanninen and Popelar 1985), but the variation is small as compared to the variation of uncertainties associated with failure assessment of a real structure. The FAD concept has been extended to include the effect of thermal and residual stresses. The concept is easy to apply, because it requires only the linear elastic SIF for the case and the limit stress, which can be obtained easily (Kanninen and Popelar 1985). For the determination of load for a given crack size, the limit stress and K_C are required. This may require iterations using Eq. (8.35). Alternatively, for a given load and crack size, by knowing the limit stress and K_C, the factor of safety can be calculated.

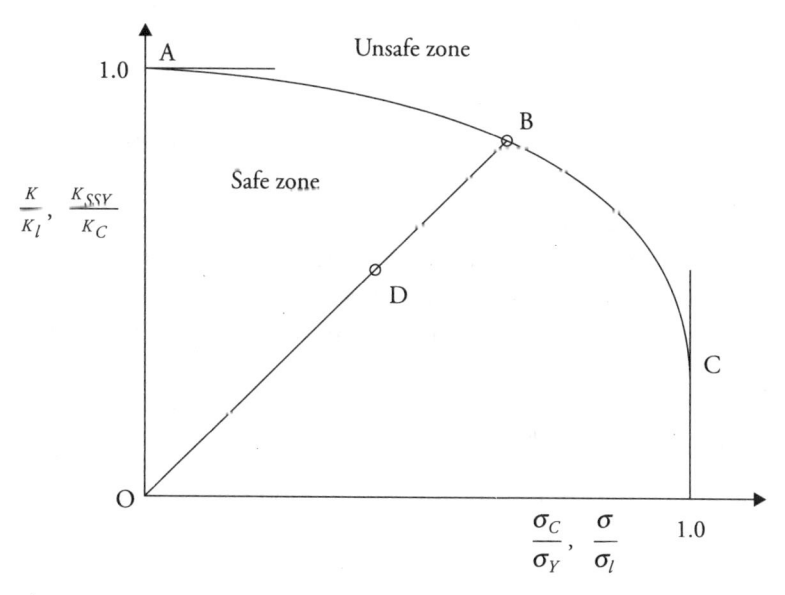

Figure 8.9 Fracture assessment diagram.

8.7 Mode I Crack-Tip Field

8.7.1 Rice–Rosengren Analysis

For a material showing plastic deformation before fracture, the crack-tip field depends on the tensile stress–strain relationship of the material. The tensile stress–strain relationship can be represented by the Ramberg-Osgood type relationship given by

$$\frac{\epsilon}{\epsilon_Y} = \frac{\sigma}{\sigma_Y} + \alpha \left(\frac{\sigma}{\sigma_Y}\right)^n \tag{8.37}$$

where σ_Y is yield stress, ϵ_Y = yield strain = $\frac{\sigma_Y}{E}$, E is modulus of elasticity, α and n are two hardening constants. At any stage of loading, the total strain will constitute elastic and plastic strains given by

$$\epsilon_{ij} = \epsilon_{ij}^e + \int d\epsilon_{ij}^p$$

$$= \left(\frac{1-2\nu}{E}\sigma_m \delta_{ij} + \frac{1+\nu}{E}S_{ij}\right) + \int \left(\frac{3}{2}\frac{d\bar{\epsilon}^p}{\bar{\sigma}}S_{ij}\right) \tag{8.38}$$

where ν is Poisson's ratio, σ_m is hydrostatic stress, δ_{ij} is Kronecker's delta, S_{ij} is deviatoric stress, and equivalent stress $\bar{\sigma}$ is given below.

$$\bar{\sigma} = \sqrt{3J_2} = \left(\frac{3}{2}S_{ij}S_{ij}\right)^{\frac{1}{2}} = \frac{1}{\sqrt{2}}\left[(\sigma_x - \sigma_y)^2 + (\sigma_y - \sigma_z)^2 + (\sigma_z - \sigma_x)^2 \right.$$
$$\left. +6(\tau_{xy}^2 + \tau_{xy}^2 + \tau_{xy}^2)\right]^{\frac{1}{2}} \tag{8.39}$$

Equivalent plastic strain increment $d\bar{\epsilon}^p$ is given by Eq. (8.9b). J_2 is the second invariant of deviatoric stress tensor. In the face of monotonically increasing loading till fracture, the loading is proportional, and the deformation theory of plasticity, which is the same as nonlinear elasticity, can be employed. The plastic strains can be related to the final state of stress.

Consequently, the stress–strain relations can be written as follows.

$$\epsilon_{ij} = \left(\frac{1-2\nu}{E}\sigma_m \delta_{ij} + \frac{1+\nu}{E}S_{ij}\right) + \left(\frac{3}{2}\frac{\bar{\epsilon}^p}{\bar{\sigma}}S_{ij}\right) \tag{8.40}$$

Neglecting the elastic part of the strains, the relation can be simplified in the form

$$\epsilon_{ij} = \left(\frac{3}{2}\frac{\bar{\epsilon}^p}{\bar{\sigma}}S_{ij}\right) = \frac{3}{2}\alpha \epsilon_Y \left(\frac{\bar{\sigma}}{\sigma_Y}\right)^n \frac{S_{ij}}{\bar{\sigma}} = \frac{3}{2}\alpha \epsilon_Y \left(\frac{\bar{\sigma}}{\sigma_Y}\right)^{n-1} \frac{S_{ij}}{\sigma_Y} \tag{8.41}$$

Considering crack-tip to be surrounded by a plastic zone, the stress–strain variation will be governed by Eq. (8.40). The exact form of spatial variation of this field was obtained by Hutchinson (1968a,b) and Rice and Rosengren (1968) independently and they showed the existence of different singularities in stresses and strains.

Considering a circular path for J integral around the crack-tip (Fig. 8.10),

$$J = \int_S \left[W dy - T_i \frac{\partial u_i}{\partial x} dS \right] = \int_{-\pi}^{\pi} \left[W(r,\theta) \cos\theta - T_i(r,\theta) \frac{\partial u_i(r,\theta)}{\partial x} \right] r \, d\theta \quad (8.42)$$

If this integral is to be path independent, the integrand must be independent of radius r. That is, the expression within the square brackets of the right most expression of Eq. (8.42), which is roughly proportional to the product of stress and strain, should be equal to a function of θ divided by r. That is

$$\sigma_{ij}\epsilon_{ij} \alpha \frac{1}{r} \quad (8.43)$$

Since strain is proportional to σ^n, as per the above relation, stress $\sigma_{ij} \alpha \left(\frac{1}{r}\right)^{\frac{1}{n+1}}$ and $\epsilon_{ij} \alpha \left(\frac{1}{r}\right)^{\frac{n}{n+1}}$. This means that there is stress singularity of order $1/(n+1)$ and strain singularity of order $n/(n+1)$. If the material is elastic-perfectly-plastic, that is, $n = \infty$, there is no stress singularity but there is a strain singularity of order 1, which is similar to the case of Mode III discussed earlier.

Considering Airy stress function approach (Rice and Rosengren 1968), and noting that stress

$$\sigma_{ij} \alpha \, r^{-\frac{N}{N+1}} \text{ and } \epsilon_{ij} \alpha \, r^{-\frac{1}{N+1}}, \text{ where } N = 1/n,$$

it is possible to write a stress function

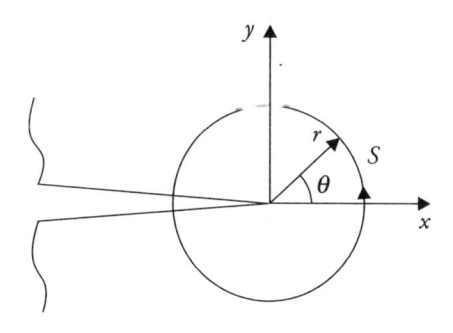

Figure 8.10 Crack-tip polar coordinates.

$$\Phi = \frac{(1+N)^2}{2+N} r^{\frac{2+N}{N+1}} f(\theta) \tag{8.44}$$

where $f(\theta)$ is an arbitrary function of θ. This function satisfies the equilibrium equations. Therefore,

$$\sigma_{\theta\theta} = \frac{\partial^2 \Phi}{\partial r^2} = r^{\frac{-N}{N+1}} f(\theta) \tag{8.45}$$

$$\sigma_{rr} = \frac{1}{r}\frac{\partial \Phi}{\partial r} + \frac{1}{r^2}\frac{\partial^2 \Phi}{\partial \theta^2} = (N+1) r^{\frac{-N}{N+1}} f(\theta) + \frac{(1+N)^2}{2+N} r^{\frac{-N}{N+1}} f''(\theta) \tag{8.46}$$

$$\tau_{r\theta} = -\frac{\partial}{\partial r}\left(\frac{1}{r}\frac{\partial \Phi}{\partial \theta}\right) = -\frac{1+N}{2+N} r^{\frac{-N}{N+1}} f'(\theta) \tag{8.47}$$

To ensure strain compatibility condition, the displacements were represented in the form of another function,

$$\psi = r^{\frac{1+2N}{N+1}} g(\theta) \tag{8.48}$$

where $g(\theta)$ is another function of θ. The two displacements, radial and circumferential, were defined by

$$u = -\frac{1}{r}\frac{\partial \psi}{\partial \theta} = -r^{\frac{N}{N+1}} g'(\theta), \text{ and } v = \frac{\partial \psi}{\partial r} = \frac{1+2N}{N+1} r^{\frac{N}{N+1}} g(\theta) \tag{8.49}$$

This gives

$$\epsilon_{rr} = \frac{1}{r}\frac{\partial u}{\partial \theta} = -\frac{N}{N+1} r^{\frac{-1}{N+1}} g'(\theta) \tag{8.50}$$

$$\epsilon_{\theta\theta} = \frac{u}{r} + \frac{1}{r}\frac{\partial u}{\partial \theta} = \frac{N}{N+1} r^{\frac{-1}{N+1}} g'(\theta) \tag{8.51}$$

$$\gamma_{r\theta} = 2\epsilon_{r\theta} = \frac{1}{r}\frac{\partial u}{\partial \theta} + \frac{\partial v}{\partial r} - \frac{v}{r} = -r^{\frac{-1}{N+1}}\left[\frac{1+2N}{(N+1)^2} g(\theta) + g''(\theta)\right] \tag{8.52}$$

Although the superscript p has been dropped from the above strains, they all indicate total plastic strains. Under plane strain condition, $\epsilon_{zz} = 0$, assuming z axis is perpendicular to the plane of the body. The above strains ensure incompressibilty condition, $\epsilon_{rr} + \epsilon_{\theta\theta} + \epsilon_{zz} = 0$. Considering a power law hardening material expressed in the following form

$$\gamma = \gamma_Y \left(\frac{\tau}{\tau_Y}\right)^{\frac{1}{N}} \tag{8.53}$$

where τ_Y is the yield stress in shear and shear yield strain $\gamma_Y = \dfrac{\tau_Y}{\mu}$, μ = rigidity modulus. Hence,

$$\tau^2 = \tau_Y^2 \left(\dfrac{\gamma}{\gamma_Y}\right)^{2N} \tag{8.54}$$

From the Mohr circle relationship, the maximum shear stress and strain are given by

$$\tau^2 = \dfrac{1}{4}(\sigma_{rr} - \sigma_{\theta\theta})^2 + \tau_{r\theta}^2 \tag{8.55}$$

$$\left(\dfrac{\gamma}{2}\right)^2 = \dfrac{1}{4}(\epsilon_{rr} - \epsilon_{\theta\theta})^2 + \left(\dfrac{\gamma_{r\theta}}{2}\right)^2 \tag{8.56}$$

Substituting τ and γ in Eq. (8.54) in terms of the three polar components of stresses and strains using Eqs. (8.55) and (8.56), the following relation is obtained.

$$\left[f''(\theta) + \dfrac{N(2+N)}{(N+1)^2} f(\theta)\right]^2 + \dfrac{4}{(N+1)^2}[f'(\theta)]^2$$

$$= \left(\dfrac{1}{\gamma_Y}\right)^{2N} \tau_Y^2 \dfrac{4(2+N)^2}{(N+1)^4} \left[\dfrac{4N^2}{(N+1)^2}\{g'(\theta)\}^2 + \left\{g''(\theta) + \dfrac{(2N+1)}{(N+1)^2}g(\theta)\right\}^2\right]^N \tag{8.57}$$

Further, according to Levy–Mises rule

$$\dfrac{\epsilon_{rr}}{S_{rr}} = \dfrac{\epsilon_{\theta\theta}}{S_{\theta\theta}} = \dfrac{\epsilon_{r\theta}}{S_{r\theta}} \tag{8.58}$$

This gives,

$$\epsilon_{rr} S_{\theta\theta} = \epsilon_{\theta\theta} S_{rr} \tag{8.59a}$$

$$\epsilon_{r\theta} S_{rr} = \epsilon_{rr} S_{r\theta} \tag{8.59b}$$

Noting that

$$\sigma_{zz} = \dfrac{1}{2}(\sigma_{rr} + \sigma_{\theta\theta}), \tag{8.60a}$$

$$S_{\theta\theta} = \dfrac{1}{2}(\sigma_{\theta\theta} - \sigma_{rr}) = -S_{rr}, \tag{8.60b}$$

Eq. (8.59b) gives

$$-\frac{1}{2}r^{-\frac{1}{N+1}}\left[\frac{(2N+1)}{(N+1)^2}g(\theta)+g''(\theta)\right]\left[\frac{1}{2}r^{-\frac{N}{(N+1)}}\left\{(N+1)f(\theta)+\frac{(N+1)^2}{(2+N)}f''(\theta)-f(\theta)\right\}\right]$$

$$=-\frac{N}{(N+1)}r^{-\frac{1}{N+1}}g'(\theta)\left[-\frac{1+N}{2+N}r^{\frac{-N}{N+1}}f'(\theta)\right] \tag{8.61}$$

After simplification

$$\left[f''(\theta)+\frac{N(2+N)}{(N+1)^2}f(\theta)\right]\left[g''(\theta)+\frac{(2N+1)}{(N+1)^2}g(\theta)\right]=-\frac{4N}{(N+1)^2}g'(\theta)f'(\theta) \tag{8.62}$$

Rice and Rosengren (1968) obtained a fourth order differential equation in $f(\theta)$ and solved it by applying the Runge–Kutta method and presented the solution in terms of plastic zone radius as follows.

$$\tau=\tau_Y\left[\frac{R(\theta)}{r}\right]^{\frac{N}{N+1}},\ \gamma=\gamma_Y\left[\frac{R(\theta)}{r}\right]^{\frac{1}{N+1}} \tag{8.63}$$

where $R(\theta)$ = distance up to the elastic plastic boundary (Fig. 8.11). For a perfectly plastic material (i.e., N = 0),

$$\tau=\tau_Y,\ \gamma\alpha\frac{1}{r} \tag{8.64}$$

The similar stress-strain variation within the plastic zone was obtained by Hult and McClintock (1957) for Mode III problem. The COD for Mode I was obtained by Rice and Rosengren as $0.58\frac{J}{\tau_Y}\approx\frac{J}{2\tau_Y}$.

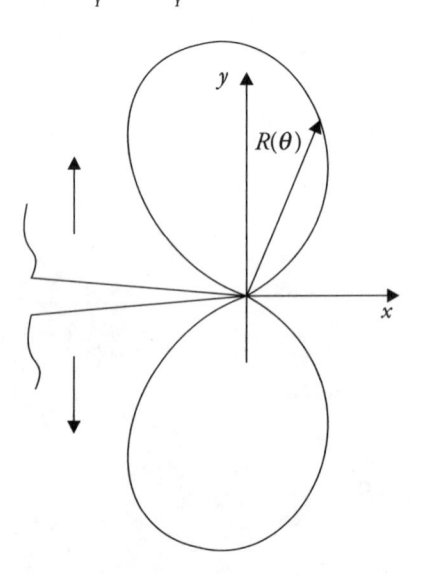

Figure 8.11 Radial variation of plastic zone size.

8.7.2 Hutchinson's Analysis

Hutchinson (1968a,b) solved the problem through the eigenfunction type approach of Williams (1957) for the linear elastic case. The material was considered to have the same Ramberg–Osgood type stress–strain property [Eq. (8.37)] in tensile test. Assuming proportional loading until initiation of crack growth, the stress–strain relations are given by

$$\epsilon_{ij} = \frac{1-2\nu}{3E}\sigma_{kk}\delta_{ij} + \frac{1+\nu}{E}S_{ij} + \frac{3}{2}\alpha\epsilon_Y \frac{1}{\bar{\sigma}}\left(\frac{\bar{\sigma}}{\sigma_Y}\right)^n S_{ij}$$

$$= \frac{1-2\nu}{3E}\sigma_{kk}\delta_{ij} + \frac{1+\nu}{E}S_{ij} + \frac{3}{2}\alpha\frac{1}{E}\left(\frac{\bar{\sigma}}{\sigma_Y}\right)^{n-1} S_{ij} \qquad (8.65)$$

where $\bar{\sigma}$ is the equivalent stress and is given under plane stress by

$$\bar{\sigma}^2 = \sigma_r^2 + \sigma_\theta^2 - \sigma_r\sigma_\theta + 3\tau_{r\theta}^2 \qquad (8.66)$$

Noting that

$$S_r = \frac{1}{3}(2\sigma_r - \sigma_\theta), \quad S_\theta = \frac{1}{3}(2\sigma_\theta - \sigma_r), \quad S_{r\theta} = \tau_{r\theta} \qquad (8.67)$$

The strains are given by

$$\epsilon_r = \frac{\sigma_r - \nu\sigma_\theta}{E} + \frac{\alpha}{E}\left(\frac{\bar{\sigma}}{\sigma_Y}\right)^{n-1}\left(\sigma_r - \frac{1}{2}\sigma_\theta\right) \qquad (8.68)$$

$$\epsilon_\theta = \frac{\sigma_\theta - \nu\sigma_r}{E} + \frac{\alpha}{E}\left(\frac{\bar{\sigma}}{\sigma_Y}\right)^{n-1}\left(\sigma_\theta - \frac{1}{2}\sigma_r\right) \qquad (8.69)$$

$$\epsilon_{r\theta} = \frac{1+\nu}{E}\tau_{r\theta} + \frac{3}{2}\frac{\alpha}{E}\left(\frac{\bar{\sigma}}{\sigma_Y}\right)^{n-1}\tau_{r\theta} \qquad (8.70)$$

The compatibility equation for the case has the following form.

$$\frac{1}{r}\frac{\partial^2}{\partial r^2}(r\epsilon_\theta) + \frac{1}{r^2}\frac{\partial^2\epsilon_r}{\partial r^2} - \frac{1}{r}\frac{\partial\epsilon_r}{\partial r} - \frac{2}{r^2}\frac{\partial}{\partial r}\left(r\frac{\partial\epsilon_\theta}{\partial\theta}\right) = 0 \qquad (8.71)$$

Substituting the stresses in terms of the Airy stress function Φ [see Eqs. (8.45) to (8.47)] in strain–stress relations [Eqs. (8.68) to (8.70)] and then making use of the compatibility equation [Eq. (8.71)], the following equation is obtained.

$$\nabla^4 \Phi + \frac{\alpha}{2} \left[\frac{1}{r} \frac{\partial^2}{\partial r^2} \left\{ \left(\frac{\bar{\sigma}}{\sigma_Y}\right)^{n-1} \left(2r \frac{\partial^2 \Phi}{\partial r^2} - \frac{\partial \Phi}{\partial r} - \frac{1}{r} \frac{\partial^2 \Phi}{\partial r^2}\right) \right\} + \frac{6}{r^2} \left\{ \left(\frac{\bar{\sigma}}{\sigma_Y}\right)^{n-1} r \frac{\partial}{\partial r} \left(\frac{1}{r} \frac{\partial \Phi}{\partial \theta}\right) \right\}$$

$$+ \frac{1}{r} \frac{\partial}{\partial r} \left\{ \left(\frac{\bar{\sigma}}{\sigma_Y}\right)^{n-1} \left(\frac{\partial^2 \Phi}{\partial r^2} - \frac{2}{r} \frac{\partial \Phi}{\partial r} - \frac{2}{r^2} \frac{\partial^2 \Phi}{\partial \theta^2}\right) \right\} + \frac{1}{r^2} \frac{\partial^2}{\partial \theta^2} \left\{ \left(\frac{\bar{\sigma}}{\sigma_Y}\right)^{n-1} \left(\frac{2}{r} \frac{\partial \Phi}{\partial r} - \frac{\partial^2 \Phi}{\partial r^2} + \frac{2}{r^2} \frac{\partial^2 \Phi}{\partial \theta^2}\right) \right\} \right] = 0$$

(8.72)

The stress function can be assumed in the form

$$\Phi = c r^s \tilde{\Phi}(\theta) \tag{8.73}$$

where $\tilde{\Phi}(\theta)$ is a function of θ only and c and s are real constants.

Using the stress function [Eq. (8.73)], the stresses are obtained as follows.

$$\sigma_\theta = cs(s-1) r^{s-2} \tilde{\Phi}(\theta) = c r^{s-2} \tilde{\sigma}_\theta(\theta) \tag{8.74}$$

$$\sigma_r = csr^{s-2} \tilde{\Phi}(\theta) + c r^{s-2} \tilde{\Phi}''(\theta) = c r^{s-2} \tilde{\sigma}_r(\theta) \tag{8.75}$$

$$\tau_{r\theta} = c(1-s) r^{s-2} \frac{\partial \tilde{\Phi}(\theta)}{\partial \theta} = c r^{s-2} \tilde{\tau}_\theta(\theta) \tag{8.76}$$

$$\bar{\sigma} = \left[\{cs(s-1) r^{s-2} \tilde{\Phi}(\theta)\}^2 + \left\{ csr^{s-2} \tilde{\Phi}(\theta) + c r^{s-2} \frac{\partial^2 \tilde{\Phi}(\theta)}{\partial \theta^2} \right\}^2 - \{cs(s-1)$$

$$r^{s-2} \tilde{\Phi}(\theta)\} \left\{ csr^{s-2} \tilde{\Phi}(\theta) + c r^{s-2} \frac{\partial^2 \tilde{\Phi}(\theta)}{\partial \theta^2} \right\} + 3 \left\{ c(1-s) r^{s-2} \frac{\partial \tilde{\Phi}(\theta)}{\partial \theta} \right\}^2 \right]^{\frac{1}{2}}$$

$$= csr^{s-2} \tilde{\bar{\sigma}}(\theta), \text{ say} \tag{8.77}$$

The dominant term is determined by the nonlinear part, that is, the second term after $\nabla^4 \Phi$, of the governing equation [Eq. (8.72)]. Since stresses σ_{ij} are proportional to r^{s-2}, strains ϵ_{ij} around the crack-tip region are proportional $r^{n(s-2)}$. The strain energy density near the tip (i.e., $r \to 0$) will dominate, provided $(n+1)(s-2) < 0$. That is, $s < 2$.

The nonlinear part of Eq. (8.72) gives rise to the eigenvalue equation in s in the following form.

$$\left[n(s-2) - \frac{\partial^2}{\partial \theta^2}\right] \left[\{\tilde{\sigma}(\theta)\}^{n-1} \left\{s(s-3)\tilde{\Phi}(\theta) - 2\frac{\partial^2 \tilde{\Phi}(\theta)}{\partial \theta^2}\right\}\right]$$

$$+ \left[\{n(s-2)+1\}n(s-2)\{\tilde{\sigma}(\theta)\}^{n-1} \left\{s(2s-3)\tilde{\Phi}(\theta) - \frac{\partial^2 \tilde{\Phi}(\theta)}{\partial \theta^2}\right\}\right]$$

$$+ \left[6\{n(s-2)+1\}(s-1)\frac{\partial}{\partial \theta}\left\{\frac{\partial \tilde{\Phi}(\theta)}{\partial \theta}\{\tilde{\sigma}(\theta)\}^{n-1}\right\}\right] = 0 \qquad (8.78)$$

This is a nonlinear equation. It has been solved numerically noting that the stress field is symmetric about x axis ($\theta = 0$) and crack edges are stress-free, that is, $\sigma_\theta = \tau_{r\theta} = 0$ for $\theta = \pm\pi$. The symmetry requires $\frac{\partial \tilde{\Phi}(\theta)}{\partial \theta} = \frac{\partial^3 \tilde{\Phi}(\theta)}{\partial \theta^3} = 0$ for $\theta = 0$, and the stress-free condition demands $\tilde{\Phi}(\theta) = \frac{\partial \tilde{\Phi}(\theta)}{\partial \theta} = 0$ for $\theta = \pm\pi$. The numerical solution gives the approximate solution $s = \frac{2n+1}{n+1}$ for both the plane stress and plane strain conditions.

The solution for s indicates that there is $r^{-\frac{1}{n+1}}$ singularity in stress and $r^{-\frac{n}{n+1}}$ singularity in strain. The complete stress–strain field around the crack-tip is given by

$$\sigma_{ij} = \sigma_Y \left(\frac{J}{\alpha \sigma_Y \epsilon_Y I_n r}\right)^{\frac{1}{n+1}} \tilde{\sigma}_{ij}(\theta, n) \qquad (8.79)$$

$$\epsilon_{ij} = \alpha \epsilon_Y \left(\frac{J}{\alpha \sigma_Y \epsilon_Y I_n r}\right)^{\frac{n}{n+1}} \tilde{\epsilon}_{ij}(\theta, n) \qquad (8.80)$$

$$COD = \delta = \frac{J}{\sigma_Y I_n} \left(\frac{\alpha \sigma_Y \epsilon_Y I_n r}{J}\right)^{\frac{1}{n+1}} \tilde{v}(\theta, n) \qquad (8.81)$$

where I_n is a constant dependent on the state of stress and material constant n. Typical values are shown in Table 8.1 (Hutchinson 1968a).

The θ-dependent variation of the three dimensionless functions $\tilde{\sigma}_{ij}(\theta, n)$, $\tilde{\epsilon}_{ij}(\theta, n)$, and $\tilde{v}(\theta, n)$ depends on the state of stress, plane stress or plane strain condition, and the material constant n.

Table 8.1 Dependance of I_n on n and state of stress.

State of stress	Values of I_n			
	$n = 3$	$n = 5$	$n = 9$	$n = 13$
Plane stress	3.86	3.41	3.03	2.87
Plane strain	5.33	5.01	4.60	4.40

Source: Hutchinson (1968a)

Furthermore,

$$u_i = \alpha \epsilon_Y \left(\frac{J}{\alpha \sigma_Y \epsilon_Y I_n} \right)^{\frac{n}{n+1}} r^{\frac{1}{n+1}} \tilde{u}_i(\theta, n) \tag{8.82}$$

$$\sigma_{33} = \frac{1}{2}(\sigma_{11} + \sigma_{22}), \quad \epsilon_{33} = 0 \quad \text{for plane strain} \tag{8.83a}$$

$$\epsilon_{33} = -(\epsilon_{11} + \epsilon_{22}), \quad \sigma_{33} = 0 \quad \text{for plane stress} \tag{8.83b}$$

$$I_n = \int_{-\pi}^{\pi} \left\{ \frac{n}{n+1} \left(\frac{\bar{\sigma}}{\sigma_Y} \right)^{n+1} \cos\theta - \frac{1}{\sigma_Y} \left[\tilde{\sigma}_{rr} \left(\tilde{u}_\theta - \frac{\partial \tilde{u}_r}{\partial \theta} \right) - \tilde{\sigma}_{r\theta} \left(\tilde{u}_r + \frac{\partial \tilde{u}_\theta}{\partial \theta} \right) \right] \sin\theta \right.$$

$$\left. - \frac{1}{n+1} \frac{1}{\sigma_Y} [\tilde{\sigma}_{rr} \tilde{u}_r + \tilde{\sigma}_{r\theta} \tilde{u}_\theta] \cos\theta \right\} d\theta \tag{8.84}$$

$$\tilde{u}_r = (n+1) \left(\frac{\bar{\sigma}}{\sigma_Y} \right)^{n-1} \left[\tilde{\Phi}'' + \frac{s(3-s)\tilde{\Phi}}{2} \right] \tag{8.85}$$

$$\frac{\partial \tilde{u}_\theta}{\partial \theta} = -\frac{1}{2} \left(\frac{\bar{\sigma}}{\sigma_Y} \right)^{n-1} [\tilde{\Phi}'' - s(2s-3)\tilde{\Phi}] - \tilde{u}_r, \quad s = \frac{2n+1}{n+1} \tag{8.86}$$

According to Tracy's (1976) definition of CTOD, δ is equal to the opening BC between the two intersection points B and C of the crack profile and 45° intercepts OB and OC from the crack-tip O (Fig. 8.12).

$$\delta = \text{CTOD} = 2v(r, \pi) = 2[r - u(r, \pi)] \tag{8.87}$$

Noting that $u_1 = u(r, \pi)$ and $u_2 = v(r, \pi)$. $u(r, \pi)$ and $v(r, \pi)$ indicate the displacements of a point A at a distance r behind the crack-tip.

$$r = u(r, \pi) + v(r, \pi) = (\alpha \epsilon_Y)^{\frac{1}{n}} [\tilde{u}(\pi) + \tilde{v}(\pi)]^{\frac{n+1}{n}} \frac{J}{\sigma_Y I_n} \tag{8.88}$$

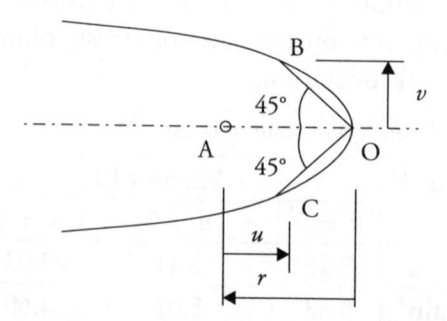

Figure 8.12 Crack-tip opening displacement according to Tracy (1976).

where $\tilde{u}(\pi) = \tilde{u}(\pi, n)$ and $\tilde{v}(\pi) = \tilde{v}(\pi, n)$.

$$\delta = d_n \frac{J}{\sigma_Y}, \quad d_n = 2\left\{\alpha \epsilon_Y [\tilde{u}(\pi) + \tilde{v}(\pi)]\right\}^{\frac{1}{n}} \tilde{v}(\pi)/I_n \qquad (8.89)$$

For a perfectly plastic material ($n = \infty$), the strain field only exhibits a r^{-1} singularity, which was seen in the case of Mode III as well. In this perfectly plastic situation, the crack-tip fields can be written in the following form.

$$\sigma_{ij} = \sigma_Y \tilde{\sigma}_{ij}(\theta, \infty) \qquad (8.90)$$

$$\epsilon_{ij} = \frac{J}{\sigma_Y I_n r} \tilde{\epsilon}_{ij}(\theta, \infty) = \frac{\delta}{d\, I_n\, r} \tilde{\epsilon}_{ij}(\theta, \infty) \qquad (8.91)$$

where $J = \frac{\sigma_Y \delta}{d}$, $d = 0.8$ for large n and 0.3 for $n = 3$.

Equations (8.79) and (8.80), and Eqs. (8.90) and (8.91) are known as Hutchinson-Rice–Rosengren (HRR) singularity fields, and the associated singularity is known as HRR singularity. Either J or δ can be treated as field parameters. When HRR field embeds the fracture process zone, the field parameters J or δ are likely to characterize the fracture process or the crack growth.

Under monotonic loading, yielding behaviour of metallic materials is the same as given by the deformation theory till fracture. If the region of finite strain, where void nucleation, coalescence, and crack growth, called process zone (Fig. 8.13), is small compared to the zone where small strain formulation is valid, the initiation of crack growth can be given in terms of J or δ. The finite strain based finite element calculations have shown that over a distance of about 2δ to 3δ, results based on small strain and finite strain calculations differ. The process zone can be taken less than 3δ. Therefore, if the radius R of the HRR field is large enough compared to this size of the process zone, J or δ can serve as the fracture characterizing parameter in elastic–plastic or yielding fracture mechanics in close parallel to K, the SIF, in the LEFM. In the case of hardening materials, under small- or large-scale yielding, the HRR singularity dominates over a significant distance R. This distance is always greater than four to six times δ ahead of the crack-tip when the uncracked ligament b is mostly subjected to primarily bending load as in the case of single edge notched bend (SENB), or three point bend (TPB) test, specimen. If the ligament is primarily in tension, however, as in the case of centre cracked tension (CCT) specimen, the size R singularity zone is not that significant. It has been established (Hutchinson 1983) that under fully yielded condition of the ligament b, in the case of TPB specimen, acceptable size of R is guaranteed,

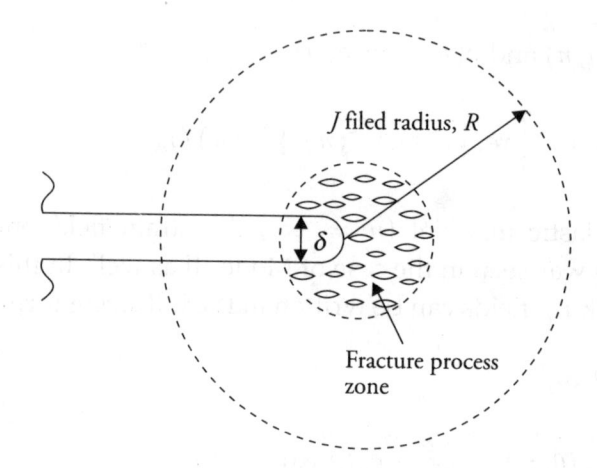

Figure 8.13 Near-tip fracture process zone and extent of J field.

provided dimension b is greater than $25 \dfrac{J}{\sigma_Y}$. Similarly, in the case of CCT specimen, the similar condition is guaranteed, provided b is greater than $175 \dfrac{J}{\sigma_Y}$.

The size R of the region of J dominance depends on geometry of specimen and the hardening exponent n. The geometry has a strong influence in the case of low hardening (i.e., n $\to$ ∞, or perfectly plastic) material. The hardening solution for large n does not converge to the slip-line field solution. This is because the governing equation for the hardening material is elliptic, and the equation for a non-hardening material is hyperbolic. In the case of non-hardening material, there is no unique solution independent of geometry.

Stress state at the crack-tip is not unique under large-scale yielding and perfect plasticity. Under fully yielded conditions, stress–strain field and J are strong functions of specimen boundary geometry even under plane strain condition. J is a unique configuration-independent parameter associated with the crack-tip field only when there is some strain hardening.

8.8 Experimental Determination of J

Begley and Landes (1972) and Landes and Begley (1972) gave a procedure for experimental determination of J. Specimens can be prepared with different crack sizes (Fig. 8.14(a)). By loading them gradually, variation of load with load point displacement can be plotted (Fig. 8.14(b)). Using these data variation of strain energy, U with crack size a up to different displacement levels (Fig. 8.14(c)) can obtained from Fig. 8.14(b). By determining the slope of these plots for a particular crack size a, variation of $J = \dfrac{1}{B}\dfrac{dU}{da}$, where B is specimen thickness, with

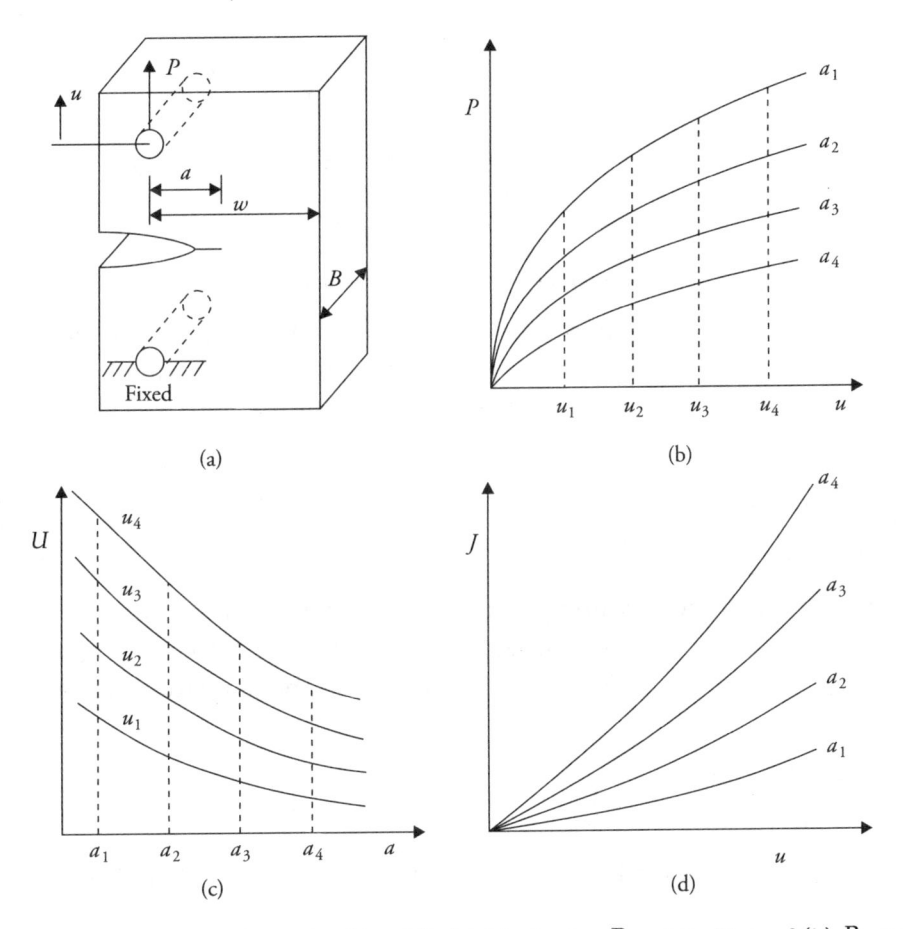

Figure 8.14 Experimental determination of J. (a) Monitoring P and u. Plots of (b) P versus u for constant crack sizes, (c) U versus a for constant displacements and (d) J versus u for constant crack sizes.

displacement (Fig. 8.14(d)) for constant crack length can be determined. By noting the experimental displacement at the onset of crack extension corresponding to a specified crack size a, the fracture toughness can be determined.

8.9 Alternative Methods for Measuring J

The above procedure implies testing of a number of specimens. Later, Rice, Paris and Merkle (1973) and Merkle and Corten (1974) have given some basis whereby the same data can be obtained through testing of less number of specimens. In the case of bend specimens, if the whole ligament undergoes plastic deformation (Fig. 8.15) it is possible to write

$$\theta = \theta_{el} + \theta_{pl} \tag{8.92}$$

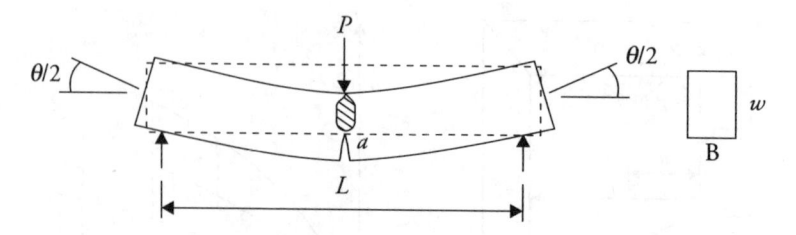

Figure 8.15 Bending deformation of three point bend specimen.

Considering Ramberg–Osgood type of material, when the plastic deformation dominates, the elastic part can be neglected and $\theta \approx \theta_{pl}$. The angle can then be written in terms of bending moment acting on the section, $\epsilon_Y = \sigma_Y/E$, and hardening exponent n

$$\theta_{pl} = f\left(\frac{M}{Bb^2\sigma_Y}, \frac{\sigma_Y}{E}, n\right), \tag{8.93}$$

where $b = w - a$ is the remaining ligament dimension and f is a function of the three dimensionless variables. Noting that for a TPB specimen bending moment at the centre of the span $M = PL/4$ and inverting the above relation it is possible to write

$$P = \frac{4Bb^2\sigma_Y}{L} h\left(\theta_{pl}, \frac{\sigma_Y}{E}, n\right) \tag{8.94}$$

where h is a function of the three dimensionless variables. By definition

$$J = -\frac{1}{B}\frac{\partial}{\partial a}\left\{\int P du\right\}_{u=\text{constant}} = -\frac{1}{B}\frac{\partial}{\partial a}\left\{\int \frac{4Bb^2\sigma_Y}{L} h\left(\theta_{pl}, \frac{\sigma_Y}{E}, n\right) du\right\}_{u=\text{constant}}$$

$$= \frac{1}{B}\frac{\partial}{\partial b}\left\{\int \frac{4Bb^2\sigma_Y}{L} h\left(\theta_{pl}, \frac{\sigma_Y}{E}, n\right) du\right\}_{u=\text{constant}} = \frac{2}{Bb}\int_0^{u^*} P du \tag{8.95}$$

noting the form of P above [Eq. (8.94)] and b is equal to $(w - a)$. Finally

$$J = \frac{2A}{Bb} \tag{8.96}$$

where A is the area under the load–load point displacement diagram up to the displacement u^* (Fig. 8.16). Equation (8.96) has been derived considering only the plastic part of the deformation. A similar expression can also be derived considering the elastic part of the deformation. This gives

$$J = \frac{2}{Bb}\left\{\int_0^{\theta_{el}} P du\right\}_{u=u^*} + \frac{2}{Bb}\left\{\int_0^{\theta_{pl}} P du\right\}_{u=u^*} = \frac{K^2}{E'} + \frac{2}{Bb}\left\{\int_0^{\theta_{pl}} P du\right\}_{u=u^*} \tag{8.97}$$

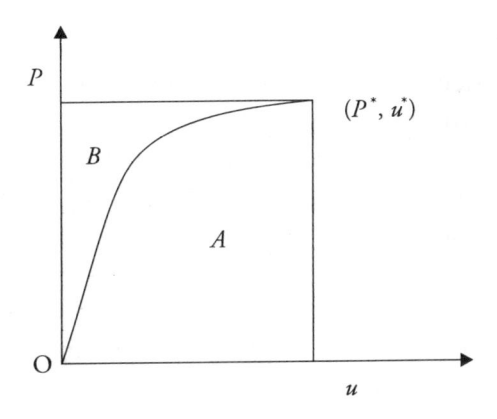

Figure 8.16 Areas involved in J calculation.

where K is the SIF and $E' = E$, the modulus of elasticity, under plane stress and $E' = E/(1-v^2)$ under plane strain. Strictly Eq. (8.96) should include only the plastic part of the displacement. This relation is accurate for crack size $a/w > 0.5$ for TPB specimen. According to Landes, Walker, and Clarke (1979), J can be obtained through Eq. (8.96) considering the total of elastic and plastic displacements.

For compact tension (CT) specimen, Eq. (8.96) is not accurate because the specimen is subjected to both tensile and bending load on the crack plane. Merkle and Corten (1974) suggested the following relation for the CT specimen.

$$J = \frac{2}{Bb}\frac{1+\beta}{1+\beta^2}\int_0^{u^*} P\,du + \frac{2\beta}{Bb}\frac{1-2\beta-\beta^2}{(1+\beta^2)^2}\int_0^{P^*} u\,dP \qquad (8.98)$$

where β is given by

$$\beta = \left[4r^2 + 4r + 2\right]^{1/2} - 2r - 1, \quad r - \frac{a}{b} \qquad (8.99)$$

and a is crack size, and P^* and u^* are load and displacements at fracture. Indeed, the second integral of Eq. (8.98) indicates the complementary strain energy and is given by the area B (Fig. 8.16) between the load and load–load line displacement diagram and the load axis up to level P^*. When area A is large compared to area B,

$$J = \frac{2}{Bb}\frac{1+\beta}{1+\beta^2}\int_0^{u^*} P\,du \qquad (8.100)$$

The above relation is recommended for the calculation of J (Kanninen and Popelar 1985) from experimental load-displacement data. Equations (8.96) and (8.98) or (8.100) permit determination of J through testing of single specimen.

Problem 8.1

During TPB testing of steel, the following data was obtained: Displacement varies linearly from 0 to 2.0 mm with a slope 25 MN/m. Further variation of load from this displacement level to 4.0 mm is $P = 0.05 \times 10^6 + 1.5 \times 10^6 [(u-2)10^{-3}]^{\frac{1}{2}}$ N. The specimen dimensions are: $B = a = b = 40$ mm. Calculate J_c if the crack extension begins from the given crack size 40 mm.

Solution

Area A under the load displacement diagram up to 4 mm consists of both the linear and nonlinear parts. Through integration $A = 239.425$ Nm.

$$J_C = \frac{2A}{Bb} = \frac{2 \times 238.425}{0.04 \times 0.04} = 0.30 \text{ MPam (Ans.)}.$$

Problem 8.2

For the same crack-section dimensions, when a CT specimen of the same material was tested, the area under the load displacement record reduced by a factor of 0.9 from that obtained by testing a TPB specimen (earlier solved example). Calculate the fracture toughness J_C.

Solution

Area under the load displacement record up the point of fracture, $A = 861.3$ Nm.
Using Eq. (8.99), for this case $\beta = [4r^2 + 4r + 2]^{1/2} - 2r - 1$.
Since $r = \frac{a}{b} = 1$ and $\beta = 0.16227$, J_C is given by

$$J_C = \frac{1+\beta}{1+\beta^2} \frac{2A}{Bb} = 0.304 \text{ MPam (Ans.)}.$$

8.10 Crack-Tip Constraints: T Stress and Q Factor

The elastic stress field in the neighbourhood of the crack-tip is influenced by the boundary loading and component in-plane dimensions. The stress field as radius $r \to 0$ is given by the first term of the Williams' eigenfunction expansion. The first non-singular term contributes to the stress in the direction parallel to the crack. This is generally referred to as T stress. The crack-tip stress field can be then represented by

$$\sigma_{ij} = (\sigma_{ij})_{SST} + T\delta_{ij} \qquad (8.101)$$

where $(\sigma_{ij})_{SST}$ is the contribution due to the stress singularity term and δ_{ij} is Kronecker's delta.

$\delta_{ij} = 0$ for $i = j = 2$.

T stress affects the plastic zone around the crack-tip, stability of crack growth, and, in the case of mixed mode problems, it influences the direction of crack extension. In case there is plastic deformation around the crack-tip, the stress field consists of the HRR field and the contributions of the non-singular terms. The field can be represented by

$$\sigma_{ij} = (\sigma_{ij})_{HRR} + (\sigma_{ij})_{Diff} \qquad (8.102)$$

where $(\sigma_{ij})_{Diff}$ is the contributions due to non-singular terms. It is shown (O'Dowd and Shih 1991) that the difference field does not vary appreciably over the angular span $\theta = -45°$ to $\theta = +45°$ irrespective of r. Further, $(\sigma_x)_{Diff} \approx (\sigma_y)_{Diff} \gg (\sigma_{xy})_{Diff}$. Therefore, in the presence of plastic deformation around the tip, plane strain condition, and plastic incompressibilty, $(\sigma_z)_{Diff} = \frac{1}{2}[(\sigma_x)_{Diff} + (\sigma_y)_{Diff}] = (\sigma_x)_{Diff} = (\sigma_y)_{Diff}$. Hence, the difference in each of the normal stresses is equal to the shift in hydrostatic stress. The existence of triaxiality at the crack-tip can be expressed by a parameter Q given by

$$Q = \frac{\sigma_y - (\sigma_y)_{HRR}}{\sigma_Y} = \frac{\sigma_x - (\sigma_x)_{HRR}}{\sigma_Y} \qquad (8.103)$$

at a finite distance from the crack-tip for $\theta = 0°$. This distance r can be taken as $2J/\sigma_Y$. This parameter Q is known as plastic constraint. It depends on the in-plane dimensions of the specimen and its thickness. In the case of uniaxial loading, there are two-dimensional constraints: in-plane and out-of-plane directions. The in-plane constraint will depend on the position of the outer boundary with respect to the crack-tip. The out-of-plane constraint will depend on the specimen thickness. Hence, in general, the fracture toughness J_C measured using a particular specimen configuration will vary with specimen dimensions. Thereby, it is possible to obtain a variation of fracture toughness with Q. The parameter Q can be determined through finite element analysis of the specimen. Q is given by

$$\sigma_y = (\sigma_y)_{HRR\, or\, T=0} + Q\sigma_Y \qquad (8.104)$$

If Q is zero, the stresses are given by the HRR field. If Q is positive, the difference field increases. If Q is negative, there is loss of constraint in both the in-plane and out-of-plane directions, and the stresses fall below the level given by the HRR field. Through modified boundary layer analysis, it has been shown that T and Q are directly related over a wide range of hardening exponent (O'Dowd and

Shih 1991, 1992). However, T has relevance in predominantly elastic situations and cases involving small-scale plasticity; Q has relevance under both small- and large-scale plastic deformations. In general, if Q is positive, there is constraint on the plastic deformation at the crack-tip. As Q becomes negative, the constraint reduces, and the plastic deformation occurs on a larger scale. As a result, the fracture toughness increases (Fig. 8.17). It has been shown by the same authors that Q remains almost zero in the case of TPB specimen over a large span ahead of the crack-tip. This is not so in the case of CCT specimen; Q becomes negative after a small distance ahead of the crack-tip. Therefore, there is more constraint over a larger span ahead of the crack-tip in the case of TPB specimen compared to the CCT specimen. A position similar to TPB specimen exists in the case of deeply cracked CT specimen.

The plastic constraint Q is dependent on the geometry of the specimen or the component. Therefore, it is not possible to get a single value of J_C by testing different specimens. Through testing specimens of different sizes and geometry, it is possible to obtain the variation of J_C with Q for the material. To apply this material data in design, it is necessary to ensure that $J_{Design} = J_C$, and the constraint factor in the design example has the same value as the one associated with J_C. This is explained in Fig. 8.17. If the toughness J_C of a material varies with Q as given by the curve OAB (Fig. 8.17), and the loading on a component of the same material leads to the variation of applied J with Q along the curve OAC, the fracture load will correspond to the point A, where both the cases have the same Q. This shows how laboratory test data can be transferred to practice.

Anderson and Dodds Jr. (1991) have shown that to employ a microscopic fracture model like that of Ritchie, Knott and Rice (1973), which is based on stress,

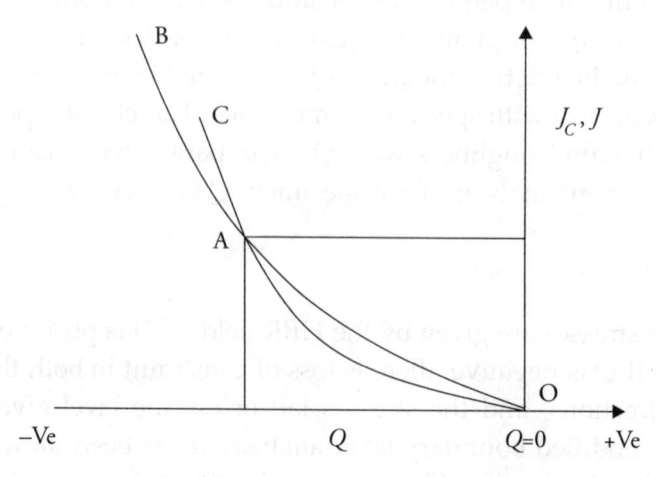

Figure 8.17 Variation of fracture toughness with constraint parameter Q.

the similarity in volume of the material undergoing the same stress level must be checked. That is, both the stress level and the volume of material subjected to the same stress level in the test case and the design problem must be the same.

8.11 Crack Propagation and Crack Growth Stability

After the onset of crack growth, further propagation in an elastic plastic material takes place with increasing load. For a growing crack, both an elastic unloading and a non-proportional loading occur near the crack-tip (Fig. 8.18) within the fracture process zone. None of these processes is adequately represented by the deformation theory of plasticity. These make the determination of crack-tip field ahead of an extending crack difficult mathematically.

Rice (1975), based on J_2 flow theory of plasticity, showed that for a fully plastic material ($n = \infty$), the incremental strains in the immediate vicinity of crack-tip are given by

$$d\epsilon_{ij} = \frac{d\delta}{r} f_{ij}(\theta) + \frac{\sigma_Y}{E} \frac{da}{r} \ln \frac{R(\theta)}{r} g_{ij}(\theta) \tag{8.105}$$

Therefore,

$$\frac{d\epsilon_{ij}}{da} = \frac{1}{r} \frac{d\delta}{da} f_{ij}(\theta) + \frac{\sigma_Y}{E} \frac{1}{r} \ln \frac{R(\theta)}{r} g_{ij}(\theta) \tag{8.106}$$

where $d\delta$ = increase in COD, da = increment in crack length due to extension, $R(\theta)$ is the distance of the elastic plastic boundary from the crack-tip, and $f_{ij}(\theta)$ and $g_{ij}(\theta)$ are the functions of θ.

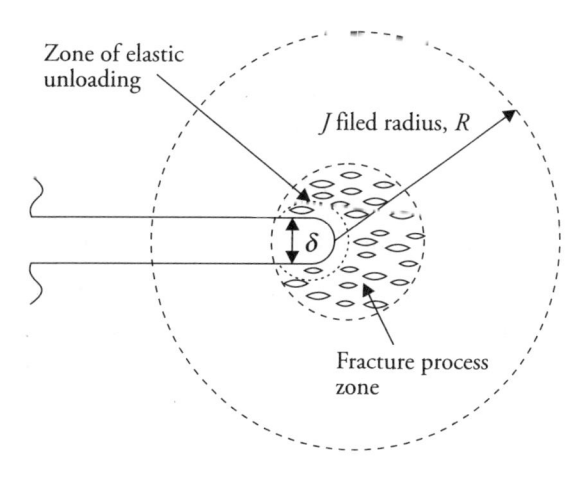

Figure 8.18 Elastic unloading and fracture process zones embedded in J field.

$g_{ij}(\theta)$ has value of the order of unity. The first term represents additional strain due to crack-tip blunting if the crack did not advance during load/displacement increments. The second term represents the additional plastic strains caused by the advance of the stress field through the material. In many cases, the first term dominates over a significant interval, several times CTOD δ_t, except right at the crack-tip, where the singularity $\frac{1}{r} \ln \frac{R(\theta)}{r}$ dominates. In other words, the strains in the close neighbourhood of crack-tip are uniquely characterized by the crack-tip opening angle $\frac{d\delta}{da}$, if $\frac{d\delta}{da} \gg \frac{\sigma_Y}{E} \ln \frac{R(\theta)}{r}$.

Hutchinson and Paris (1979), based on J_2 deformation theory of plasticity, showed that for a fully plastic material ($n = \infty$),

$$d\epsilon_{ij} = \frac{1}{\alpha\sigma_Y} \frac{dJ}{r} f_{ij}(\theta) + \frac{J}{\alpha\sigma_Y} \frac{da}{r^2} h_{ij}(\theta) \qquad (8.107)$$

That is,

$$\frac{d\epsilon_{ij}}{da} = \frac{1}{\alpha\sigma_Y} \frac{dJ}{da} \frac{1}{r} f_{ij}(\theta) + \frac{J}{\alpha\sigma_Y} \frac{1}{r^2} h_{ij}(\theta) \qquad (8.108)$$

where $h_{ij}(\theta)$ is the dimensionless function of θ of magnitude about unity. This relation indicates that the first term dominates the field if $\frac{dJ}{da} \gg \frac{J}{r}$. The similarity in the structure of the two relations [Eqs. (8.105) and (8.107)] is notable though they are derived based on J_2 flow and J_2 deformation theories, respectively. The first term in Eq. (8.107) indicates proportional increments in the strain fields due to an increase in the strength J of the HRR singularity, while the second term shows the nonproportional strain increments due to an advance of the HRR field with the extending crack. Therefore, if the HRR field increases in strength more rapidly than it advances, the crack-tip opening angle $\frac{d\delta}{da}$ and $\frac{dJ}{da}$ describe the changes in crack-tip environment. This means that when the fracture process zone is enclosed by the region dominated by $\frac{d\delta}{da}$ or $\frac{dJ}{da}$, Eqs. (8.105) and (8.107) together with the constraints, $\frac{d\delta}{da} \gg \frac{\sigma_Y}{E} \ln \frac{R(\theta)}{r}$ and $\frac{dJ}{da} \gg \frac{J}{r}$, provide the basis for COD-based and J-based resistance curve approach to stable crack growth respectively.

Similar results are obtainable through a slightly different approach. According to the HRR solution, the strain field for a hardening material (Hutchinson and Paris 1979) is given by

$$\epsilon_{ij} = b_n J^{\frac{n}{n+1}} r^{-\frac{n}{n+1}} \tilde{\epsilon}_{ij} \qquad (8.109)$$

where b_n is a dimensionless constant and $\tilde{\epsilon}_{ij} = \tilde{\epsilon}_{ij}(\theta, n)$. θ-variation of $\tilde{\epsilon}_{ij}$ depends on state of stress and hardening exponent n.

$$d\epsilon_{ij} = \frac{\partial \epsilon_{ij}}{\partial J} dJ + \frac{\partial \epsilon_{ij}}{\partial x} dx \qquad (8.110)$$

Assuming crack grows along the x axis under Mode I loading, $dx = -da$ and

$$\frac{\partial \epsilon_{ij}}{\partial x} = \frac{\partial \epsilon_{ij}}{\partial r} \cos\theta + \frac{\partial \epsilon_{ij}}{\partial \theta}\left(-\frac{\sin\theta}{r}\right) \qquad (8.111)$$

$$d\epsilon_{ij} = b_n \frac{n}{n+1} J^{-\frac{1}{n+1}} dJ\, r^{-\frac{n}{n+1}} \tilde{\epsilon}_{ij} + b_n J^{\frac{n}{n+1}} \left[\frac{n}{n+1} r^{-\frac{2n+1}{n+1}} \tilde{\epsilon}_{ij} \cos\theta + r^{-\frac{n}{n+1}} \frac{\sin\theta}{r} \frac{\partial \tilde{\epsilon}_{ij}}{\partial \theta}\right] da$$

$$= b_n J^{\frac{n}{n+1}} r^{-\frac{n}{n+1}} \left[\frac{dJ}{J} \frac{n}{n+1} \tilde{\epsilon}_{ij} + \frac{da}{r}\left(\frac{n}{n+1}\cos\theta\, \tilde{\epsilon}_{ij} + \frac{\partial \tilde{\epsilon}_{ij}}{\partial \theta}\sin\theta\right)\right] \qquad (8.112)$$

Finally,

$$d\epsilon_{ij} = b_n J^{\frac{n}{n+1}} r^{-\frac{n}{n+1}} \left[\frac{dJ}{J} \frac{n}{n+1} \tilde{\epsilon}_{ij} + \frac{da}{r} \tilde{\beta}_{ij}\right], \qquad (8.113)$$

where $\tilde{\beta}_{ij}$ stands for the expression within () in Eq. (8.112). Since J is directly related to the external load, the first term corresponds to the proportional part of strain increment and proportional loading. The second part, arising from da, is non-proportional. Since $\tilde{\epsilon}_{ij}$ and $\tilde{\beta}_{ij}$ are comparable, nearly proportional loading will occur, provided

$$\frac{dJ}{J} \gg \frac{da}{r} \qquad (8.114)$$

Paris et al. (1979) introduced a non-dimensional parameter

$$T_{mc} = \frac{E}{\sigma_Y^2} \frac{dJ_R}{da} \qquad (8.115)$$

called tearing modulus. This signifies material's resistance to crack growth. In the above relationship, $\frac{dJ_R}{da}$ is the slope of the resistance curve J_R versus Δa (Fig. 8.19). The condition for crack growth instability is given by

$$T_m \geq T_{mc} \qquad (8.116)$$

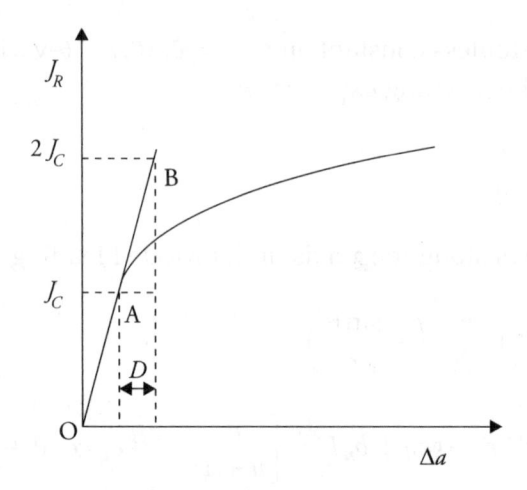

Figure 8.19 J-resistance curve.

where $T_m = \dfrac{E}{\sigma_Y^2}\dfrac{dJ}{da}$ is the slope of the loading curve or the crack driving force. During the stable crack growth, the slope of the J-loading curve $\dfrac{dJ}{da}$, or crack opening angle $\dfrac{d\delta}{da}$, remains constant and $T_m < T_{mc}$.

According to Paris et al., during stable crack extension, T_m remains constant and the value of T_{mc} is dependent on specimen geometry, loading, etc. The tearing modulus can also be defined in terms of δ-resistance curve:

$$T_{mc(\delta)} = \dfrac{E}{\sigma_Y^2}\dfrac{d\delta_R}{da} \tag{8.117}$$

where $\dfrac{d\delta_R}{da}$ is the slope of the resistance curve δ_R versus a. Under J-controlled crack extension, Shih (1981) has shown that $T_{mc(\delta)}$ and T_{mc} are related by a constant.

From J-resistance curve, J_R versus Δa (Fig. 8.19), an approximate crack growth D that corresponds to an increase in the fracture resistance from J_C to $2J_C$ is given by

$$D = \dfrac{J_C}{\left(\dfrac{dJ_R}{da}\right)_A} \tag{8.118}$$

where $\left(\dfrac{dJ_R}{da}\right)_A$ is approximately the slope of resistance curve at the initiation point A. For the crack growth D to occur under J dominance, $D \ll r \ll R$ must be satisfied. r is the radius referred from the crack-tip and R is the radius of the J field. For a specimen that can ensure J-controlled crack growth, the ligament dimension

$b(= w - a)$, where w is total width, must be much greater than R. Therefore, the ratio $\omega = \dfrac{b}{D}$ must be a very large quantity. That is,

$$\omega = \frac{b}{J_C}\frac{dJ_R}{da} \gg 1 \tag{8.119}$$

For fixing specimen dimensions, ω has been recommended to be about 40.

8.12 Engineering Estimates of J

SIF handbooks facilitated application of the principles of LEFM extensively. Similar source for the evaluation of J under the condition of elastic plastic deformation was missing. Kumar, German, and Shih (1981) of the Electric Power Research Institute (EPRI) first presented the solutions for its evaluation for a number of common geometries. The basis for deriving the relation for J is discussed subsequently.

According to Il'yushin (Kanninen and Popelar 1985), when a component made up of a material with nonlinear stress-strain relationship is loaded gradually by a point load, or displacement, on its boundary, the stress and strain within the body display two important trends. The stress increases in direct proportion to the load P, and the strain increases in proportion to P^n. Further, since the stresses increase in proportion to the load, the results based on the deformation and incremental theory of plasticity are the same. Therefore

$$\epsilon = \epsilon_Y \alpha \left(\frac{\sigma}{\sigma_Y}\right)^n, \tag{8.120}$$

$$\frac{\sigma}{\sigma_Y} \propto \frac{P}{P_Y}, \quad \frac{\epsilon}{\epsilon_Y} \propto \left(\frac{P}{P_Y}\right)^n \tag{8.121}$$

where P_Y is the reference load. Kumar, German, and Shih termed P_Y as the limit load, and it is proportional to σ_Y. Since J depends on the product of stress and strain, it must be proportional to $\left(\frac{P}{P_Y}\right)^{n+1}$.

The same results can also be obtained from the HRR solution for Mode I crack-tip stresses and strains given by Eqs. (8.79) and (8.80), respectively, based on the deformation theory of plasticity. Inverting the first equation,

$$J = \alpha \epsilon_Y \sigma_Y I_n r \left(\frac{\sigma_{ij}}{\sigma_Y}\right)^{n+1} [\tilde{\sigma}_{ij}(\theta)]^{-(n+1)} \tag{8.122}$$

J given by Eqs. (8.96) and (8.98) or (8.100) can be considered to be due to the plastic part of the deformation only. Since σ_{ij} is proportional to external load P and σ_Y is proportional to P_Y, the above relation can be rewritten to give J_p in the following form.

$$J_p = \alpha \epsilon_Y \sigma_Y b \; h_1\left(\frac{a}{w}, n\right) \left(\frac{P}{P_Y}\right)^{n+1} \tag{8.123}$$

where h_1 is a dimensionless function of $\frac{a}{w}$ and n, and b is related to the specimen dimensions. The function h_1 depends on the aspect ratio a/w of the crack, stress exponent n in Ramberg–Osgood relation, and the state of stress. For the CT specimen, the plastic parts of crack mouth displacement δ_{mthp} and load line displacement δ_{lldp} are given by

$$\delta_{mthp} = \alpha \, \epsilon_Y a \, h_2\left(\frac{a}{w}, n\right) \left(\frac{P}{P_Y}\right)^{n+1} \tag{8.124}$$

$$\delta_{lldp} = \alpha \, \epsilon_Y a \, h_3\left(\frac{a}{w}, n\right) \left(\frac{P}{P_Y}\right)^{n+1} \tag{8.125}$$

where h_2 and h_3 are functions of $\frac{a}{w}$ and n.

For a single edge cracked panel (SECP) under uniform remote uniform tension, the elastic part J_e of the J integral is obtained from the following relation.

$$J_e = \frac{K^2}{E'} \tag{8.126}$$

where $K = \sigma \sqrt{\pi a_e} \; f\left(\frac{a_e}{w}\right)$, $f\left(\frac{a_e}{w}\right)$ is the SIF correction factor, effective crack length a_e is calculated from (Kumar, German, and Shih 1981)

$$a_e = a + r_y = a + \frac{1}{1 + \left(\frac{P}{P_Y}\right)^2} \left[\frac{n-1}{n+1}\right] \frac{1}{k\pi} \left(\frac{K}{\sigma_Y}\right)^2 \tag{8.127}$$

$k = 2$ for plane stress and 6 for plane strain. $E' = E$ for plane stress and $\frac{E}{1-v^2}$ for plane strain. For calculating a_e, SIF K in Eq. (8.126) can be calculated using given crack size, that is, without the correction for plastic zone. No further iteration is required.

For the single edge crack plate under uniform tension at the ends, the plastic part of relative rotation θ_p between the two ends of the specimen is given by

$$\theta_p = \alpha \, \epsilon_Y \, h_5\left(\frac{a}{w}, n\right) \left(\frac{P}{P_Y}\right)^n \tag{8.128}$$

Elastic plastic fracture mechanics

In general, the total J can be obtained by adding the elastic and plastic parts, $J = J_e + J_p$. Data for the functions, h_1, h_2, h_3, and h_5, wherever appropriate, for four specimens, CT, CCT specimen under remote loading, TPB, and SECP under remote uniform tension, under the condition of both plane strain and plane stress, are presented in Tables 8.2 to 8.9, respectively, at the end of this chapter. The appropriate expressions for calculation of various quantities (J_p, δ_{mthp}, and δ_{lldp}) and the limit load P_Y for the four specimens are reproduced below from Kumar, German, and Shih (1981). Note that P_Y stands for load per unit specimen thickness.

CT geometry (Fig. 8.20):

$$J_p = \alpha \sigma_Y \epsilon_Y b\, h_1\left(\frac{a}{w}, n\right) \left(\frac{P}{P_Y}\right)^{n+1} \quad (8.129)$$

$$\delta_{mthp} = \alpha \epsilon_Y a\, h_2\left(\frac{a}{w}, n\right) \left(\frac{P}{P_Y}\right)^{n} \quad (8.130)$$

This opening is associated with the outer edges of the crack.

$$\delta_{lldp} = \alpha \epsilon_Y a\, h_3\left(\frac{a}{w}, n\right) \left(\frac{P}{P_Y}\right)^{n} \quad (8.131)$$

$$P_Y = 1.455\, \eta\, b\, \sigma_Y \quad \text{for plane strain} \quad (8.132a)$$

$$ = 1.071\, \eta\, b\, \sigma_Y \quad \text{for plane stress} \quad (8.132b)$$

where $\eta = \left[4r^2 + 4r + 2\right]^{\frac{1}{2}} - (2r + 1)$, and $r = a/b$. (8.133)

CCT geometry (Fig. 8.21):

$$J_p = \alpha \sigma_Y a \frac{b}{w} h_1\left(\frac{a}{w}, n\right) \left(\frac{P}{P_Y}\right)^{n+1} \quad (8.134)$$

$$\delta_{mthp} = \alpha \epsilon_Y a\, h_2\left(\frac{a}{w}, n\right) \left(\frac{P}{P_Y}\right)^{n} = \text{crack opening at the centre line} \quad (8.135)$$

$$\Delta_c = \alpha \epsilon_Y a\, h_3\left(\frac{a}{w}, n\right) \left(\frac{P}{P_Y}\right)^{n} \quad (8.136)$$

Assuming one load point fixed and the other moving, Δ_c is the difference between the moving load point displacements corresponding to the plate with and without crack.

$$P_Y = 4 b \sigma_Y / \sqrt{3} \quad \text{for plane strain} \quad (8.137a)$$

$$ = 2 b \sigma_Y \quad \text{for plane stress} \quad (8.137b)$$

TPB geometry (Fig. 8.22):

$$J_p = \alpha \sigma_Y \epsilon_Y b\, h_1\left(\frac{a}{w}, n\right) \left(\frac{P}{P_Y}\right)^{n+1} \tag{8.138}$$

$$\delta_{mthp} = \alpha \epsilon_Y a\, h_2\left(\frac{a}{w}, n\right) \left(\frac{P}{P_Y}\right)^{n} \tag{8.139}$$

$$\Delta_c = \alpha \epsilon_Y a\, h_3\left(\frac{a}{w}, n\right) \left(\frac{P}{P_Y}\right)^{n} \tag{8.140}$$

Δ_c is equal to the difference between the displacements of the load points corresponding to the specimen geometry with and without crack.

$$P_Y = 0.728\, \sigma_Y\, b^2/L \quad \text{for plane strain} \tag{8.141a}$$

$$= 0.536\, \sigma_Y\, b^2/L \quad \text{for plane stress} \tag{8.141b}$$

SECP under remote uniform tension (Fig. 8.23):

$$J_p = \alpha \sigma_Y \epsilon_Y b \frac{a}{w} h_1\left(\frac{a}{w}, n\right) \left(\frac{P}{P_Y}\right)^{n+1} \tag{8.142}$$

$$\delta_{mthp} = \alpha \epsilon_Y a\, h_2\left(\frac{a}{w}, n\right) \left(\frac{P}{P_Y}\right)^{n} \tag{8.143}$$

$$\Delta_C = \alpha \epsilon_Y a\, h_3\left(\frac{a}{w}, n\right) \left(\frac{P}{P_Y}\right)^{n} \tag{8.144}$$

Δ_C is equal to the difference between the displacements of the loaded edges at the plate centreline with and without crack.

$$P_Y = 1.455\, \eta\, b\, \sigma_Y \quad \text{for plane strain} \tag{8.145a}$$

$$= 1.072\, \eta\, b\, \sigma_Y \quad \text{for plane stress} \tag{8.145b}$$

where $\eta = \left[1 + r^2\right]^{\frac{1}{2}} - r$, and $r = a/b$. \hfill (8.146)

Hereafter, h_1, h_2, h_3 and h_5, which are all functions of $\frac{a}{w}$ and n, are written dropping the arguements.

Problem 8.3

During testing of a component in the form of a CT specimen under plane strain, load recorded is 0.80 MN. The specimen material is A533 B steel and test temperature is 90°C (Kumar, German, and Shih 1981). The material parameters are: yield stress = 413.7 MPa, yield strain $\epsilon_Y = \dfrac{\sigma_Y}{E} = 0.002$, modulus of elasticity = 207 GPa, Poisson's ratio $\nu = 0.25$, Ramberg–Osgood material constants in $\left[\dfrac{\epsilon}{\epsilon_Y} = \dfrac{\sigma}{\sigma_Y} + \alpha \left(\dfrac{\sigma}{\sigma_Y} \right)^n \right]$, $\alpha = 1.12$, and $n = 9.71$. The specimen details are: crack size $a = 117.22$ mm, thickness $B = 101.6$ mm, and width $w = 203.2$ mm. Calculate total J.

Solution

To calculate the yield load P_Y using Eq. (8.132a), η parameter required is calculated using

$$\eta = [4r^2 + 4r + 2]^{\tfrac{1}{2}} - (2r+1) = 0.1318, \text{ since } r = \dfrac{a}{b} = \dfrac{a}{w-a} = \dfrac{117.22}{85.98} = 1.363.$$

The yield load $P_Y = 1.455\, \eta\, b\, B\, \sigma_Y = 1.455 \times 0.1318 \times 0.08598 \times 0.1016 \times 413.7$

$= 0.694$ MN, as b = 85.98 mm.

Collecting appropriate formula for the SIF K_I from second chapter (Fig. A2.1.5),

$$K_I = \dfrac{P}{B\sqrt{w}} f(t) = \dfrac{0.80 \times 12.315}{0.1016 \times \sqrt{0.2032}} = 215.112\ \mathrm{MPa}\sqrt{m},$$

$$t = \dfrac{a}{w} = \dfrac{0.11722}{0.2032} = 0.5768 \text{ and}$$

$$f(t) = [(2+t)(0.866 + 4.64t - 13.32t^2 + 14.72t^3 - 5.6t^4)] / (1-t)^{1.5} = 12.315.$$

Elastic part of J integral

$$J_e = (1 - \nu^2) \dfrac{K_I^2}{E} = 0.2095\ \mathrm{MPam}, \text{ since } \nu = 0.25.$$

The plastic part of J integral

$$J_p = \alpha\, \epsilon_Y\, \sigma_Y\, b\, h_1 \left(\dfrac{P}{P_Y} \right)^{n+1} = 1.12 \times 0.002 \times 413.7 \times 0.08598 \times 0.565 \times 1.1527^{9.71+1}$$

$= 0.2063$ MPam.

h_1 (for $a/w = 0.5768$) is obtained through interpolation of data given in Table 8.2, corresponding to $n = 9.71$ and a/w in the span 0.5 and 0.625.

Total $J = J_e + J_p = 0.4158$ MPam (Ans.).

In the above calculation of J_e, correction to given crack length due to crack-tip plastic zone [Eqs. (8.126) and (8.127)] has not been done. Upon applying the corrections the following results are obtained: $r_y = 5$ mm, $a_e = 122.22$ mm, $t_e = 0.60147$, $K_I = 236.282$ MPa$\sqrt{m}$, and total $J = 0.45915$ MPam. Note that J_p remains unchanged.

Problem 8.4
A component in the form of a CT specimen, made of SS304 steel, was tested at room temperature. The material parameters are: yield stress = 207 MPa, yield strain $\epsilon_Y = \frac{\sigma_Y}{E} = 0.001$, modulus of elasticity = 207 GPa, Poisson's ratio $\nu = 0.25$, Ramberg–Osgood material constants in $\left[\frac{\epsilon}{\epsilon_Y} = \frac{\sigma}{\sigma_Y} + \alpha \left(\frac{\sigma}{\sigma_Y}\right)^n\right]$, $\alpha = 1.691$ and $n = 5.421$ [Kumar, German, and Shih 1981]. The specimen details are: crack size $a = 63.5$ mm, thickness $B = 50.8$ mm, and width $w = 101.6$ mm. Calculate total J for a load 1.5 times the yield load assuming plane strain condition.

Solution
Proceeding in the same manner as in the case of earlier example,

$$\eta = [4r^2 + 4r + 2]^{\frac{1}{2}} - (2r + 1) = 0.113, \text{ since } r = \frac{a}{b} = \frac{a}{w-a} = \frac{63.5}{38.1} = 1.667.$$

The yield load $P_Y = 1.455 \, \eta \, b \, B \, \sigma_Y = 1.455 \times 0.113 \times 0.0381 \times 0.0508 \times 207$

$= 0.0658$ MN, as $b = 38.1$ mm.

Therefore, the applied load $P = 0.0987$ MN.

Collecting appropriate formula for the SIF K_I from the end of second chapter,

$$K_I = \frac{P}{B\sqrt{w}} f(t) = \frac{0.0987 \times 14.885}{0.0508 \times \sqrt{0.1016}} = 90.73 \text{ MPa}\sqrt{m},$$

since $t = \frac{a}{w} = \frac{63.5}{101.6} = 0.625$,

$$f(t) = \left[(2+t)(0.866 + 4.64t - 13.32t^2 + 14.72t^3 - 5.6t^4)\right] / (1-t)^{1.5} = 14.885.$$

The elastic part of J integral

$$J_e = (1 - \nu^2) \frac{K_I^2}{E} = 0.0372 \text{ MPam, since } \nu = 0.25.$$

Note that no correction for plasticity is applied to calculate K_I and J_e.

The plastic part of J integral

$$J_p = \alpha\, \epsilon_Y\, \sigma_Y b h_1 \left(\frac{P}{P_Y}\right)^{n+1} = 1.691 \times 0.001 \times 207 \times 0.0381 \times 0.927 \times 1.5^{5.421+1}$$

$= 0.167$ MPam, since $h_1 = 0.927$ for $a/w = 0.625$ (Table 8.2).

Total $J = J_e + J_p = 0.204$ MPam (Ans.)

Problem 8.5
A three point beam of A533B steel with the material data as given in Problem 8.3 was loaded up to J reaching a value 0.45 MPam. The specimen dimensions are: thickness $B = 50.8$ mm, depth $w = 2B$, span length $L = 4w$, and crack size $a = 38.1$ mm. Determine the load level (a) neglecting the elastic part of J and (b) without neglecting it. Assume plane strain condition and $\nu = 0.25$.

Solution
Given $b = w - a = 101.6 - 38.1 = 63.5$ mm, $L = 406.4$ mm and $\sigma_Y = 413.7$ MPa.

Yield load P_Y [using Eq. (8.141a)] $= 0.728\, \sigma_Y B \dfrac{b^2}{L} = 0.1518$ MN. Representing the load by P,

$$K_I = \frac{PL}{BW^{1.5}} \frac{3\sqrt{r}\left[1.99 - r(1-r)(2.15 - 3.93r + 2.7r^2)\right]}{2(1+2r)(1-r)^{1.5}}, \quad r = \frac{a}{w}$$

$$= \frac{P \times 0.4064}{0.0508 \times 0.1016^{1.5}} \times 1.8511 \quad \text{as } r = 0.375$$

$= 457.28P$.

$$J_e = (1-\nu^2)\frac{K_I^2}{E} = (1 - 0.25^2)\frac{(457.28P)^2}{207000} = 0.9470P^2.$$

$$J_p = \alpha\, \epsilon_Y\, \sigma_Y\, b h_1 \left(\frac{P}{P_Y}\right)^{n+1}$$

$$= 1.12 \times 0.002 \times 413.7 \times 0.0635 \times 0.569 \times \left(\frac{P}{0.1518}\right)^{9.71+1}, \text{ since } h_1 = 0.569$$

(from Table 8.6). Finally $J_p = 0.03348 \times \left(\dfrac{P}{0.1518}\right)^{9.71+1} = 1.96501 \times 10^7 \times P^{10.71}$

$$J = J_e + J_p = 0.947P^2 + 1.96501 \times 10^7 \times P^{10.71}.$$

(a) Neglecting the elastic part of J and equating the applied $J = 0.45$ MPam with J_p, $P = 0.1934$ MN (Ans.).

(b) Without neglecting the elastic part, and solving the nonlinear equation

$$0.9470 P^2 + 1.96501 \times 10^7 \times P^{10.71} = 0.45,$$

$P = 0.192$ MN (Ans.).

In the above calculations (b) to determine load P, correction for crack-tip plastic zone was ignored. This can be done through the steps as follows. Based on $P = 0.192$kN, correction r_y to the specified physical crack length is given by

$$r_y = \frac{1}{1 + \left(\frac{P}{P_Y}\right)^2} \left(\frac{n-1}{n+1}\right) \frac{1}{k\pi} \left(\frac{K_I}{\sigma_Y}\right)^2 = 0.747 \text{ mm},$$

where $k = 6$, $n = 9.71$, $K_I = 457.28P$, and $P_Y = 0.1518$ MN. The effective crack length $a_e = a + r_y = 38.1 + 0.747 = 38.847$ mm. The corresponding SIF K_{Ie} is then given by

$$K_{Ie} = \frac{P L}{B w^{1.5}} f(t_e) = \frac{P \times 0.4064}{0.0508 \times 0.1016^{1.5}} \times 1.88826 = 466.456 P \text{ MPa}\sqrt{m}.$$

$f(t_e)$ is obtained as 1.88826 using $t_e = \frac{a_e}{w} = \frac{38.847}{101.6} = 0.3823$ in the formula used earlier. The corrected elastic part J_e of the total J is obtained as $0.9854 P^2$. Finally corrected load P is obtained from the equation

$$J = J_e + J_p = 0.9854 P^2 + 1.96501 \times 10^7 \times P^{10.71} = 0.45 \text{ MPam},$$

as 0.19195 MN, which is very close to the earlier load. Hence, accurate P can be obtained in this case without applying the correction for crack-tip plastic zone. This has perhaps become possible because the elastic part J_e of total J here is small compared to the plastic part J_p.

Problem 8.6
Solve Problem 8.5 assuming plane stress condition.

Solution
Yield load P_Y [using Eq. (141b)] $= 0.536\, \sigma_Y B \frac{b^2}{L} = 0.11176$ MN.
   Representing the load by P,

$$K_I = \frac{PL}{BW^{1.5}} \frac{3\sqrt{r}\left[1.99 - r(1-r)(2.15 - 3.93r + 2.7r^2)\right]}{2(1+2r)(1-r)^{1.5}}, \quad r = \frac{a}{w}$$

$$= \frac{P \times 0.4064}{0.0508 \times 0.1016^{1.5}} \times 1.8511 \text{ as } r = 0.375$$

$$= 457.28P.$$

$$J_e = \frac{K_I^2}{E} = \frac{(457.28P)^2}{207000} = 1.0102P^2$$

$$J_p = \alpha \, \epsilon_Y \, \sigma_Y \, b \, h_1 \left(\frac{P}{P_Y}\right)^{n+1}$$

$$= 1.12 \times 0.002 \times 413.7 \times 0.0635 \times 0.318 \times \left(\frac{P}{0.11176}\right)^{9.71+1},$$

$h_1 = 0.318$ (from Table 8.6).

$$J_p = 0.01871 \times \left(\frac{P}{0.11176}\right)^{9.71+1} = 2.91693 \times 10^8 \times P^{10.71}$$

$$J = J_e + J_p = 1.0102P^2 + 2.91693 \times 10^8 \times P^{10.71}$$

(a) Neglecting the elastic part of J and equating the applied $J = 0.45$ MPam with J_p, $P = 0.1504$ MN (Ans.).

(b) Without neglecting the elastic part, and solving the nonlinear equation

$$1.0102P^2 + 2.91693 \times 10^7 \times P^{10.71} = 0.45,$$

$P = 0.1497$ MN (Ans.).

8.13 Closure

The application of the principles EPFM to practice has not been possible to the same extent as in the case of LEFM. One of the reasons for this is the non-availability of solutions for COD or J for practical geometries and materials with varieties of hardening characteristics. The numerical technique like the finite element method has helped to eliminate many limitations. Simultaneously, the research and developments related to the transfer of laboratory material data to real life situations have increased the confidence in both design and safety assessments. Full-scale testing has also paved the way for practical applications of the principles of EPFM.

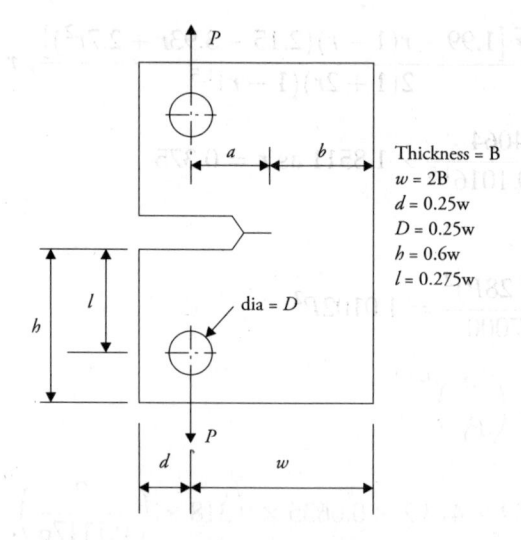

Figure 8.20 Geometry associated with Tables 8.2 and 8.3.

Table 8.2 Compact tension specimen in plane strain.

a/w	h_i	\multicolumn{9}{c}{n}
		1	2	3	5	7	10	13	16	20
1/4	h_1	2.23	2.05	1.78	1.48	1.33	1.26	1.25	1.32	1.57
	h_2	17.9	12.5	11.7	10.8	10.5	10.7	11.5	12.6	14.6
	h_3	9.85	8.51	8.17	7.77	7.71	7.92	8.52	9.31	10.9
3/8	h_1	2.15	1.72	1.39	0.97	0.693	0.443	0.276	0.176	0.098
	h_2	12.6	8.18	6.52	4.32	2.97	1.79	1.10	0.686	0.370
	h_3	7.94	5.76	4.64	3.10	2.14	1.29	0.793	0.494	0.266
1/2	h_1	1.94	1.51	1.24	0.919	0.685	0.461	0.314	0.216	0.132
	h_2	9.33	5.85	4.30	2.75	1.91	1.20	0.788	0.530	0.317
	h_3	6.41	4.27	3.16	2.02	1.41	0.888	0.585	0.393	0.236
5/8	h_1	1.76	1.45	1.24	0.974	0.752	0.602	0.459	0.347	0.248
	h_2	7.61	4.57	3.42	2.36	1.81	1.32	0.983	0.749	0.485
	h_3	5.52	3.43	2.58	1.79	1.37	1.00	0.746	0.568	0.368
3/4	h_1	1.71	1.42	1.26	1.033	0.864	0.717	0.575	0.448	0.345
	h_2	6.37	3.95	3.18	2.34	1.88	1.44	1.12	0.887	0.665
	h_3	4.86	3.05	2.46	1.81	1.45	1.11	0.869	0.686	0.514
1	h_1	1.57	1.45	1.35	1.18	1.08	0.95	0.85	0.73	0.63
	h_2	5.39	3.74	3.09	2.43	2.12	1.80	1.57	1.33	1.14
	h_3	4.31	2.99	2.47	1.95	1.79	1.44	1.26	1.07	0.909

Source: Kumar, German, and Shih (1981)

Table 8.3 Compact tension specimen in plane stress.

a/w	h_i	\multicolumn{9}{c}{n}
		1	2	3	5	7	10	13	16	20
1/4	h_1	1.61	1.46	1.28	1.06	0.903	0.729	0.601	0.511	0.395
	h_2	17.6	12.0	10.7	8.74	7.32	5.74	4.63	3.75	2.92
	h_3	9.67	8.00	7.21	5.94	5.00	3.95	3.19	2.59	2.023
3/8	h_1	1.55	1.25	1.05	0.801	0.647	0.484	0.377	0.284	0.220
	h_2	12.4	8.20	6.54	4.56	3.45	2.44	1.83	1.36	1.02
	h_3	7.80	5.73	4.62	3.25	2.48	1.77	1.33	0.99	0.746
1/2	h_1	1.40	1.08	0.901	0.686	0.558	0.436	0.356	0.298	0.238
	h_2	9.16	5.67	4.21	2.80	2.12	1.57	1.25	1.03	0.814
	h_3	6.29	4.15	3.11	2.09	1.59	1.18	0.938	0.774	0.614
5/8	h_1	1.27	1.03	0.875	0.695	0.593	0.494	0.423	0.37	0.31
	h_2	7.47	4.48	3.35	2.37	1.92	1.54	1.29	1.12	0.928
	h_3	5.42	3.38	2.54	1.80	1.47	1.18	0.988	0.853	0.71
3/4	h_1	1.23	0.977	0.833	0.683	0.598	0.506	0.431	0.373	0.314
	h_2	6.25	3.78	2.89	2.14	1.78	1.44	1.20	1.03	0.857
	h_3	4.77	2.92	2.24	1.66	1.38	1.12	0.936	0.80	0.666
1	h_1	1.13	1.01	0.775	0.68	0.65	0.62	0.49	0.47	0.42
	h_2	5.29	3.54	2.41	1.91	1.73	1.59	1.23	1.17	1.03
	h_3	4.23	2.83	1.93	1.52	1.39	1.27	0.985	0.933	0.824

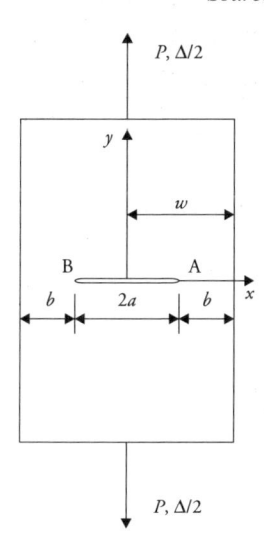

Source: Kumar, German, and Shih (1981)

Figure 8.21 Geometry associated with Tables 8.4 and 8.5

Table 8.4 Centre crack tension specimen under remote tension and in plane strain.

a/w	h_i	1	2	3	5	7	10	13	16	20
1/8	h_1	2.80	3.61	4.06	4.35	4.33	4.02	3.56	3.06	2.46
	h_2	3.05	3.62	3.91	4.06	3.93	3.54	3.07	2.60	2.06
	h_3	0.303	0.574	0.84	1.30	1.63	1.95	2.03	1.96	1.77
1/4	h_1	2.54	3.01	3.21	3.29	3.18	2.92	2.63	2.34	2.03
	h_2	2.68	2.99	3.01	2.85	2.61	2.30	1.97	1.71	1.45
	h_3	0.536	0.911	1.22	1.64	1.84	1.85	1.80	1.64	1.43
3/8	h_1	2.34	2.62	2.65	2.51	2.28	1.97	1.71	1.46	1.19
	h_2	2.35	2.39	2.23	1.88	1.58	1.28	1.07	0.89	0.715
	h_3	0.699	1.06	1.28	1.44	1.40	1.23	1.05	0.888	0.719
1/2	h_1	2.21	2.29	2.20	1.97	1.76	1.52	1.32	1.16	0.978
	h_2	2.03	1.86	1.60	1.23	1.00	0.799	0.664	0.564	0.466
	h_3	0.803	1.07	1.16	1.10	0.968	0.796	0.665	0.565	0.469
5/8	h_1	2.12	1.96	1.76	1.43	1.17	0.863	0.628	0.458	0.30
	h_2	1.71	1.32	1.04	0.707	0.524	0.358	0.250	0.178	0.114
	h_3	0.844	0.937	0.879	0.701	0.522	0.361	0.251	0.178	0.115
3/4	h_1	2.07	1.73	1.47	1.11	0.895	0.642	0.461	0.337	0.216
	h_2	1.35	0.857	0.596	0.361	0.254	0.167	0.114	0.081	0.0511
	h_3	0.805	0.70	0.555	0.359	0.254	0.168	0.114	0.0813	0.0516
7/8	h_1	2.08	1.64	1.40	1.14	0.987	0.814	0.688	0.573	0.461
	h_2	0.889	0.428	0.287	0.181	0.139	0.105	0.0837	0.0682	0.0533
	h_3	0.632	0.40	0.291	0.182	0.140	0.106	0.0839	0.0683	0.0535

Source: Kumar, German, and Shih (1981)

Table 8.5 Centre crack tension specimen under remote tension and in plane stress.

a/w	h_i	n=1	2	3	5	7	10	13	16	20
1/8	h_1	2.80	3.57	4.01	4.47	4.65	4.62	4.41	4.13	3.72
	h_2	3.53	4.09	4.43	4.74	4.79	4.63	4.33	4.00	3.55
	h_3	0.35	0.661	0.997	1.55	2.05	2.56	2.83	2.95	2.92
1/4	h_1	2.54	2.97	3.14	3.20	3.11	2.86	2.65	2.47	2.20
	h_2	3.10	3.29	3.30	3.15	2.93	2.56	2.29	2.08	1.81
	h_3	0.619	1.01	1.35	1.83	2.08	2.19	2.12	2.01	1.79
3/8	h_1	2.34	2.53	2.52	2.35	2.17	1.95	1.77	1.61	1.43
	h_2	2.71	2.62	2.41	2.03	1.75	1.47	1.28	1.13	0.988
	h_3	0.807	1.20	1.43	1.59	1.57	1.43	1.27	1.13	0.994
1/2	h_1	2.21	2.20	2.06	1.81	1.63	1.43	1.30	1.17	1.00
	h_2	2.34	2.01	1.70	1.30	1.07	0.871	0.757	0.666	0.557
	h_3	0.927	1.19	1.26	1.18	1.04	0.867	0.758	0.668	0.56
5/8	h_1	2.12	1.91	1.69	1.41	1.22	1.01	0.853	0.712	0.573
	h_2	1.97	1.46	1.13	0.785	0.617	0.474	0.383	0.313	0.256
	h_3	0.975	1.05	0.97	0.763	0.62	0.478	0.386	0.318	0.273
3/4	h_1	2.07	1.71	1.46	1.21	1.08	0.867	0.745	0.646	0.532
	h_2	1.55	0.97	0.685	0.452	0.361	0.262	0.216	0.183	0.148
	h_3	0.929	0.802	0.642	0.45	0.361	0.263	0.216	0.183	0.149
7/8	h_1	2.08	1.57	1.31	1.08	0.972	0.862	0.778	0.715	0.63
	h_2	1.03	0.485	0.31	0.196	0.157	0.127	0.109	0.0971	0.0842
	h_3	0.73	0.452	0.313	0.198	0.157	0.127	0.109	0.0973	0.0842

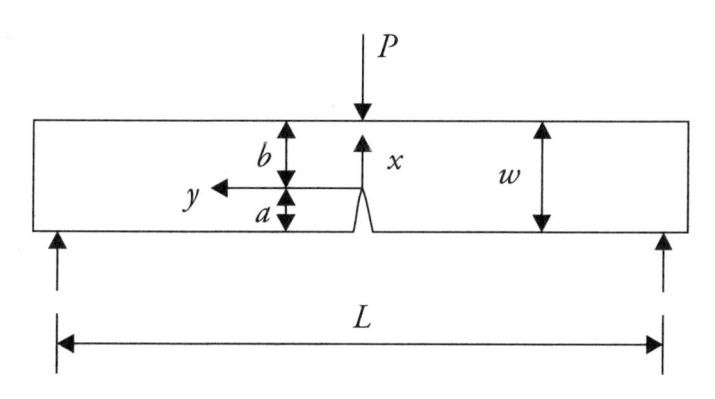

Source: Kumar, German, and Shih (1981).

Figure 8.22 Geometry associated with Tables 8.6 and 8.7.

250 Fracture mechanics

Table 8.6 h_1, h_2, and h_3 for SECP in plane strain under three point bending.

a/w	h_i	n=1	2	3	5	7	10	13	16	20
1/8	h_1	0.936	0.869	0.805	0.687	0.58	0.437	0.329	0.245	0.165
	h_2	6.97	6.77	6.29	5.29	4.38	3.24	2.40	1.78	1.19
	h_3	3	22.1	20	15	11.7	8.39	6.14	4.54	3.01
1/4	h_1	1.2	1.034	0.93	0.762	0.633	0.523	0.396	0.303	0.215
	h_2	5.8	4.67	4.01	3.08	2.45	1.93	1.45	1.09	0.758
	h_3	4.08	9.72	8.36	5.86	4.47	3.42	2.54	1.9	1.32
3/8	h_1	1.33	1.15	1.02	0.084	0.695	0.556	0.442	0.36	0.265
	h_2	5.18	3.93	3.2	2.38	1.93	1.47	1.15	0.928	0.684
	h_3	4.51	6.01	5.03	3.74	3.02	2.3	1.8	1.45	1.07
1/2	h_1	1.41	1.09	0.922	0.675	0.495	0.331	0.211	0.135	0.0741
	h_2	4.87	3.28	2.53	1.69	1.19	0.773	0.48	0.304	0.165
	h_3	4.69	4.33	3.49	2.35	1.66	1.08	0.669	0.424	0.23
5/8	h_1	1.46	1.07	0.896	0.631	0.436	0.255	0.142	0.084	0.0411
	h_2	4.64	2.86	2.16	1.37	0.907	0.518	0.287	0.166	0.0806
	h_3	4.71	3.49	2.7	1.72	1.14	0.652	0.361	0.209	0.102
3/4	h_1	1.48	1.15	0.974	0.693	0.5	0.348	0.223	0.14	0.0745
	h_2	4.47	2.75	2.1	1.36	0.936	0.618	0.388	0.239	0.127
	h_3	4.49	3.14	2.4	1.56	1.07	0.704	0.441	0.272	0.144
7/8	h_1	1.5	1.35	1.2	1.02	0.855	0.69	0.551	0.44	0.321
	h_2	4.36	2.9	2.31	1.7	1.33	1	0.782	0.613	0.459
	h_3	4.15	3.08	2.45	1.81	1.41	1.06	0.828	0.649	0.486

Source: Kumar, German, and Shih (1981)

Elastic plastic fracture mechanics 251

Table 8.7 h_1, h_2, and h_3 for SECP in plane stress under three point bending.

a/w	h_i	1	2	3	5	7	10	13	16	20
1/8	h_1	0.676	0.6	0.548	0.459	0.383	0.297	0.238	0.192	0.148
	h_2	6.84	6.30	5.66	4.53	3.64	2.72	2.12	1.67	1.26
	h_3	2.95	20.1	14.6	12.2	9.12	6.75	5.2	4.09	3.07
1/4	h_1	0.869	0.731	0.629	0.479	0.37	0.246	0.174	0.117	0.0593
	h_2	5.69	4.5	3.68	2.61	1.95	1.29	0.897	0.603	0.307
	h_3	4.01	8.81	7.19	4.73	3.39	2.2	1.52	1.01	0.508
3/8	h_1	0.963	0.797	0.68	0.527	0.418	0.307	0.232	0.174	0.105
	h_2	5.09	3.73	2.93	2.07	1.58	1.13	0.841	0.626	0.381
	h_3	4.42	5.53	4.48	3.17	2.41	1.73	1.28	0.948	0.575
1/2	h_1	1.02	0.767	0.621	0.453	0.324	0.202	0.128	0.0813	0.0298
	h_2	4.77	3.12	2.32	1.55	1.08	0.655	0.41	0.259	0.0974
	h_3	4.6	4.09	3.09	2.08	1.44	0.874	0.545	0.344	0.129
5/8	h_1	1.05	0.786	0.649	0.494	0.357	0.235	0.173	0.105	0.0471
	h_2	4.55	2.83	2.12	1.46	1.02	0.656	0.472	0.286	0.13
	h_3	4.62	3.43	2.6	1.79	1.26	0.803	0.577	0.349	0.158
3/4	h_1	1.07	0.786	0.643	0.474	0.343	0.23	0.167	0.110	0.0442
	h_2	4.39	2.66	1.97	1.33	0.928	0.601	0.427	0.28	0.114
	h_3	4.39	3.01	2.24	1.51	1.05	0.68	0.483	0.316	0.129
7/8	h_1	1.086	0.928	0.81	6.46	0.538	0.423	0.332	0.242	0.205
	h_2	4.28	2.76	2.16	1.56	1.23	0.922	0.702	0.561	0.428
	h_3	4.07	2.93	2.29	1.65	1.3	0.975	0.742	0.592	0.452

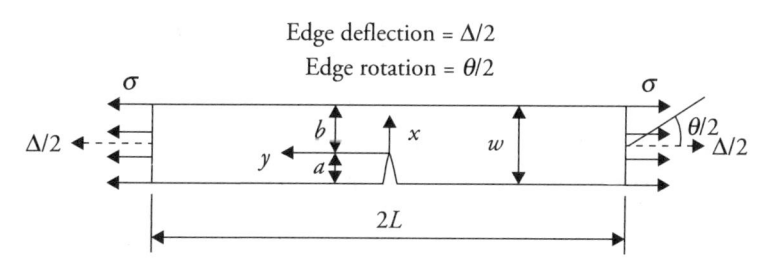

Source: Kumar, German, and Shih (1981)

Figure 8.23 Geometry associated with Tables 8.8 and 8.9.

Table 8.8 h_1, h_2, h_3 and h_5 for SECP in plane strain under remote uniform tension.

a/w	h_i	n=1	2	3	5	7	10	13	16	20
1/8	h_1	4.95	6.93	8.57	11.5	13.5	16.1	18.1	19.9	21.2
	h_2	5.25	6.47	7.56	9.46	11.1	12.9	14.4	15.7	16.8
	h_3	26.6	25.8	25.2	24.2	23.6	23.2	23.2	23.5	23.7
	h_5	0	0.558	0.807	1.26	1.62	1.98	2.19	2.33	0
1/4	h_1	4.34	4.77	4.64	3.82	3.06	2.17	1.55	1.11	0.712
	h_2	4.76	4.56	4.28	3.39	2.64	1.81	1.25	0.875	0.552
	h_3	10.3	7.64	5.87	3.7	2.48	1.5	0.97	0.654	0.404
	h_5	0	1.14	1.11	0.833	0.604	0.375	0.237	0.153	0.0894
3/8	h_1	3.88	3.25	2.63	1.68	1.06	0.539	0.276	0.142	0.0595
	h_2	4.54	3.49	2.67	1.57	0.946	0.458	0.229	0.116	0.048
	h_3	5.14	2.99	1.9	0.923	0.515	0.24	0.119	0.06	0.0246
	h_5	0	1.43	1.1	0.643	0.38	0.179	0.0879	0.0442	0.0181
1/2	h_1	3.4	2.3	1.69	0.928	0.514	0.213	0.0902	0.0385	0.0119
	h_2	4.45	2.77	1.89	0.954	0.507	0.204	0.0854	0.0356	0.011
	h_3	3.15	1.54	0.912	0.417	0.215	0.085	0.0358	0.0147	0.00448
	h_5	0	1.6	1.11	0.562	0.3	0.121	0.0511	0.0213	0.00657
5/8	h_1	2.86	1.8	1.3	0.697	0.378	0.153	0.0625	0.0256	0.0078
	h_2	4.37	2.44	1.62	0.0806	0.423	0.167	0.0671	0.0272	0.00823
	h_3	2.31	1.08	0.681	0.329	0.171	0.067	0.0268	0.0108	0.00326
	h_5	0	1.80	1.21	0.604	0.318	0.126	0.0509	0.0207	0.00626
3/4	h_1	2.34	1.61	1.25	0.769	0.477	0.233	0.116	0.059	0.0215
	h_2	4.32	2.52	1.79	1.03	0.619	0.296	0.146	0.0735	0.0267
	h_3	2.02	1.1	0.765	0.435	0.262	0.125	0.0617	0.0312	0.0113
	h_5	0	2.17	1.55	0.895	0.539	0.258	0.127	0.0639	0.0232
7/8	h_1	1.91	1.57	1.37	1.1	0.925	0.702			
	h_2	4.29	2.75	2.14	1.55	1.23	0.921			
	h_3	2.01	1.27	0.988	0.713	0.564	0.424			
	h_5	0	2.601	2.203	1.47	1.16	0.875			

Source: Kumar, German, and Shih (1981)

Table 8.9 h_1, h_2, h_3 and h_5 for SECP in plane stress under remote uniform tension.

a/w	h_i	n=1	2	3	5	7	10	13	16	20
1/8	h_1	3.58	4.55	5.06	5.3	4.96	4.14	3.29	2.6	1.92
	h_2	5.15	5.43	6.05	6.01	5.47	4.46	3.48	2.74	2.02
	h_3	26.1	21.6	18	12.7	9.24	5.98	3.94	2.72	2
	h_5	0.296	0.49	0.627	0.748	0.72	0.586	0.45	0.345	0.255
1/4	h_1	3.14	3.26	2.92	2.12	1.53	0.96	0.615	0.4	0.23
	h_2	4.67	4.3	3.7	2.53	1.76	1.05	0.656	0.419	0.237
	h_3	10.1	6.49	4.36	2.19	1.24	0.63	0.362	0.224	0.123
	h_5	0.904	1.05	0.932	0.631	0.433	0.258	0.16	0.103	0.0583
3/8	h_1	2.81	2.37	1.94	1.37	1.01	0.677	0.474	0.342	0.226
	h_2	4.47	3.43	2.63	1.69	1.18	0.762	0.524	0.372	0.244
	h_3	5.05	2.65	1.6	0.812	0.525	0.328	0.223	0.157	0.102
	h_5	1.73	1.4	1.1	0.72	0.51	0.332	0.229	0.164	0.108
1/2	h_1	2.46	1.67	1.25	0.776	0.51	0.286	0.164	0.0956	0.0469
	h_2	4.37	2.73	1.91	1.09	0.694	0.38	0.216	0.124	0.0607
	h_3	3.1	1.43	0.871	0.461	0.286	0.155	0.088	0.0506	0.0247
	h_5	2.41	1.58	1.12	0.652	0.417	0.229	0.13	0.0748	0.0395
5/8	h_1	2.07	1.41	1.105	0.755	0.551	0.363	0.248	0.172	0.107
	h_2	4.3	2.55	1.84	1.16	0.816	0.523	0.353	0.242	0.15
	h_3	2.27	1.13	0.771	0.478	0.336	0.215	0.146	0.1	0.0616
	h_5	3.02	1.88	1.36	0.861	0.606	0.388	0.262	0.179	0.111
3/4	h_1	1.7	1.14	0.91	0.624	0.447	0.28	0.181	0.118	0.067
	h_2	4.24	2.47	1.81	1.15	0.798	0.49	0.314	0.203	0.115
	h_3	1.98	1.09	0.784	0.494	0.344	0.211	0.136	0.0581	0.0496
	h_5	3.44	2.12	1.56	0.986	0.686	0.421	0.27	0.174	0.0987
7/8	h_1	1.38	1.11	0.962	0.792	0.677	0.574			
	h_2	4.22	2.68	2.08	1.54	1.27	1.04			
	h_3	1.97	1.25	0.969	0.716	0.591	0.483			
	h_5	3.91	2.53	1.96	1.45	1.19	0.973			

Source: Kumar, German, and Shih (1981)

Exercise

8.1 During compact tension testing of a specimen of thickness 40 mm made of steel, the following data was obtained: load varies from 0 to 0.3 mm linearly with a slope 0.8325 MN/mm. Further variation of load from this displacement level to 1.0 mm is $P = 0.25 \times 10^6 + 0.70 \times 10^6 [(u - 0.3)\, 10^{-3}]^{\frac{1}{3}}$ N. The specimen dimensions are $a = b = 40$ mm. Calculate J_C if the crack extension begins from the given crack size 40 mm. [Ans. 0.3467 MPam]

8.2 A component with a centre crack was subjected to a tensile load 2.70 MN. It is made of SAE4340 steel with the following properties [Anand and Parks, 2004]: yield stress = 1172 MPa, modulus of elasticity = 200 GPa, Poisson's ratio $\nu = 0.25$, Ramberg–Osgood material constants in $\left[\dfrac{\varepsilon}{\varepsilon_Y} = \dfrac{\sigma}{\sigma_Y} + \alpha \left(\dfrac{\sigma}{\sigma_Y}\right)^n\right]$ $\alpha = 1.865$, and $n = 15$ (approximated). The component dimension details: crack size $a = 40$ mm $b = 60$ mm, thickness $B = 25$ mm, and semi-width $w = 100$ mm. Determine J considering plane strain condition.
[Ans. 0.4627 MPam]

8.3 Determine load for Problem 8.3 corresponding to applied $J = 0.40$ MPam.
[Ans. 0.795 MN ignoring any correction due to r_y]

8.4 Determine load for Problem 8.4 corresponding to applied $J = 0.25$ MPam and state of stress is plane stress.
[Ans. 0.0805 MN ignoring any correction due to r_y]

8.5 Determine the per cent change in load if applied $J = 0.45$ MPam in Problem 8.5 is (i) increased by 20 per cent and (ii) decreased by 15 per cent.
[Ans. (i) 0.19555 MN, (ii) 0.188595 MN; ignoring any correction due to r_y]

References

8.1 Anand L. and D.M. Parks. 2004. Mechanics and Materials II, Supplementary notes, Department of Mechanical Engineering, MIT Cambridge, Massachusetts 02139.

8.2 Anderson, T.L. and R.H. Dodds Jr. 1991. 'Specimen Size Requirements for Fracture Toughness Testing in Ductile-Brittle Transition Region.' *Journal of Testing and Evaluation* 19: 123–34.

8.3 Barenblatt, G.I. 1962. 'The Mathematical Theory of Equilibrium Cracks in Brittle Fracture.' *Advances in Applied Mechanics* VII: 55–129.

8.4 Begley, J.A. and J.D. Landes. 1972. 'The J-integral as a Fracture Criterion'. In *Fracture Toughness*, Part II, 1–23. Philadelphia: American Society for Testing and Materials [ASTM STP 514].

8.5 Chakrabarty, J. 1987. *Theory of Plasticity.* New York: McGraw-Hill Book Company.

8.6 Dugdale, D.S. 1960. 'Yielding of Steel Sheets Containing Slits.' *Journal of Mechanics and Physics of Solids* 8: 100–08.

8.7 Harrison, R.P., K. Loosemoore and I. Milne. 1976. *Assessment of Integrity of Structures Containing Crack.* Central Electricity Generating Board Report No. R/H/R6, UK.

8.8 Hult, J.A.H. and F. A. McClintock. 1957. 'Elastic Plastic Stress and Strain Distributions Around Sharp Notches under Repeated Shear', Vol. 8, 51–58. *Proceedings of the 9th International Congress of Applied Mechanics,* University of Brussels.

8.9 Hutchinson, J.W. 1968a. 'Singular Behavior at the End of a Tensile Crack in a Hardening Material.' *Journal of Mechanics and Physics of Solids* 16: 13–31.

8.10 ———. 1968b. 'Plastic Stress and Strain Fields at a Crack Tip.' *Journal of Mechanics and Physics of Solids* 16: 337–47.

8.11 ———. 1983. 'Fundamentals of the Phenomenological Theory of Nonlinear Fracture Mechanics.' *Journal of Applied Mechanics, Transactions of ASME* 50: 1042–51.

8.12 Hutchinson, J.W. and P. C. Paris. 1979. 'Stability Analysis of J-controlled Crack Growth'. In *Elastic - Plastic Fracture,* 37–64. Philadelphia: American Society for Testing and Materials [ASTM STP 668].

8.13 Kanninen, M.F. and C.H. Popelar. 1985. *Advanced Fracture Mechanics.* New York: Oxford University Press.

8.14 Kumar, V., M.D. German and C. F. Shih. 1981. *An Engineering Approach for Engineering Elastic Plastic Fracture Analysis.* Electric Power Research Institute (EPRI) NP-1931, Project 1287-1, Topical Report.

8.15 Landes, J.D. and J. A. Begley. 1972. 'The Effect of Specimen Geometry on J_{IC},' 24–29. Philadelphia: *American Society for Testing and Materials* [ASTM STP 514].

8.16 Landes, J.D., H. Walker and G. A. Clarke. 1979. 'Evaluation of Estimation Procedure Used in J-integral Testing'. In *Elastic - Plastic Fracture,* 266–87. Philadelphia: American Society for Testing and Materials [ASTM STP 668].

8.17 McClintock, F.A. 1971. 'Plasticity Aspects of Fracture.' In *Fracture: An Advanced Treatise,* Vol. 3, 47–225. New York: Academic Press.

8.18 Mendelson, A. 1968. *Plasticity: Theory and Applications.* New York: The Macmillan Company.

8.19 Merkle, J.G. and H. T. Corten. 1974. 'A J-integral Analysis for the Compact Specimen, Considering Axial Force as well as Bending Effects.' *Journal of Pressure Vessel and Technology, Transactions of ASME* 96: 286–92.

8.20 O'Dowd, N.P. and C. F. Shih. 1991. 'Family of Crack-tip Fields Characterized by a Triaxiality Parameter – I. Structure of Fields.' *Journal of Mechanics and Physics of Solids* 39: 898–1015.

8.21 ———. 1992. 'Family of Crcak-tip Fields Characterized by a Triaxiality Parameter – II. Fracture Applications.' *Journal of Mechanics and Physics of Solids* 40: 939–63.

8.22 Paris, P.C., H. Tada, Z. Zahoor and H. Ernst. 1979. 'Instability of the Tearing Mode of Elastic Plastic Crack Growth'. In *Elastic - Plastic Fracture*, 5–36. Philadelphia: American Society for Testing and Materials [ASTM STP 668].

8.23 Rice, J.R. 1975. 'Elastic Plastic Model for Stable Crack Growth.' In *Proceedings of Cambridge Conference*, England, 1973, ed. May, M.J., 14–39. British Steel Corporation Physical Metallurgy Centre Publication.

8.24 Rice, J.R. and G. F. Rosengren. 1968. 'Plane Strain Deformation Near a Crack Tip in a Power-Law Hardening Material.' *Journal of Mechanics and Physics of Solids* 16: 1–12.

8.25 Rice, J.R., P. C. Paris and J. G. Merkle. 1973. 'Some Further Results on J-integral Analysis and Estimates'. In *Progress in Flaw Growth and Fracture Toughness Testing*, 231–245. Philadelphia: American Society for Testing and Materials [ASTM STP 536].

8.26 Ritchie, R.O., J. F. Knott and J. R. Rice. 1973. 'On the Relationship Between Critical Tensile Stress andFracture Toughness in Mild Steel.' *Journal of Mechanics and Physics of Solids* 21: 395–410.

8.27 Shih, C.F. 1981. 'Relationship Between the J-integral and the Crack Opening Displacement for Stationary and Extending Crack.' *Journal of Mechanics and Physics of Solids* 29: 305–26.

8.28 Tracy, D.M. 1976. 'Finite Element Solutions for Crack-tip Behavior in Small-scale Yielding.' *Journal of Engineering Materials and Technology, Transactions of ASME* 98: 146–51.

8.29 Wells, A.A. 1961. 'Unstable Crack Propagation in Metals: Cleavage and Fast Fracture', 210–30. *Proceedings of the Crack Propagation Symposium*, College of Aeronautics, Cranfield.

8.30 Williams, M.L. 1957. 'On Stress Distributions at the Base of a Stationary Crack.' *Journal of Applied Mechanics, Transactions of ASME* 24: 109–14.

9

Experimental Measurement of Fracture Toughness Data

9.1 Introduction

This chapter deals with the experimental methods related to the measurement of fracture toughness and determination of resistance curves. All tests are done as per certain standards. The standards are generally very exhaustive; they provide all relevant information that may be needed about a test. Useful information in brief about the different testing is presented in this chapter.

9.2 Measurement of Plane Strain Fracture Toughness K_{IC}

K_{IC} testing is mostly done as per ASTM E399-90 (Reapproved 1997) (2000) or its equivalent.

The standard specifies specimen geometry, procedure for preparation of specimen, testing machine requirements, the sensitivity of measurement devices, and testing fixtures, and gives guidelines for conducting tests and data collection, along with methods of calculation of toughness and reporting of the experimental data. For a rolled or forged material, there are three distinct directions of symmetry: longitudinal (L), transverse (T), and short transverse (S). The toughness of such a material will depend on the orientation of the crack and the direction of loading during testing. If the loading is in the longitudinal direction and the crack plane is in the short transverse direction, the specimen is identified as L–S specimen; if the loading is in the transverse direction and the crack plane is in the short transverse direction, the specimen is identified as T–S specimen. Thus there are six possible combinations: L–S, S–L, L–T, T–L, S–T, and T–S. Three such combinations are illustrated in Fig. 9.1.

The two most commonly used specimen geometries, compact tension (CT) and three point bend (TPB), are shown in Figs. 9.2 and 9.3, respectively. Arc-shaped

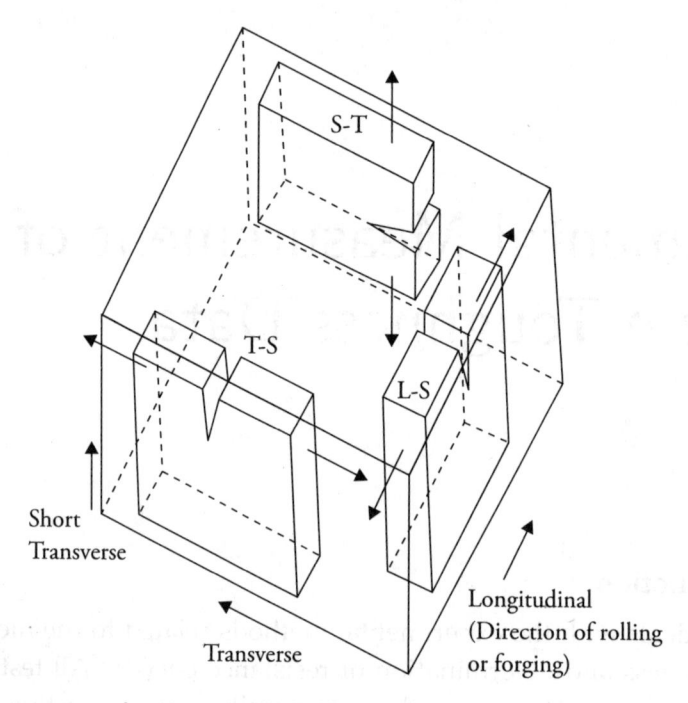

Figure 9.1 Different test specimen identifications depending on loading direction and crack plane orientation.

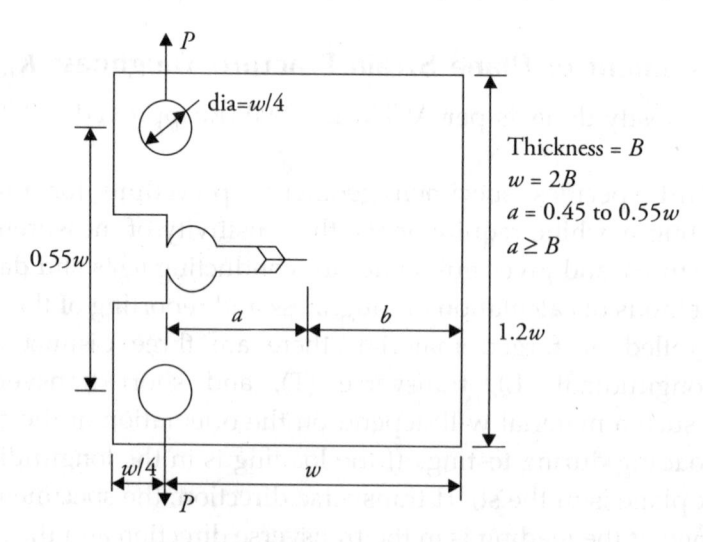

Figure 9.2 Compact tension specimen.

specimens are also recommended to facilitate their preparation from pressure vessel stock. For a measured fracture toughness K_{IC} data to be valid according to this standard, specimen thickness B and crack size a must be greater than

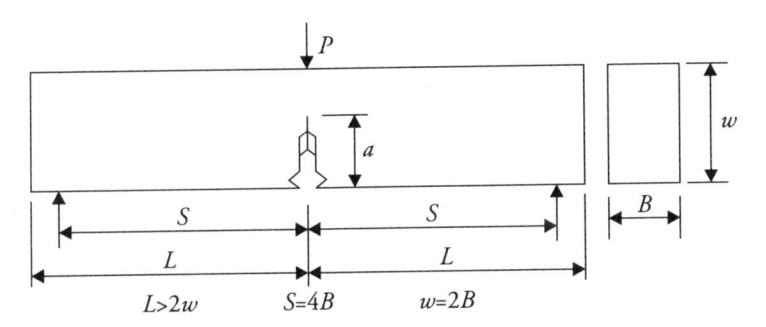

Figure 9.3 Three point bend specimen.

$2.5 \left(\dfrac{K_{IC}}{\sigma_Y}\right)^2$, where σ_Y is the yield point of the material. In the absence of a distinct yield point, 0.2% proof stress can be employed. This calls for an initial estimation of the toughness of the material. Alternatively, to fix these two parameters a table of minimum recommended values of B and a in terms of the ratio $\dfrac{\sigma_Y}{E}$, where E is the modulus of elasticity, provided by the standard can be used. Other dimensions of the specimen are dependent mostly on the thickness of the specimen. The specimens must be machined as per tolerances and surface finish specified in the standard. The machined notch is finished into a chevron notch to give rise to a crack with a straight front after fatigue pre-cracking. The fatigue pre-cracking is done under cycle amplitude $K_{I_{min}}$ to $K_{I_{max}}$, where $K_{I_{min}} = 0$ and $K_{I_{max}} < 0.65 K_{IC}$ to avoid large-scale crack-tip plastic deformation.

During the testing, the load is applied through pin and clevis in the case of CT specimen. The clevis and pin mating surfaces must be smooth so as to permit free rotation of the specimen during loading. During the bend test, the specimen is supported on rollers, which move freely on the surface of the fixture. The test load is applied via a roller. The variation of applied load with the load-line displacement (LLD) or crack opening displacement (COD) is recorded. During testing, the specimen is loaded at a rate such that the rate of increase in stress intensity factor is within 0.55 to 2.75 MPa$\sqrt{m}$/s. The COD is measured with the help of COD or clip gauge (Fig. 9.4). The specimen is loaded until the specimen can sustain no further increase in load. The standard recommends recording the maximum load from the load recorder or the dial of the testing machine. Typical records of load-displacement are shown (Fig. 9.5). Depending on the material, one of the three types of variations is obtainable. The British standard shows four types of variations. The slope $\tan \alpha$ of secant line corresponds to 95% slope of the initial linear part of a load-displacement record.

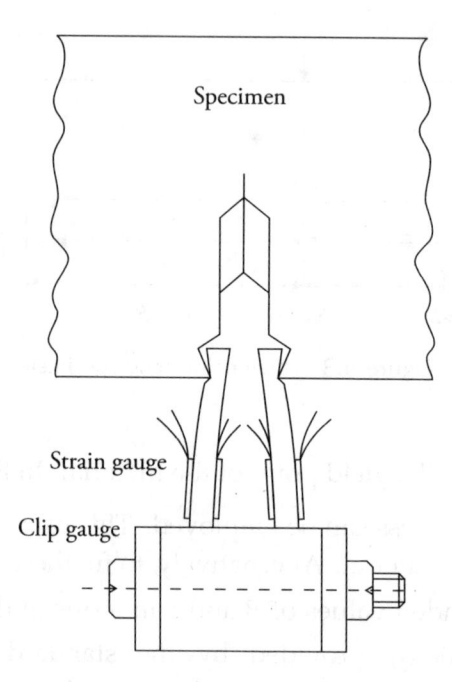

Figure 9.4 Crack opening displacement or clip gauge.

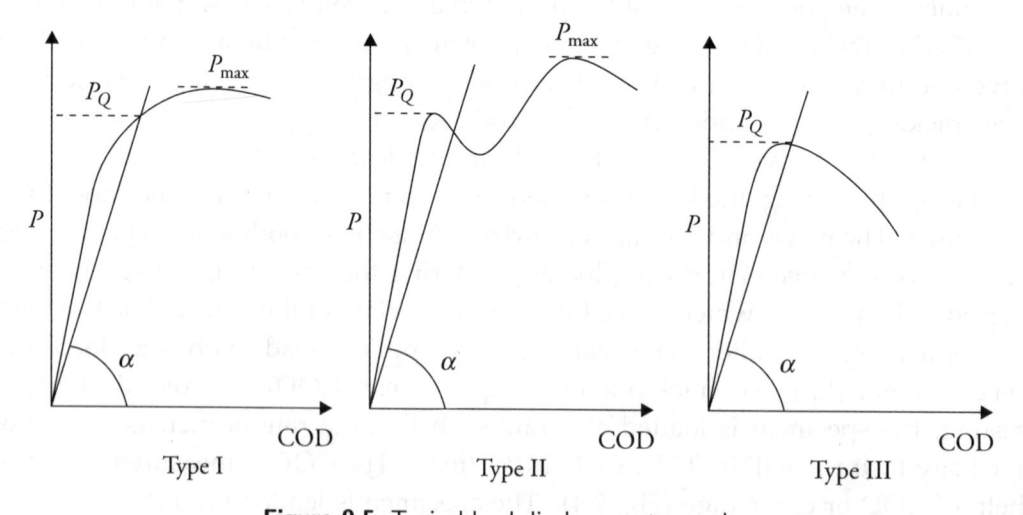

Figure 9.5 Typical load-displacement records.

For the Type I record, fracture load is equal to P_Q, provided $\dfrac{P_{max}}{P_Q} < 1.1$. If $\dfrac{P_{max}}{P_Q} > 1.1$, the test is invalid. For the Types II and III, fracture load is equal to P_Q. The 5% reduction in the slope of the linear part, or 95% secant method, in conjunction with the above specimen dimensions ensures a crack growth of 2% a, where a is the crack size used for the fracture toughness K_{IC} calculation. When the test is completed, the specimen is broke open after heat-tinting. The actual crack growth due to test-loading and fatigue crack growth can thereby be clearly distinguished (Fig. 9.6). In the case of some aluminium alloys, the latter portion shows striation or beach marks corresponding to each cycle. Their spacing indicates the amount of crack growth per cycle. Crack size a to be associated with fracture toughness calculations is $\frac{1}{3}(a_1 + a_2 + a_3)$, where a_1, a_2, and a_3, as per old standard, are measured at the one-third, middle, and two-thirds locations from one of the specimens face. According to 1997 reapproved standard a_1 and a_3 are measured at the surfaces. Test is invalid, if difference between any two of the three measurements exceeds $0.10a$. It is also invalid, if $\frac{|a_s - a|}{a} > 0.10$. The difference between the two surface crack length measurements should not differ by more than 10% of average crack length measured. For straight through thickness starter notch, no part of the crack front shall be closer to the machined starter notch than 2.5% w or 1.3 mm minimum, nor shall the surface crack length measurements differ from the average by more than 15%, and the difference between these two measurements shall not exceed 10% of the average crack length.

The tentative fracture toughness K_Q is calculated using the following relations.

$$K_Q = \frac{P}{B\sqrt{w}} f(r), \quad r = \frac{a}{w} \qquad (9.1)$$

$$f(r) = [(2+r)(0.866 + 4.64r - 13.32r^2 + 14.72r^3 - 5.6r)]/(1-r)^{1.5} \qquad (9.2)$$

for CT specimen (ASTM 1987).

$$K_Q = \frac{Pl_\varepsilon}{Bw^{1.5}} \frac{3\sqrt{r}\,[1.99 - r(1-r)(2.15 - 3.93r + 2.7r^2)]}{2(1+2r)(1-r)^{1.5}}, \quad r = \frac{a}{w}, \qquad (9.3)$$

$L_s = 2S = 4w$ for TPB specimen.

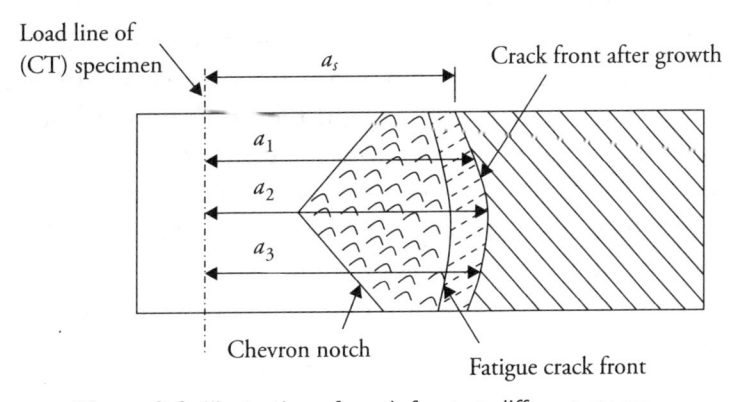

Figure 9.6 Illustration of crack front at different stages.

Standard Checks

If both crack size a and specimen thickness B satisfy the following condition

$$a, B \geq 2.5 \left(\frac{K_Q}{\sigma_Y}\right)^2, \tag{9.4}$$

and crack size satisfies the checks mentioned earlier, then the tentative fracture toughness K_Q is specified as the plane strain fracture toughness K_{IC}.

According to BS 5447 (1977), the formula for calculation of K_Q is slightly different. For the CT specimen

$$K_Q = \frac{P}{B\sqrt{w}} f(r), \quad r = \frac{a}{w} \tag{9.5}$$

$$f(r) = 29.6 r^{0.5} - 185.5 r^{1.5} + 655.7 r^{2.5} - 1017 r^{3.5} + 638.9 r^{4.5} \tag{9.6}$$

and for TPB specimen

$$K_Q = \frac{3\, PL_s}{Bw^{1.5}} f(r), \quad r = \frac{a}{w} \tag{9.7}$$

$$f(r) = 1.93 r^{0.5} - 3.07 r^{1.5} + 14.53 r^{2.5} - 25.11 r^{3.5} + 25.80 r^{4.5} \tag{9.8}$$

The British Standard provides for a validity check list. Before fatigue pre-cracking, check that the test piece dimensions and tolerances are as per the specifications. Before the testing, it should be ensured that the surface fatigue crack length is at least $0.45w$. Further, both ends of the fatigue crack have extended at least 1.25 mm or $2.5\% w$ from the root of the machined notch, whichever is greater. The two surface crack dimensions do not differ more than $5\% w$. The plane of the crack does not have slope more than $10°$ with the notch plane. After the fracture testing, it is necessary to ensure that multi-plane fracture is not present at the fatigue crack front. The average of the three crack length measurements (at $25\% B$, $50\% B$, and $75\% B$) is such that a/w ratio lies within 0.45 to 0.55. No two of three crack length measurements differ more than $2.5\% w$, and the maximum and minimum crack lengths measured do not differ more the $5\% w$. During the analysis of the data, it is necessary to check that force-displacement record has an angular disposition of $40°$ to $65°$ with respect to vertical axis. Further, the deviation from the linear variation at the load level $0.8 P_Q$ is less than one-quarter of the deviation at the level P_Q (for Type I record only). Additionally, it is required that P_{max}/P_Q is less than 1.10. Lastly, it is necessary to ensure that the specimen thickness and crack length exceed $2.5\, (K_Q/\sigma_Y)^2$.

9.3 Measurement of J_{IC}

This test is done as per the standard ASTM E1820-99a (2000). The standard E1820 deals with procedures and guidelines for the determination of fracture toughness of metallic materials in terms of K, J, and CTOD δ. Toughness can be measured in the form of resistance curve or as a point value. This standard covers only the opening mode (Mode I) loading. The recommended specimen geometries include TPB or single edge (SE(B)), compact tension (CT) (Fig. 9.7), and disk-shaped compact (DC(T)) specimens. All specimens contain notches followed by fatigue cracks. The specimen dimensions vary depending on the fracture toughness data to be collected. The guidelines are established through the consideration of material toughness and yield strength and individual qualification requirements associated with the toughness measure.

The test method requires continuous measurement of load versus LLD and crack mouth displacement. If any stable tearing occurs, then an R-curve is developed, and the amount of slow stable crack extension is measured. Two alternative procedures, the basic procedure and the resistance curve procedure, for measuring crack extension are recommended. The basic procedure is directed

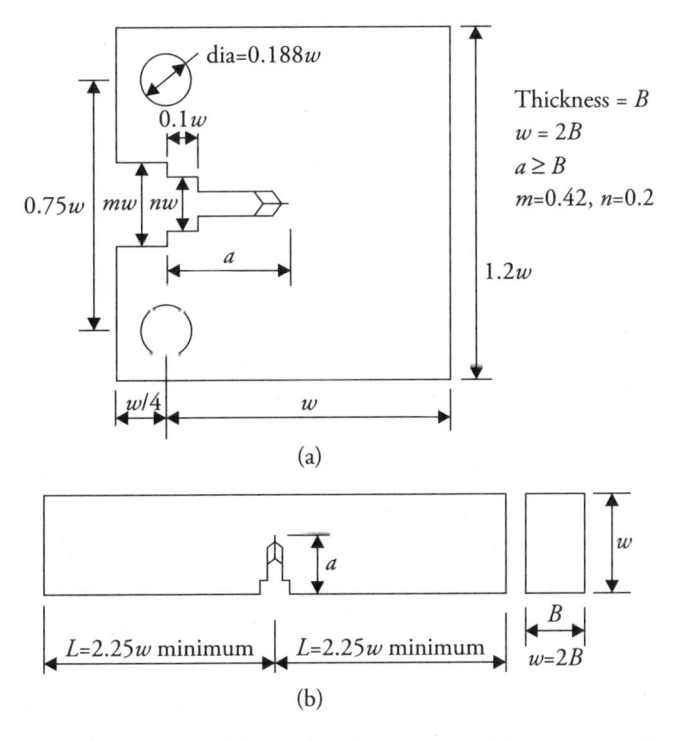

Figure 9.7 Standard geometries for (a) compact tension and (b) three point bend specimens for J_{IC} testing.

towards obtaining the value such as K_{IC}, J_C, or δ_C. The basic procedure involves physical marking of the crack advance and multiple specimens to develop a plot from which a single point initiation toughness value can be determined. The resistance curve procedure is based on the elastic compliance method and multiple point data are collected from a single specimen. The basic procedure is elaborated subsequently.

The specimen dimensions are recommended. For an evaluation of the initiation toughness data, w/B is 2. Suggested alternative proportions for w/B are 1 to 4 for TPB and 2 to 4 for CT specimens. The specimen is chosen keeping in mind the size restrictions in terms of the fracture toughness and machine capacity available. Specimens are to be made with the recommended machining tolerances. All requirements of fixture alignments, loading rate during experiment, and temperature stability and accuracy have to be ensured during the test. The specimens are initially made with sharp notches and then fatigue cracked to give a pre-crack. The pre-crack is produced by cyclically loading between 10^4 to 10^6 cycles. The fatigue load limits are

$$P_f = \frac{0.5 B b_0^2 \sigma_{Ye}}{S} \quad \text{for TPB specimen} \tag{9.9}$$

$$P_f = \frac{0.4 B b_0^2 \sigma_{Ye}}{(2w + a_0)} \quad \text{for CT specimen} \tag{9.10}$$

where S is the distance between the supports, and b_0 and a_0 are the initial crack and ligament dimensions. σ_{Ye} is effective yield stress; it is the average of the yield stress σ_Y (or 0.2% yield stress) and the ultimate tensile strength of the material.

The specimens are loaded under displacement control. In the basic procedure, each specimen is loaded to a selected level and the amount of crack growth is determined. During the test load versus displacement variation is recorded. Both the initial and final physical crack sizes are measured by optical methods. Multiple specimens can also be used to evaluate J at the initiation of ductile cracking, J_{IC} or δ_{IC}. The optical method of physical crack length measurement involves heat tinting, breaking open the specimen, and measurement. Alternatively, the specimen can be fatigue loaded after the test, and then measurement can be done after breaking open the specimen. Use of liquid penetrants is not recommended. The heat tinting is recommended for steels and titanium alloys. For other materials, the fatigue loading after the test is suggested. In the first case, the starting crack size will be clearly demarcated by the end of fatigue region and the final crack size will be separated by the heat-tinted region. In the second case, both will be demarcated by the starting and end fatigue regions. The electrical potential drop method is also allowed for measuring the crack sizes. Crack sizes are measured at nine equally spaced points on the starting or end crack fronts. The points are located over a region leaving a span of $0.005w$ from both the surfaces. None of the nine physical measurements of the initial crack size should differ by more than 5% from the average a_0. The similar measurement and restrictions apply for the final crack size. From the final crack size and the initial crack size, the extent of crack extension can be obtained. J is calculated through the following relation.

$$J = J_{el} + J_{pl} \tag{9.11}$$

where $J_{el} = (1 - \nu^2) \dfrac{K_I^2}{E}$,

$$K_I = \dfrac{P}{B\sqrt{w}} f\left(\dfrac{a_0}{w}\right) \quad \text{for the CT specimen,} \tag{9.12}$$

$$K_I = \dfrac{P}{B\, w^{3/2}} f\left(\dfrac{a_0}{w}\right) \quad \text{for the bend specimen,} \tag{9.13}$$

and a_0 is the initial actual crack length. $f\left(\dfrac{a_0}{w}\right)$ is taken from the handbook of stress intensity factors (SIFs) or the standard. If side-grooved specimens are employed in the testing to ensure plane-strain condition, then for the CT specimen

$$K_I = \dfrac{P}{\sqrt{B\, B_N w}} f\left(\dfrac{a}{w}\right) \tag{9.14}$$

and for the bend specimen

$$K_I = \dfrac{P}{\sqrt{B B_N} w^{3/2}} f\left(\dfrac{a_0}{w}\right) \tag{9.15}$$

where B_N is the reduced thickness and B is the original thickness. J_{pl} is given (Fig. 9.8) by the following relation.

$$J_{pl} = \dfrac{\eta\, A_{pl}}{B_N h_0} \tag{9.16}$$

where $\eta = 2 + 0.522 \left(\dfrac{b_0}{w}\right)$, $b_0 = w - a_0$, for CT specimen, and $\eta = 2$ for TPB specimen.

The variation of J versus Δa (Fig. 9.9) is plotted. Draw a construction line with $M = 2$ or some other value, which can be determined from the test data. The details of its determination are given in the standard. Draw the exclusion line parallel to the construction line at 0.15 mm and 1.5 mm. The limiting crack sizes and the J_{limit}

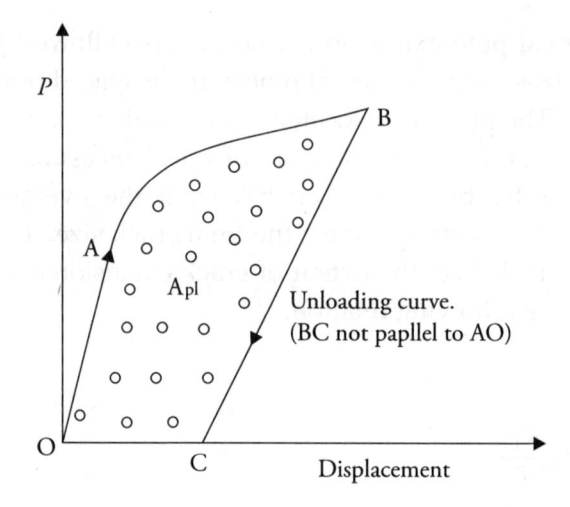

Figure 9.8 Loading and unloading curve associated with measurement of J.

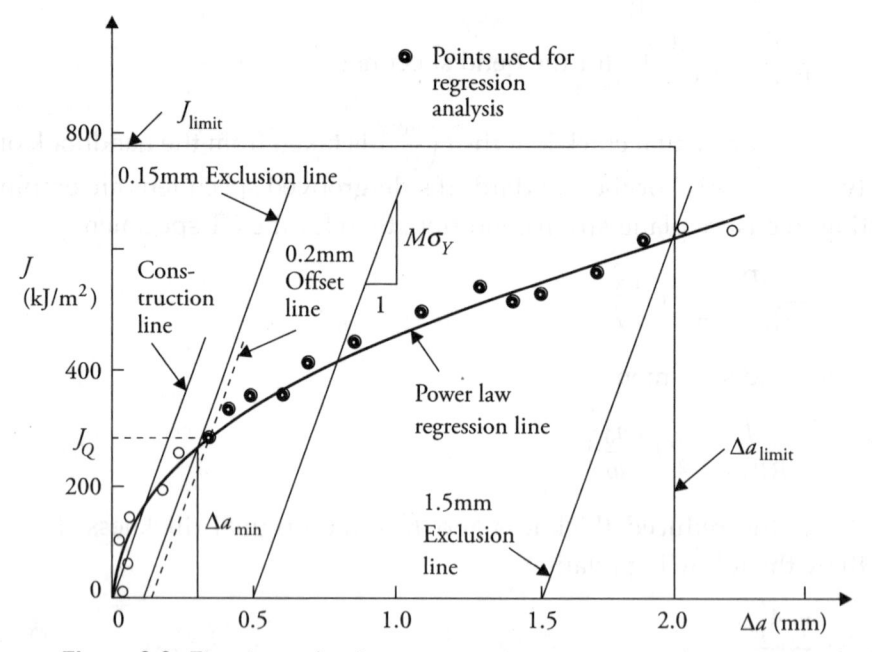

Figure 9.9 Experimental resistance curve. [Source: ASTM E1820-99a(2000)]

are drawn. J_{limit} is equal to $\dfrac{b_0 \sigma_{Ye}}{15}$. All data points lying within the space marked by the axes and the Δa_{limit} and J_{limit} lines are important and qualified for further procedure.

Draw a line parallel to the construction line passing through 0.5mm offset. It is necessary to ensure that there is at least one data point lying between 0.15 mm exclusion line and 0.5 mm offset line. The similar position must exist between 0.5

mm offset line and 1.5 mm exclusion line. All the data points that lie between the two exclusion lines are only considered for regression. Using the method of least squares, a power law regression line of the following form passing through all the qualified points is drawn.

$$\ln J = \ln C_1 + C_2 \frac{\Delta a}{p} \tag{9.17}$$

where $p = 1.0$ mm. The intersection of the regression line with 0.2 mm offset line defines the tentative fracture toughness J_Q. To confirm this tentative toughness as the plane strain fracture toughness J_{IC}, the following checks must be satisfied.

Thickness $B > 25 J_Q / \sigma_{Ye}$
Initial ligament $b_0 > 25 J_Q / \sigma_{Ye}$
The regression line slope dJ/da at Δa_Q is less than σ_{Ye}.

9.4 Measurement of Critical COD δ_C

The COD tests are carried out following either the standard ASTM E1290-99 (2000) or BS 5762 (1979). The ASTM standard recommends the use of both CT and TPB specimens with $\frac{a_0}{w} = 0.45$ to 0.55. The BS standard recommends only TPB specimens with $\frac{a_0}{w} = 0.15$ to 0.70. a_0 is the initial crack size after fatigue pre-cracking and w is specimen width. The specimen geometries are given. After getting the specimen ready through machining, they are fatigue pre-cracked with loading rate $\frac{dK_I}{dt} = 0.5$ to 2.5 MPa$\sqrt{m}$/s. Upon pre-cracking, the specimen is loaded gradually till the maximum load. The variation of load with crack mouth opening displacement (CMOD) v_g is recorded during the test. CMOD is usually measured through a clip gauge.

The total COD δ_t consists of the elastic part δ_e and the plastic component δ_p. The plastic part v_g^p of the total crack mouth opening, $v_g = v_g^e + v_g^p$ (Fig. 9.10), is calculated by considering rigid body rotation of the crack flanks/edges about point A (Fig. 9.11). This point is considered at a distance from the crack-tip, which is equal to a certain fraction of the ligament depth $(w - a)$. That is, A is the point where plastic hinge is formed. The two parts are given by

$$\delta_e = \frac{1}{2} \frac{K_I^2}{E \sigma_Y} (1 - v^2) \tag{9.18}$$

$$\delta_p = v_g^p \frac{r(w - a_0)}{r(w - a_0) + a_0 + z} \tag{9.19}$$

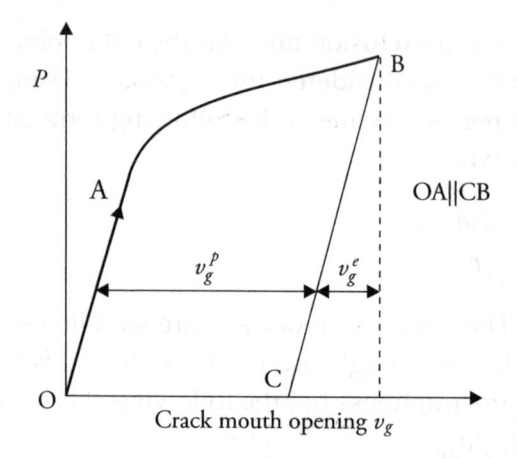

Figure 9.10 Components of total displacement.

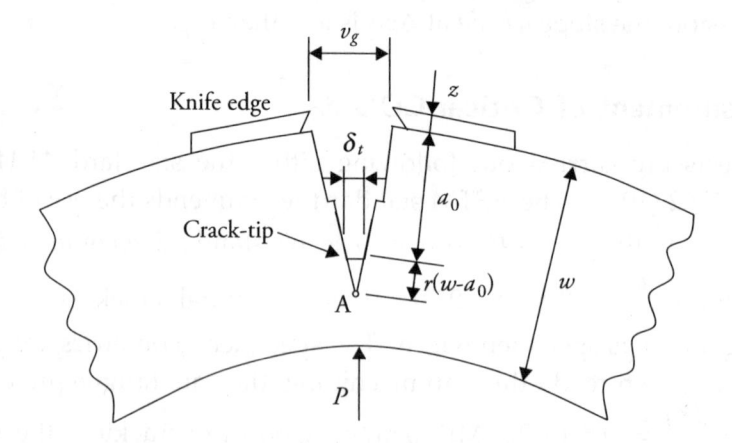

Figure 9.11 Displacements associated with crack opening displacement measurement.

K_I is the standard SIF corresponding to the load P, initial crack size a_0, and TPB specimen geometry. r is taken as 0.4 as per the BS and 0.44 as per the ASTM E1290-99 for the TPB specimen. CT specimen is permitted by the ASTM standard for the COD test. In this case

$$r = 0.4 \left[1 + 2\left\{\alpha^2 + \alpha + 0.5\right\}^{\frac{1}{2}} - 2\left\{\alpha + 0.5\right\} \right], \ \alpha = \frac{a_0}{b_0}, \tag{9.20}$$

a_0 is the initial crack size, and $b_0 = w - a_0$, the initial ligament size.

The variations of load P with CMOD can have different forms (Fig. 9.12). To cater for this type of situations, the standard BS 5762 indicates the following data for collection. The critical COD δ_c at the onset of unstable fracture with less than 0.2 mm of stable crack growth, is relevant for the type of P-CMOD curve shown in Fig. 9.12(a). δ_i and δ_u are relevant for the type shown in Fig. 9.12(b). δ_i corresponds to initiation of stable crack growth, which is analogous to K_{IC} or J_{IC}. δ_u is the critical COD that signals the onset of unstable fracture, which has been preceded by more than 0.2 mm stable crack growth. δ_m is the COD that corresponds to the maximum of the load plateau (Fig. 9.12(c)). P_c, $P_{\max}$, and P_u may not be seen in the same test. P_i may be seen along with P_u or $P_{\max}$.

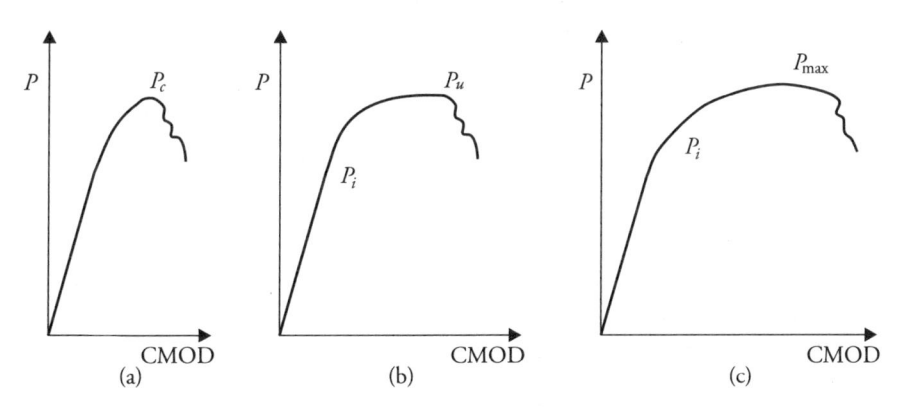

Figure 9.12 Three different forms of load versus CMOD records. (a) Load reaches level P_c then drops suddenly. (b) Load reaches slowly to highest level P_u then drops suddenly. (c) Load reaches maximum level $P_{\max}$ followed by slow drop and then sudden drop.

9.5 Measurement of K-Resistance Curve

Experimentally, R-curve can be determined following the ASTM E561-98 (2000) standard. For testing centre cracked tensile panel, CT specimen and wedge loaded specimen can be employed (Fig. 9.13). The specimen can have non-standard dimensions. The handling of data in the case of linear elastic and elastic plastic materials is diffcrent.

9.5.1 Linear Elastic Material

Generally, thin specimens are used for testing. This may lead to out-of-plane buckling. For prevention of such a situation, antibuckling guide plates are employed (Fig. 9.13(b)). To reduce friction at the contact surfaces, some lubricant (e.g., Teflon sheets) are helpful.

In the case of non-standard specimens, the instantaneous crack size a can be obtained using the secant line slope (Fig. 9.14) and the experimentally generated calibration curves. In the case of standard specimens, this can be obtained using the relationship or data provided in the standard. The corresponding resistance R is determined through the following relation.

$$R = K = \frac{P}{B\sqrt{w}} f\left(\frac{a}{w}\right) \tag{9.21}$$

Finally, variation of R with crack growth $\Delta a = a - a_0$, where a_0 is starting crack size, can be plotted to obtain the experimental resistance curve.

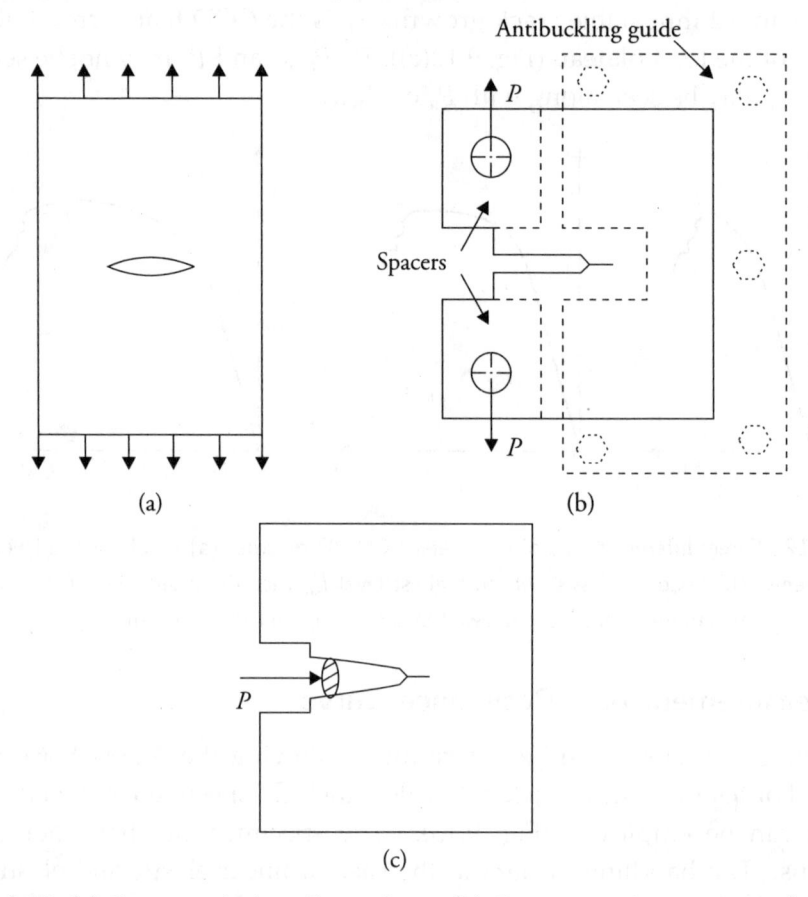

Figure 9.13 (a) Centre-cracked tensile specimen. (b) Compact tension specimen. (c) Wedge loaded specimen.

9.5.2 Elastic Plastic Material

The specimen and testing procedures are the same as in Section 9.5.1. The instantaneous crack size is obtained through the determination of physical crack size through optical method, or unloading compliance or secant compliance (Fig. 9.15). The resistance curve is plotted against the crack growth, where the effective crack size calculation is based on the Irwin plastic zone correction. That is,

$$\Delta a = a_{\text{eff}} - a_0 \tag{9.22a}$$

$$= a_{\text{measured}} + r_p - a_0, r_p = \frac{1}{2\pi}\left(\frac{K}{\sigma_Y}\right)^2 \tag{9.22b}$$

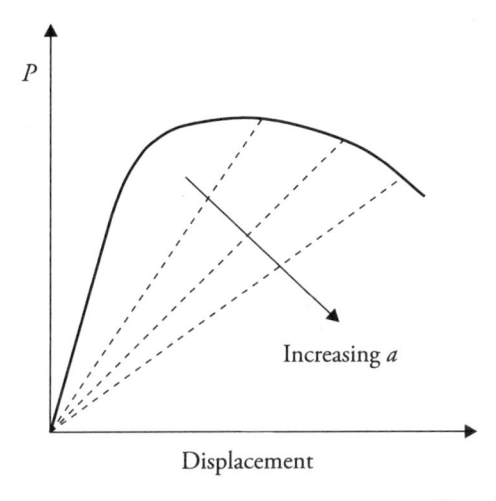

Figure 9.14 Typical load-displacement curve of elastic material.

Further,

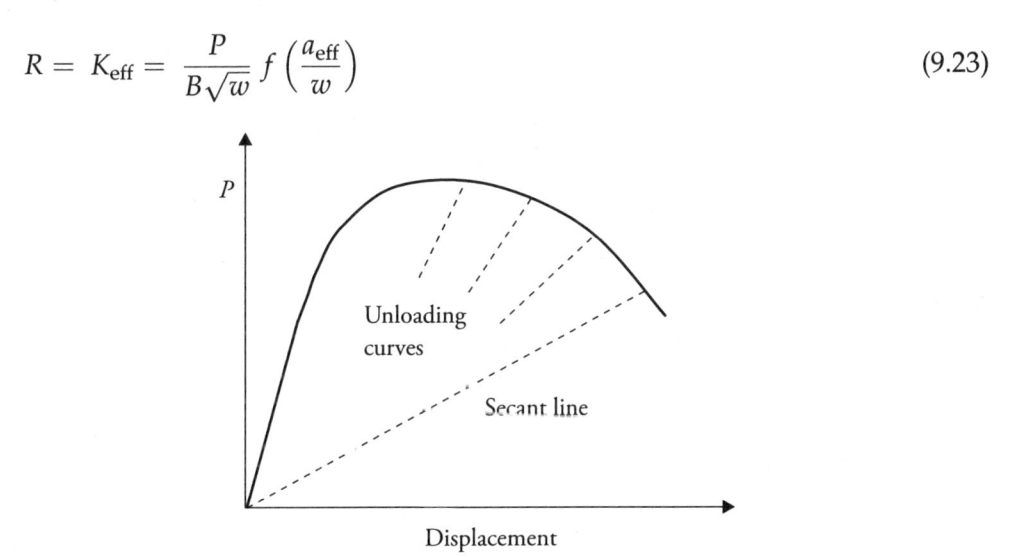

$$R = K_{\text{eff}} = \frac{P}{B\sqrt{w}} f\left(\frac{a_{\text{eff}}}{w}\right) \tag{9.23}$$

Figure 9.15 Typical load-displacement curve of elastic plastic material.

When the plastic zone is very extensive, the standard recommends the use of the secant method for the determination of the experimental instantaneous crack size. In the case of non-standard specimens, the instantaneous crack size a can be obtained using the secant line slope (Fig. 9.14) and the experimentally generated calibration curves. In the case of standard specimens, this can be obtained using the relationship or data provided in the standard.

In the first step of calculation, a_{eff} can be taken equal to the measured crack size a, neglecting the correction due to plastic zone. This allows calculation of K_{eff}

using Eq. 9.23. In the next step, a_{eff} is calculated using the plastic zone correction factor r_p through Eq. (9.22b). Fresh calculation of K_{eff} is then done. This process is repeated till satisfactory convergence is achieved. This process has been illustrated in Problem 4.3 of chapter 4.

References

9.1 ASTM E399-90 (Reapproved 1997). 2000. 'Standard Test Method for Plane-Strain Fracture Toughness of Metallic Materials. 'In *Annual Book of Standards*, Section 3, Vol. 03.01, *Metals Test Methods and Analytical Procedures*, 431–61. Philadelphia: American Society for Testing and Materials.

9.2 ASTM E561-98. 2000. 'Standard Practice for *R*-curve Determination. 'In *Annual Book of Standards 2000*, Section 3, Vol. 03.01, *Metals Test Methods and Analytical Procedures*, 522–34. Philadelphia: American Society for Testing and Materials.

9.3 ASTM E1290-99. 2000. 'Standard Test Method for Crack-tip Opening Displacement (CTOD) FractureToughness Measurement. 'In *Annual Book of Standards*, Section 3, Vol. 03.01, *Metals Test Methods and Analytical Procedures*, 856–67. Phila-delphia: American Society for Testing and Materials.

9.4 ASTM E1820-99a. 2000. 'Standard Test Method for Measurement of Fracture Toughness. 'In *Annual Book of Standards*, Section 3, Vol. 03.01, *Metals Test Methods and Analytical Procedures*, 1000–33. Philadelphia: American Society for Testing and Materials.

9.5 BS 5447. 1977. *Methods of Testing Plane Strain Fracture Toughness (K_{IC}) of Metallic Materials*. London: British Standards Institution.

9.6 BS 5762. 1979. *Methods of Crack Opening Displacement Testing*. London: British Standards Institution.

Index

Airy stress function, 65, 68, 140, 217, 221
Analytic function, 68–69, 78, 104, 108
   Examples of, 68
ASTM E 1290–99 [2000], 267–268
   Measurement of δ_C, 267–269
ASTM E 1820–99a [2000], 263
   Measurement of J_{IC}, 263–267
ASTM E 399–90 [2000], 49–50, 257
   Measurement of K_{IC}, 257–262
ASTM E 561–98 [2000], 269
   Measurement of K-resistance curve, 269

Basic testing for J_{IC}, 263–267
Beach marks, 261
   (see also striation marks)
Boundary collocation method, 102, 108
Boundary element method (BEM), 4, 102, 139
Boundary loading in terms of Airy stress function, 67–68
British standard BS 5762 [1979], 267
   Measurement of critical COD δ_C, 267–269
   (see also ASTM E 1290–99 [2000])
   Test procedure, 267
Brittle fracture, 2, 6, 10–12, 152,
   Energy balance theory of, 10–12
   Irwin–Orowan modification, 12–13
   Griffith theory of, 10–12

Calculation of:
   K_{IC}, 261–262
   J_{IC}, 265
   δ_C, 267–269
   K_{eff}, 271
Calculation of toughness, 257
   CT specimen, 261–262, 265, 267–269
   Three point bend (TPB) specimen, 261–262, 265, 267–269
Cauchy–Riemann condition, 68

(see also analytic functions)
Centre crack tension (CCT), 225
   specimen configuration, 247
Characterisation of crack growth:
   at maximum load, 269
   initiation, 225, 264, 269
   instability, 87, 95–97, 203, 235
Chevron notch, 259
Cleavage, 6–7
   fracture, 6
   theoretical strength, 6–7
COD, 2, 86–89, 208–209, 211–213, 220, 223, 233–234, 245, 259, 267–269
   measurement of, 267–269
   (see BS 5762 [1979] and ASTM E 1290–99 [2000])
   Tracy's definition for, 224
   Wells's definition for, 86–88, 208–209
Compact tension (CT), 89, 229, 246–247, 258–265, 266–270
   correction due to plastic zone, 270
   SIF relation for, 261–262
   specimen configuration, 258, 263
Compatibility condition/equation, 65–67, 114, 218, 221
   for FEM analysis, 114
   for stress analysis, 65
Condition of brittle fracture in individual modes:
   Mode I, 19
   Mode II, 19
   Mode III, 19
Crack closure during fatigue, 174
   due to debris, 175
   due to roughness, 175
   due to plastic deformation, 178
Crack closure integral (CCI), 18, 118, 130
   Irwin's definition of, 18
Crack driving force, 95–97, 236
   at instability, 96–97
   variation with crack extension, 95–96

R-curve, 95–96, 263, 269
Crack growth under fatigue loading, 169
 constant amplitude cycling,170–174
 random amplitude cycling, 169
 variable amplitude cycling, 182
Crack growth under quasistatic loading:
 condition for instability, 87, 97
 condition for stable growth, 236
 resistance at initiation, 264
 unstable crack propagation, 10
Crack path, 153
 (*see* mixed mode fracture and criteria of mixed mode fracture)
 direction of initial crack growth θ_C, 153–154, 158, 160
 prediction of, 153–154, 158–160
 under mixed mode, 153-154, 158–160
Crack tip elastic field in individual modes: determination of:
  Westergaard stress function approach, 69, 71–77
  Williams' eigen function expansion approach, 80–83
  Mode I, 71, 216
  Mode II, 75
  Mode III, 77-80, 209
CTOD, 209, 224, 234, 263
 measurement of critical, 267
 (*see* also experimental measurement of δ_C),
 Tracy's definition for, 224
 Wells's definition for, 209

Damage tolerant design, 13
Debris on crack flanks, 175
 (*see* also crack closure during fatigue crack growth)
δ resistance curve, 234
Determination of SIF, 102
 Methods of, 102–143
Displacement based determination of SIF, 118
Displacement controlled loading (hard loading), 23, 97
 Crack extension in brittle materials, 97
 Crack extension in ductile materials, 202
Displacement FE formulation, 113

Domain integral based evaluation of J, 122–129
 for mixed mode, 125
 separation into symmetric and anti-symmetric parts, 125
Double cantilever beam (DCB) geometry, 57
Ductile fracture, 12, 86–89
 COD approach for, 209
 Irwin–Orowan modification of energy balance theory, 12
 Irwin correction for small scale plastic deformation, 24, 26, 30
 J based approach for, 225
 Large scale plasticity at crack tip, 212–213
 (*see* Elastic plastic fracture mechanics (EPFM))

Elastic plastic zone around crack tip, 27–31
 Dugdale–Barenblatt model, 212–213
 (*see* strip yield model)
 Irwin's plastic zone correction, 270
EPFM, 202
(*see* also ductile fracture)
 Condition of crack initiation:
  in terms of J, 225
  in terms of δ, 225
 Crack tip field in terms of J, 225–226
  J dominated region in TPB specimen, 225, 232
  J dominated region in CCT specimen, 225–226, 232
 Crack tip field in terms of δ, 225
 Crack growth instability, 235
  tearing modulus, 235–236
EPRI (Electric Power Research Institute), 237
EPRI method of estimation of J, 237
Evaluation of CCI though FEM, 130–136
 accuracy, 133
 local smoothing, 133
Evaluation of J integral though FEM:
 Contour integral, 120–122
 Domain integral, 122–129
Evaluation of SIF experimentally:
 strain gauge based, 102–103, 140–142
  Mode I, 141

Index

Mixed mode, 142
photoelasticity, 142–143
Evaluation of SIFs through FEM:
  Crack closure integral (CCI) method, 130–136
  Displacement method, 118–119
  Energy method, 120–136
    Direct, 120
    Variant of, 120–136
    J integral method, 120–129
    Stiffness derivative procedure (SDP), 118, 120, 129–130
    (see also CCI method)
  Extrapolation method, 119–120
  Stress method, 119–120
Extended FEM (XFEM), 139–140

Failure locus, 156–157, 215
Fatigue pre-cracking load limits for:
  K_{IC} testing, 259
  J_{IC} testing, 264
Fatigue threshold, 170
FCGR, 169, 172, 174, 176, 178, 180–181
FEM, 113–118
  Displacement formulation:
    Element displacement variation, 113
    Stress-strain relation, 117
    Element stiffness matrix, 117
    Element equations, 115-117
    Global stiffenss matrix, 118
    Equilibrium equations, 118
    Displacement calculations, 118
    Element stress calculations, 118
Flat fracture, 32–34
(see also slant fracture)
Fracture toughness, 33–34, 41–47, 257, 260–262, 267
  K_{IC}, 33–34, 41–47
  Table of values at room temperature, 41–47
  Experimental measurement, 257
    as per ASTM E 399–90 [2000], 263
    checks for confirmation of, 262
  J_{IC} measurement, 263–267
    as per ASTM E 1820–99a [2000], 263
    checks for confirmation, 267

δ_C measurement as per BS 5762 [1979], 267
  Thickness dependence of, 33–34
Fracture, brittle, 2, 6, 10–12, 16, 152
  Griffith theory of brittle fracture, 10–12
  Irwin–Orowan modification of, 12–13
  Plastic zone shape in mode I, 28–29
    plane stress, 28–29, 87–89,
    plane strain, 28–29
    plastic zone correction, 26, 30, 270
  Plastic zone shape in mode II, 30
    in mode III, 30
  Mixed mode fracture, 2–4
  Criterion of fracture based on:
    Energy, 153
    Maximum tangential stress (MTS), 154–157
    Maximum tangential principal stress (MTPS), 158–159
    Strain energy density (SED), 90, 121, 128, 159–163
    Maximum volumetric energy density, 161
Fracture, ductile, 12
  Crack tip field in Mode I, 220, 223
    HRR singularity, 225
      elastic plastic material, 223
      perfectly plastic material, 225
    characterisation in terms of J, 225–226
    region of J dominance in, 226
      TPB, 225, 232
      CCT, 225-226, 232
    stress singularity, 223
    strain singularity, 223
  Crack tip field in mode III, 209–212
  Propagating crack, 233–237
    stability/instability, 95–97, 235–236
    tearing modulus, 235–236

G and K relationship, 17–18
  for Mode I, 18
  for Mode II, 19
  for Mode III, 19
G as crack driving force, 95–97
G evaluation:
  under hard loading, 23–24
  (see also crack growth under displacement control)

276 Index

under soft loading, 22–23
(see also crack growth under load control)
by FEM, 120–136
experimental, 22–24
Gauss quadrature, 118, 127–128
G_{IC}, fracture resistance, 19
Green's function method, 102, 108
Griffith theory of brittle fracture, 10–12
Griffith's explanation based on stress concentration, 8–9

Hard loading, 23–24, 97
Hardening of materials:
  Bauschinger effect, 205
  Isotropic hardening, 204–206
  Kinematic hardening, 190, 205–206
  Work hardening, 204
Heat-tinting, 261
High cycle fatigue, 169
  Factors affecting crack propagation, 174
  Fatigue crack growth rate, 170
  Life calculations, 176, 182–183
    Constant amplitude loading, 176
    Random cyclic amplitude loading, 3, 182–183, 190
    Crack growth retardation, 178
    Under cyclic interactions, 182–183, 190
    Rainflow cycle counting, 190–191
  Paris law, 170–171
HRR singularity, 225
(see also EPFM)
Hutchinson's analysis (for nonlinar material), 221
Hydrostatic stress, 216, 231

Improving accuracy of CCI calculations, 130–136
  through local smoothing, 133
Incompressibility condition, 206
Incremental theory of plasticity:
  Levy–Mises theory, 207
  Prandtl–Reuss theory, 207
Inglis's solution, 10
Instability of crack extension:
  through brittle material, 11, 19
  through ductile material, 235–236
Integral transform technique, 102

Integration, numerical, 118, 127
(see also Gauss quadrature)
Irwin–Orowan modification, 12–13
Isotropic hardening, 204–206
(see workhardening)

J–Q dependance, 232
J controlled crack growth, 236
J dominated region in, 234
  TPB specimen, 225, 232
  CCT specimen, 225–226, 232
J integral, 2, 91–92, 120, 212, 225
  as potential energy release rate, 92–95
    linear elastic material, 92–94
  characterising crack tip field, 2, 225
  elastic plastic material, 212–213, 263
    (see also HRR singularity)
  graphical representation for non-linear elastic material, 94–95
J resistance curve, 236

K characterisation of crack tip field, 14–19
  Influence of non-singular terms, 157, 230–232
  Square-root singularity, 16
  Relation between G and K, 17–19
K resistance curve, 97, 269–272
K_C:
  Definition, 33
  Thickness dependence, 33
K_{op} during fatigue cycling, 176
Kinematic hardening, 190, 205–206
(see work hardening)
Kolosoff–Muskhelishvili potential formulation, 65, 68, 80

Laplace equation, 68
LEFM, 2–4, 6, 86, 102, 192, 202
  Energy balance theory, 10
    (see Griffith theory)
    (see Irwin–Orowan modification)
  Instability, 87, 95–98, 203–204, 235
  Three modes of crack loading, 15
  Crack tip stress field, 20–21, 28, 65 103, 140, 143, 154, 157, 230
  Plastic zone size, 12, 24–27, 30–32, 34, 87, 96, 181, 220

Irwin's calculation of, 12
  plastic zone correction, 26, 30, 88, 100, 270, 272
    plane strain, 30,
    plane strain constraint, 30
    plane stress, 26
    equivalent elastic crack, 26
Strip-yield model, 26
(see also Dugdale–Barenblatt model)
Plastic zone shape:
  Mode I, 28–29, 220
    by Mises criterion, 29
    by Tresca criterion, 28
  Mode II, 30
  Mode III, 30
Instability condition using resistance curve, 97
resistance to brittle fracture, 12
(see also K_{IC})
G calculation:
  for in-plane extension, 23–24
  for kinking, 138–139
Levy–Mises theory, 207
Life estimation under fatigue, 176, 183
Load control, 22, 97
Load line displacement (LLD), 229, 238, 259
Local smoothing, 133–136
Low cycle fatigue, 168, 192
Liquid penetrant, 264

Maximum principal stress theory, 154
  (same as Rankine princpal stress theory)
Measurement of critical COD δ_C, 267–268
  (see also ASTM E 1290 99 [2000])
  (see also BS 5762 [1979])
  Calculation of δ_C at
    initiation, 268
    instability, 269
    maximum load, 269
  Test procedure, 270
Measurement of fracture toughness, 257–269,
  K_{IC}, 257–262
  J_{IC}, 263–267
  δ_C, 267–269
  specimen geometry, 257–259, 263, 267–268
Meshless method, 1, 4, 102, 140
Microscopic fracture model, 232

Ritchie–Knott–Rice model, 232–233
Mises criterion, 27–30, 205
Mixed mode fracture, 2–3
  criterion of, 153–163
Mode I, 13–14, 65, 71–75, 80, 118–120, 122, 125, 140, 155, 161, 216–226, 230–237, 257, 263, 267
Mode II, 13, 15, 75, 80, 83, 119, 120, 122, 132, 135–136, 152, 156, 162
Mode III, 16, 19, 30, 77–80, 209–212 209–210, 217, 220, 225

Non-proportional loading, 233
Numerical integration, 118, 121, 125, 177
(see Gauss quadrature)
Numerical methods, 1, 4, 102, 113–136
(see also FEM)

Overload effects on FCGR, 178

Paris law, 4, 171–174, 176, 192
Path dependence of plastic deformation, 206
Plane strain, 14–16, 18–19, 28–34, 65, 68, 75, 87, 117, 141, 176, 180, 181, 218, 223–224, 226, 231, 238–240, 246, 248, 250, 252, 257–262, 265, 267
Plane stress, 14–16, 18–19, 28–32, 65, 68, 70, 74, 88, 96, 117, 140, 176, 180, 221, 223–224, 238–240, 246–247, 249, 251, 253
Plastic constraint Q, 231–232
Plastic deformation energy per unit crack extension, 12–13
Plain strain constraint, 30
Plastic zone correction, 26, 30, 88, 209, 270, 272
Plasticity induced crack closure, 175, 178
Point of instability:
  of crack growth, 87, 96–97, 233–236, 268–269
    through elastic-plastic material, 268–269
    through linear elastic material, 87, 96
Potential energy release rate G, 23–24, 120–136
(see under G evaluation)
  experimental evaluation of critical, 22–24
  FE based evaluation of, 120
  under hard loading, 22–23, 97
  under soft loading, 23, 97

Potential energy, 10, 92
Power law hardening, 218
  Ramberg–Osgood material, 216
Prandtl–Reuss theory, 207
Process zone, 225–226, 233–234
Proportional loading, 221, 235
  deformation theory of plasticity, 86, 208, 216, 233–234, 237

Q factor 230

Ramberg–Osgood relation, 216, 254
Rankine theory, 159
Region of finite strain, 225
Residual stress induced retardation, 178
Resistance curve approach for measurement of J_{IC}, 263
Resistance curve, 86–87, 95–98, 257, 263 269–271
  in terms of K, 96
  in terms of J, 236
Retardation effects during fatigue cycling, 178
Retardation of crack growth, 178–182
Retarded fatigue crack growth rate:
  Wheeler model, 179, 182
  Willenborg et al. model, 178, 181–182
Reversal of load during fatigue cycling, 175
  Bauschinger effect, 205
Roughness induced closure, 175

Short cracks, 192
Side-grooved specimens, 265
SIF K, 13–17, 48–54, 81, 83, 103–111, 119 (see under K)
Single edge crack panel (SECP), 240
Single edge notched bend (SENB), 225
Single specimen testing for J_{IC}, 264
Singularity elements, 114, 132–133, 136, 139
Slant fracture, 33
Soft loading, 22–23
Specimen configurations for measurement of
  K_{IC}, 258–259
  J_{IC}, 263–264
  δ_C, 267–268
Stable crack extension, 236, 263

Stiffness derivative procedure, 118, 120, 129–130, 139
Strain (or potential) energy release rate, 6, 19, 22, 120, 130–136, 138
Strain singularity, 16, 211, 217, 225
  Elastic material, 16
  Material showing plastic deformation, 211, 217, 223, 225
Stress method, 119
Stress relaxation, 24, 25, 30, 32, 34
Stresses in terms of stress function, 67
Striation or beach marks, 261
Strip yield model, 26, 88, 213
Surface energy, 8, 10–12
Symmetric and anti-symmetric parts of displacements and stresses, 126

T criterion, 161
T stress, 230
Tearing modulus, 235–236
Test procedure for measurement of:
  K_{IC}, 257–262
  J_{IC}, 263–267
  δ_C, 267–269
  K-Resistance curve:
    Elastic plastic material, 270
    Linear elastic material, 269
Testing fixtures, 257
Theoretical strength:
  Atomic level modelling for, 6–8
  Explanation based on stress concentration, 8–9
Thickness dependence of fracture toughness, 33–34
Three modes of crack extension:
  Condition of fracture, 19
  Crack-tip stress field, 14–16
  Onset of crack extension in a particular mode, 19
Three point bend (TPB) geometry, 257
Total or deformation theory of plasticity, 208
Tracy's definition of CTOD, 224
Tresca criterion, 28–29, 205
Triaxiality at crack front, 31–33
  Plastic zone shape around crack tip, 31

Unloading
  during crack extension, 233
  (*see* also non-proportional loading)
  during tensile test, 204
Underload cycle, 178

Weight function method, 110–111
Williams' eigenfunction expansion, 80, 83
Work hardening, 204

XFEM, 139–140

Yield condition, 28, 205, 211
Yield criterion of :
  Mises, 28, 205
  Tresca, 28, 205
Yield surface, 205
Yielding fracture mechanics (YFM), 2, 202, 225